BOOK 2

CORE MATHEMATICS

Revision tests for first examinations

JEFFREY STEAD and PETER WRIGHT

Cambridge University Press

CAMBRIDGE

LONDON NEW YORK NEW ROCHELLE

MELBOURNE SYDNEY

Published by the Press Syndicate of the University of Cambridge
The Pitt Building, Trumpington Street, Cambridge CB2 1RP
32 East 57th Street, New York, NY 10022, USA
296 Beaconsfield Parade, Middle Park, Melbourne 3206, Australia

First published 1981

Printed in Hong Kong by Wing King Tong

British Library Cataloguing in Publication Data
Stead, Jeffrey
Core mathematics
Book 2
1. Mathematics - 1961 -
1. Title II. Wright, Peter, *b. 1940*
510 QA39.2 80-41766

ISBN 0 521 23233 3

Contents

Part 2 Notes on solutions

Section 1

Section 2

Section 3

Part 3 Answers

Section 1

Section 2

Section 3

Part 4 Reference section

Preface

The material in this course comprises two books. It is aimed at the 16+ age range and the wide variety of questions provided cover the core content of the majority of CSE and O-level syllabuses.

The content has been developed to provide a structured revision course which facilitates preparation for examinations. The material can be used either in a controlled classroom environment or for self-tuition, thus giving the conscientious student an opportunity to work at the subject independently.

The revision course as a whole provides a wealth of material for classwork, homework and testing during the 4th and 5th years of secondary school prior to the examination and, in addition, it will be useful to students in sixth forms who are working to improve their earlier performance at CSE or O-level. The content will also be suitable for students in technical colleges, and colleges of further education.

Apart from providing the student with structured sets of revision material, the books contain detailed notes on solutions to the various problems. These have been devised to aid in comprehension of method and to provide guidance and instruction for those students who experience difficulties with specific questions.

Ten Commandments

1. Know your basic facts – they will give you immense confidence.
2. Read the question slowly, carefully and then read it again.
3. Always set out your work neatly, clearly and logically.
4. Check your answers whenever possible. It may be tedious but it will pay dividends.
5. Show your answers clearly, stating the appropriate units and be sure to give the required answers.
6. Do not spend overlong on a question which you are finding difficult. Try to work at a steady rate and return to your difficulties later, if time permits.
7. Programme your revision for regular study periods. Do not be a prevaricator and leave yourself with an enormous work load at the last moment.
8. Remember that a calculator does not necessarily make a mathematician. However good it is, simply pressing the wrong key may produce a ridiculous answer. If you are allowed to use a calculator, use it sensibly. By means of a rough check, make sure that your answer makes sense and if you have time, double check.
9. Cultivate the habit of taking a pride in your work.
10. CONCENTRATE!

Part 1 Questions

Section 1

Set 1

1 A survey was taken of the listening habits of adults. The Venn diagram shows the results of the survey.

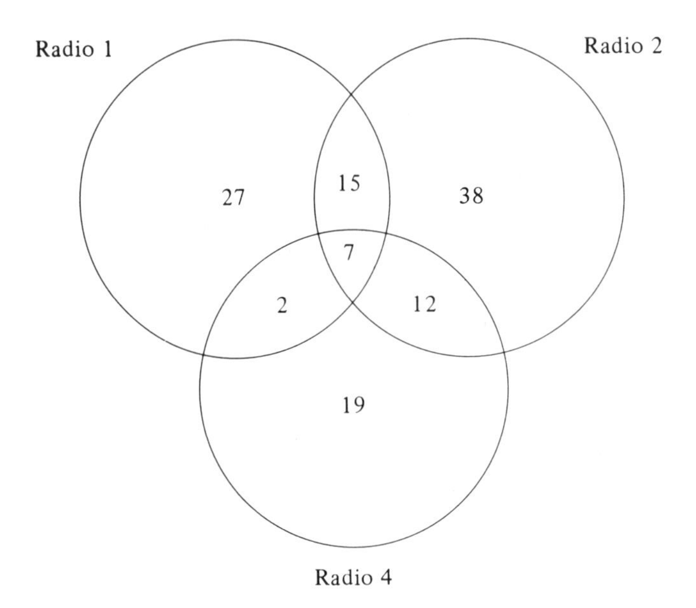

(i) How many adults were questioned in the survey?
(ii) How many people listened to both Radio 1 and Radio 4?
(iii) How many people listened to Radio 2 only?
(iv) How many people never listened to Radio 1?
(v) What percentage of adults listened to Radio 4?

2 (i) Write $10\,001_2$ in base 10.
(ii) Write 65_{10} as a binary number.
(iii) Evaluate $453_8 + 76_8$ and give your answer in base 8.
(iv) Write 100_8 in base 2.
(v) If $32_x = 14_{10}$, find the value of x.

3 (i) If petrol costs 28p a litre, find the cost of $9\frac{1}{2}$ litres.
(ii) 8 text books cost £10.24, find the cost of one book.
(iii) A shopping bill totalled £17.87, what change should you get from a £20 note?
(iv) Calculate the volume of a cube whose side is 8 cm.
(v) How many cubes of side 2 cm would exactly fill an 8 cm cube?

4 The following scores were recorded in a competition by 10 competitors: 7, 2, 3, 8, 5, 1, 7, 4, 9, 7

(i) Write down the mode.
(ii) Find the median score.
(iii) Calculate the mean.
(iv) An eleventh competitor scored x points and this lowered the mean to exactly 5. Find the value of x.
(v) What percentage of the original 10 competitors scored more than 7?

5 Solve the equations:

(i) $3a = 21$
(ii) $x - 5 = 6$
(iii) $2a = a - 9$
(iv) $x^2 = 81$
(v) $(b + 3)(b - 4) = 0$

6 (i) Write £29.78 correct to the nearest pound.
(ii) Write 38 900 correct to 2 significant figures.
(iii) Write 3.476 correct to 2 decimal places.
(iv) Write 56 000 000 in Standard Form notation.
(v) Simplify $(7.0 \times 10^4) \times (8.0 \times 10^3)$ leaving your answer in Standard Form notation.

7 Given that $A = \frac{1}{2}bh$

(i) calculate A when $b = 8$ and $h = 7$
(ii) calculate A when $b = 6.5$ and $h = 4.6$
(iii) change the subject of the formula to b
(iv) calculate b when $A = 15$ and $h = 6$
(v) calculate h when $A = 22.5$ and $b = 6$

8 In the diagram, $AB = 20$ cm, $AD = 10$ cm, $DC = 6$ cm and angle ADB is a right angle.

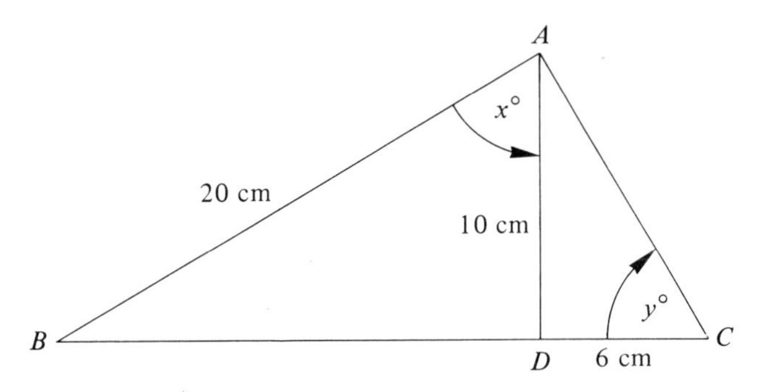

Calculate:

(i) $\cos x°$ as a decimal
(ii) angle x in degrees from your tables
(iii) angle ABD
(iv) $\tan y°$ as a decimal
(v) angle y in degrees from your tables
(vi) angle CAD
(vii) angle BAC

9 The diagram shows the plan of a garden which consists of a rectangular lawn 20 m by 12 m, in the centre of which is a circular rose bed, radius 5 m.

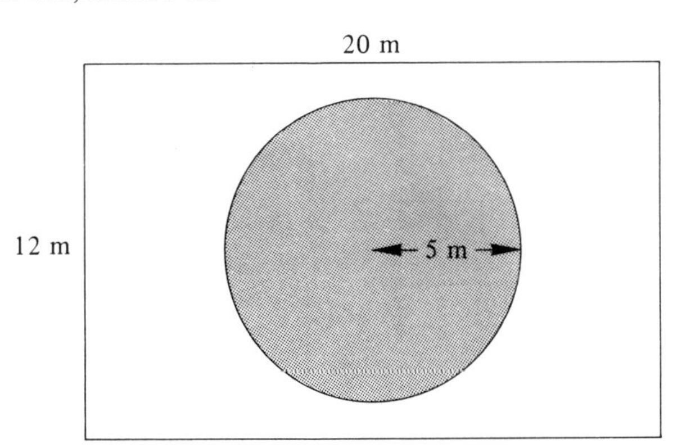

(i) Calculate:

(*a*) the total area of the garden
(*b*) the area of the rose bed ($\pi = 3.14$)
(*c*) the area of the lawn (i.e. the area of the garden which is actually put down to grass).

(ii) The rose bed is to have a low plastic fence around its circumference.

(*a*) Calculate the length of fencing needed to the nearest whole metre above.
(*b*) If the fencing costs 40p a metre, find the total cost.

10 (Squared paper) Copy the diagrams and then follow the instructions given for each one.

(i)

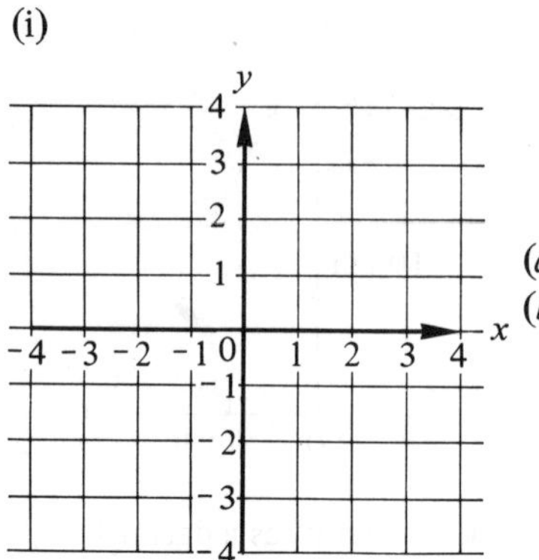

(a) Draw the graphs of $x = 1$ and $y = 3$

(b) Write down the coordinates of the point of intersection of the graphs.

(ii)

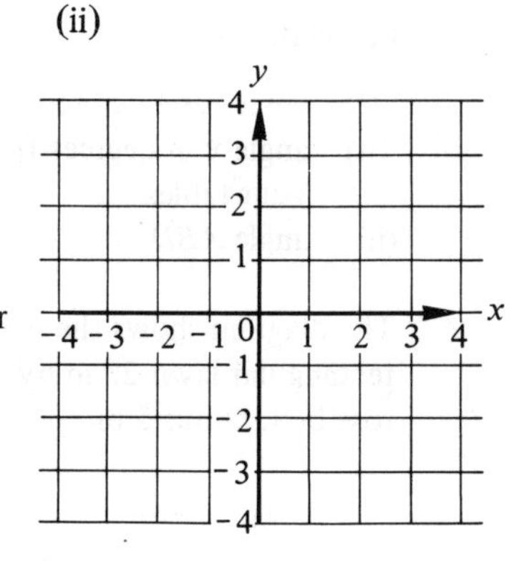

(a) Draw the graphs of $x = -2$ and $y = x$

(b) Write down the coordinates of their point of intersection.

(iii)

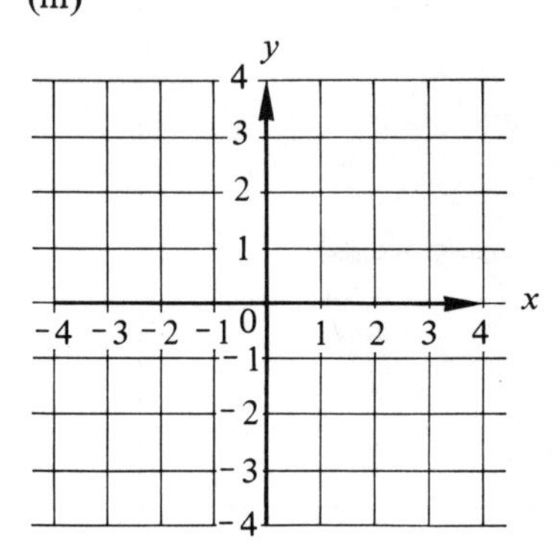

(a) Draw the graph of $x = -2$

(b) Shade and label the region R, where $x \geqslant -2$

(iv)

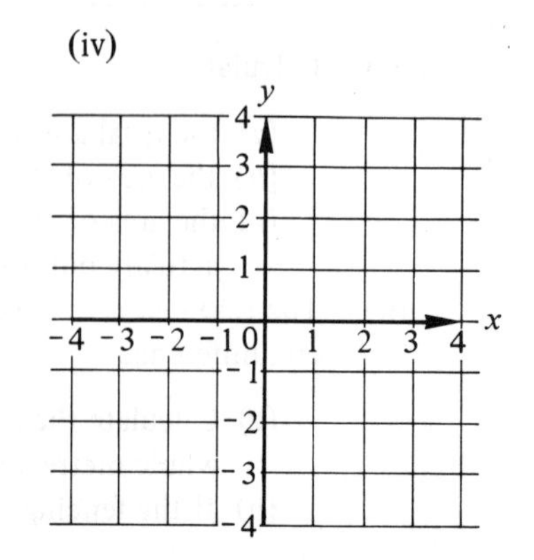

(a) Draw the graph of $y = 2$

(b) Shade and label the region Q, where $y \leqslant 2$

(v)

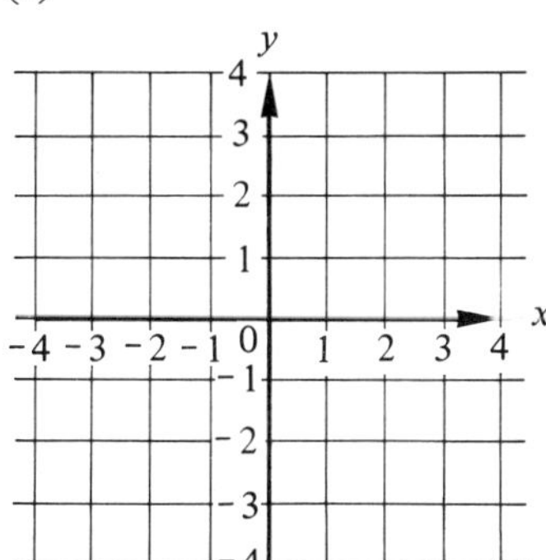

M is the region where $-3 < x < 1$.
N is the region where $-3 < y < -1$.
Shade and label the region: $M \cap N$

(vi)

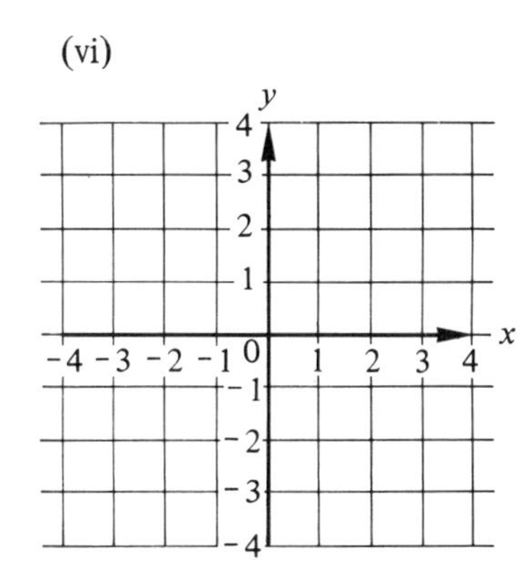

Draw the graphs of $y = x + 3$ and $x + y = 3$. Shade and label the region S where $y \geqslant 0$, $y \leqslant x + 3$ and $x + y \leqslant 3$

11 (Squared paper) The table shows the values of x for the equation: $y = x^2$

x	-4	-3	-2	-1	0	1	2	3	4
$y(x^2)$	16								

(i) Copy and complete the table of values for y; the first one has been done for you.

(ii) Draw x and y axes using the following scales:

x axis: −4 to +4, 1 cm = 1 unit
y axis: 0 to 16, 1 cm = 2 units

(iii) Plot the values from the table and join them up to give a smooth curve for the graph of $y = x^2$

(iv) Now draw on the graph the straight line whose equation is $y = 8$

(v) Write down the approximate values of x at the points of intersection of $y = x^2$ and $y = 8$

(vi) Draw on the graph a straight line from which you can read off the approximate values of the square root of 6 and write down these approximate values.

12 $A = \begin{pmatrix} 2 & -1 \\ -3 & 2 \end{pmatrix}$, $B = \begin{pmatrix} 3 & 2 \\ 4 & 0 \end{pmatrix}$, $C = \begin{pmatrix} -2 & 4 & 1 \\ 3 & -1 & 6 \end{pmatrix}$

(i) Evaluate $2C$
(ii) Evaluate $A + B$
(iii) Evaluate $B - A$
(iv) Evaluate BC
(v) Evaluate A^2
(vi) Why is it not possible to evaluate CB?

13 In the diagram the journey from A to B can be shown in vector form as $\begin{pmatrix} 4 \\ -2 \end{pmatrix}$. Copy the diagram and then follow the instructions underneath.

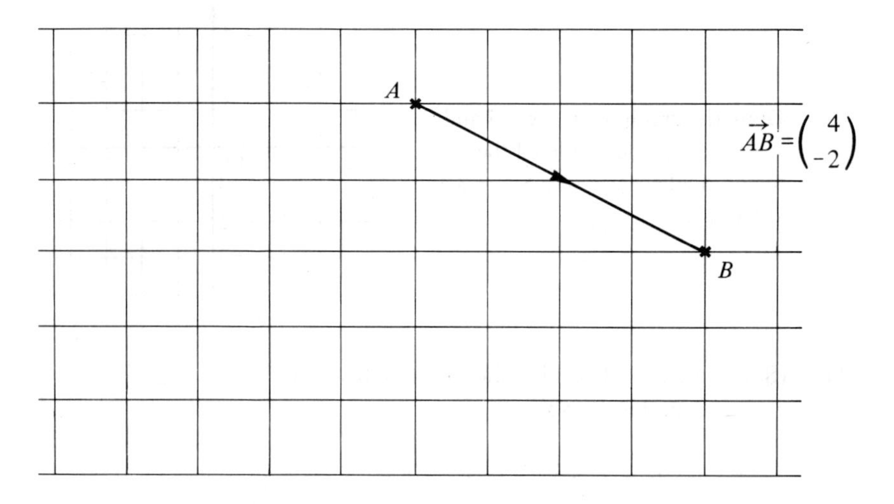

(i) Draw the vector: $\vec{BC} = \begin{pmatrix} -4 \\ -2 \end{pmatrix}$
(ii) Plot the point D, which would make $ABCD$ a rhombus.
(iii) Write down the vector $\vec{CD}$.
(iv) Write down the vector $\vec{AD}$.
(v) Write down the vector $\vec{DC}$.
(vi) Why does the vector $\vec{DC}$ = the vector $\vec{AB}$?

14 Copy the diagram which shows kite $ABCD$ and the point P (1,5)

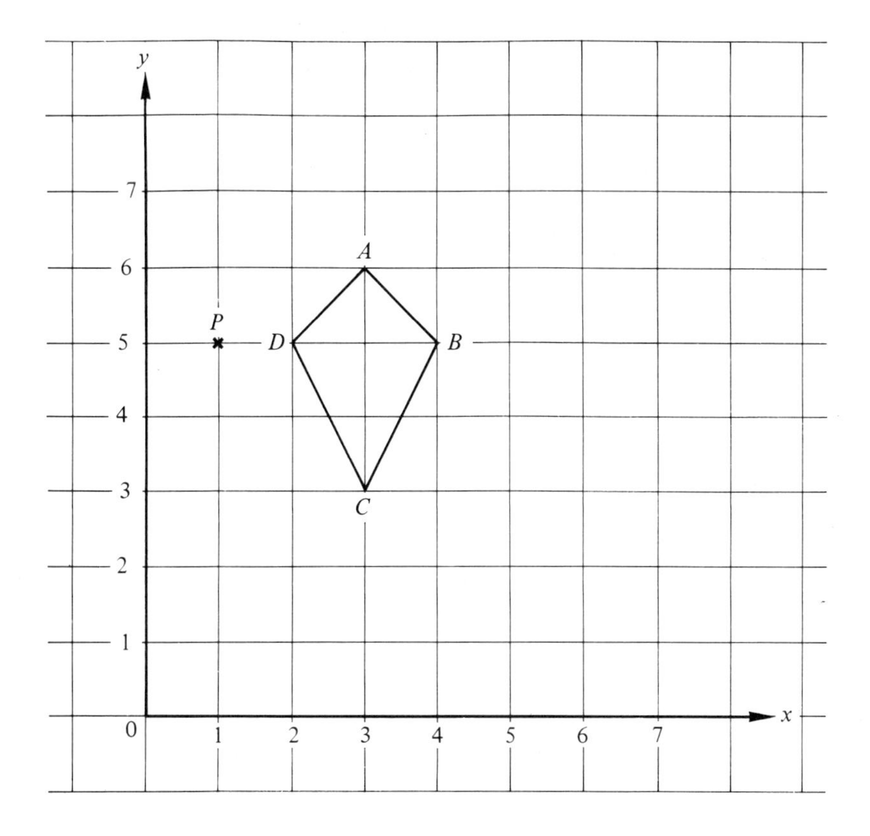

(i) Enlarge $ABCD$, Scale Factor +2, centre of enlargement P, to give $A'B'C'D'$ and write down the coordinates of the four vertices of the enlargement.

(ii) Write down the area of $ABCD$.

(iii) Write down the area of $A'B'C'D'$.

(iv) Complete and simplify the fraction: $\dfrac{\text{Area of } ABCD}{\text{Area of } A'B'C'D'}$

Set 2

1 (i) Evaluate $111_2 + 1010_2$ and leave your answer in base 2.
(ii) Evaluate $1101_2 + 1010_2$ and give your answer in base 10.
(iii) Evaluate $1001_2 - 101_2$ and leave your answer in base 2.
(iv) Evaluate $10\,111_2 - 1101_2$ and give your answer in base 10.
(v) Write $2^5 + 1$ as a binary number.
(vi) Write $8^3 + 5$ as a number in base 8.

2 (i) Find the cost of 15 litres of petrol at 35p a litre.
(ii) If a kilogram of sweets costs 96p, find the cost of 250 grams.
(iii) Calculate the perimeter of a rectangle whose dimensions are 7.8 cm by 4.9 cm.
(iv) Calculate the area of a square whose side is 0.9 m
(v) Rose trees cost 85p each. How much change should you get from a £20 note after buying 12 trees?

3 Solve the equations:

(i) $6x = 30$
(ii) $a + 15 = 0$
(iii) $4y = 16 - 6$
(iv) $3(x + 2) = 18$
(v) $2x^2 = 8$

4 The following marks were obtained from a test:

6, 3, 5, 4, 6, 3, 4, 5, 9, 7, 8, 6

(i) Arrange the marks in order, smallest first.
(ii) Write down the mode.
(iii) Calculate the median.
(iv) Calculate the mean.
(v) By how many would the total of the marks have to be increased to result in a mean of exactly 6?

5 (i) Write down 3785, correct to 2 significant figures.
(ii) Write down 0.0079 km, correct to the nearest metre.
(iii) Write down 1.469 correct to 2 decimal places.
(iv) Evaluate $(1.5)^2$ and give your answer correct to 1 decimal place.
(v) Evaluate: $\sqrt{0.04}$

6 Factorise:

(i) $2a + 4b$
(ii) $6 + 12x$
(iii) $ab + ac + ad$
(iv) $a^2 + a$
(v) $a^2 - b^2$

7

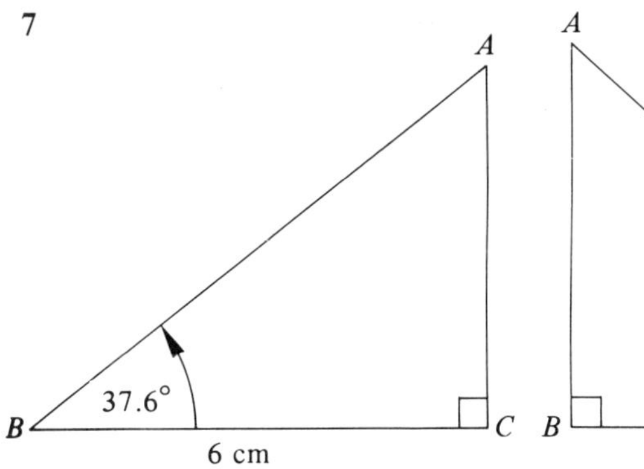

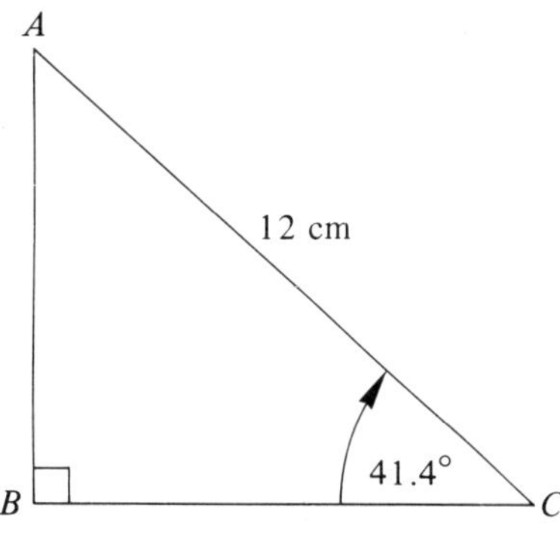

(i) Calculate: AC

(ii) Calculate: (*a*) AB
(*b*) BC

8 Given that $C = 2\pi r$,

(i) find C if $\pi = 3.14$ and $r = 2.5$
(ii) find C if $\pi = \frac{22}{7}$ and $r = 1\frac{3}{4}$
(iii) change the subject of the formula to r
(iv) find r if $\pi = 3.14$ and $C = 12.56$
(v) find r if $\pi = \frac{22}{7}$ and $C = 44$

9 For each part of this question you need to draw x and y axes on squared paper, scaled from -4 to $+4$ on both axes. Use 1 cm = 1 unit on both axes and follow the instructions for each of the six parts carefully. Label all your graphs clearly.

(i) (*a*) Draw the graphs of $x = -3$ and $y = 2$
(*b*) Write down the coordinates at the point of intersection
(ii) (*a*) Draw the graphs of $y = -x$ and $x = 3$
(*b*) Write down the coordinates at the point of intersection
(iii) (*a*) Draw the graph of $y = x$
(*b*) Shade the region R, where $y > x$
(iv) (*a*) Draw the graph of $x = -1$
(*b*) Shade the region S where $x < -1$
(v) (*a*) Draw the graphs of $y = 2$ and $y = -1$
(*b*) Shade the region T, where $-1 < y < 2$
(vi) (*a*) Draw the graphs of $x = -3$ and $x = -1$
(*b*) Shade the region V, where $-3 < x < -1$

10 $A = \begin{pmatrix} 3 & -1 \\ 2 & 3 \end{pmatrix}$, $B = \begin{pmatrix} 4 & -2 \\ 2 & -1 \end{pmatrix}$, $C = \begin{pmatrix} -3 & 4 & 2 \\ 2 & -1 & 4 \end{pmatrix}$

Evaluate:

(i) $3C$
(ii) $A + B$
(iii) $A - B$
(iv) BA
(v) $\frac{1}{4}B$

11 The journey from P to Q shown in the diagram may be written in vector form as: $\overrightarrow{PQ} = \begin{pmatrix} 2 \\ 4 \end{pmatrix}$

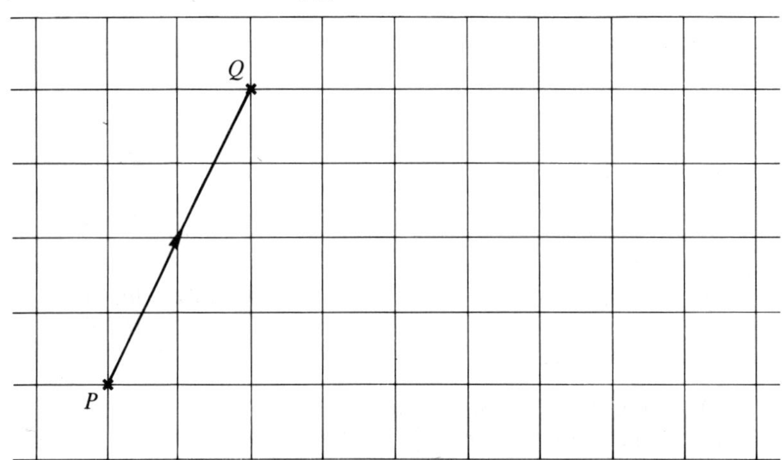

Copy the diagram and then:

(i) Draw in the vector: $\overrightarrow{QR} = \begin{pmatrix} 4 \\ 0 \end{pmatrix}$

(ii) The quadrilateral $PQRS$ is an isosceles trapezium.
Plot the point S which would complete this quadrilateral.

(iii) Draw the vector $\overrightarrow{RS}$ and complete: $\overrightarrow{RS} = \begin{pmatrix} \ \end{pmatrix}$

(iv) Draw and write down the vector $\overrightarrow{PS}$.

(v) Compare the vectors $\overrightarrow{QR}$ and $\overrightarrow{PS}$
Write down 2 relations between the lines QR and PS.

12 (i) Express $\frac{1}{4}$ as a decimal fraction.

(ii) Write down your answer to (i) as a fraction with a denominator of 100.

(iii) Write your answer to (ii) as a percentage.

(iv) Evaluate: $(\frac{1}{4})^3$

(v) Evaluate: $16^{\frac{1}{4}}$

13 (Squared paper) Draw x and y axes. Scale both axes from -4 to $+4$, using 1 cm = 1 unit on both axes.

(i) Plot the triangle whose vertices are (2,1), (4,1), (3,3).
Label it T.

(ii) Reflect triangle T in $y = 0$ to give triangle T_1. Label it clearly.

(iii) Rotate triangle T, $+90°$, centre O, the origin, to give triangle T_2. Label it.

(iv) Draw on your diagram the line whose equation is $y = -x$.
Reflect triangle T_2 in this line to give triangle T_3. Label it.

(v) Reflect triangle T_3 in $y = 0$, to give triangle T_4. Label it.

(vi) Describe a single transformation which would map triangle T on to triangle T_4.

14 (i) Express as percentage:

(*a*) $\frac{1}{5}$ (*b*) $\frac{1}{8}$ (*c*) $\frac{1}{16}$

(ii) Express as fractions in lowest terms:

(*a*) 15% (*b*) 80% (*c*) 45%

(iii) Express a score of 16 out of 25 as a percentage.

(iv) In a test out of 40 marks, a boy scored 60%.
How many marks did he obtain?

(v) An article was bought for £5.60 and sold for £7.00

Calculate:

(*a*) the profit
(*b*) the profit as a percentage of the cost price.

Set 3

1 (i) Evaluate $77_8 + 1_8$ and give your answer in base 2.
(ii) Evaluate 1001_8 as a base 10 number.
(iii) Evaluate $377_8 + 2_8 + 77_8$ and give your answer in base 10.
(iv) Evaluate $603_8 - 234_8$ and leave your answer in base 8.
(v) If $50_n = 30_{10}$, find the value of n.

2 (i) If seven similar articles cost £9.38, find the cost of one.
(ii) If $\frac{1}{2}$ a metre of wire costs $7\frac{1}{2}$p, find the cost of 20 m.
(iii) The area of a square is 1.69 m^2. Find the length of a side of the square.
(iv) A car travels 270 km in 6 hours. What is the average speed in km/h ?
(v) Find the cost of 12 bottles of wine if one bottle costs £1.85

3 Solve the equations:

(i) $2a = 1$
(ii) $3x + 9 = 0$
(iii) $\frac{1}{2}x = 12$
(iv) $3(x - 2) = 9$
(v) $x^2 + x^2 = 18$

4

(*a*) Score	(*b*) Frequency	(*c*) $S \times F$
1	15	
2	17	
3	18	
4	16	
5	19	
6	15	
Totals:		

The frequency table shows the results which were obtained when a die was rolled in a number of trials.

(i) Copy the table and add up column (*b*) to give the total number of trials.
(ii) Multiply column (*a*) by column (*b*) and put your answer in column (*c*).
(iii) Add up column (*c*) to give the total of all the scores.
(iv) Calculate the mean score.

5 (i) Write £347.85 correct to the nearest £10.
(ii) Write 534 890 correct to 3 significant figures.
(iii) Write 0.097 correct to 2 decimal places.
(iv) Write 1 000 000 in Standard Form notation.
(v) Write 0.000 001 in Standard Form notation.

6 Factorise:

(i) $3a + 3b$
(ii) $5x + 15y + 25z$
(iii) $x^3 + x^2 + x$
(iv) $a^2 - 9$
(v) $x^2 + 5x + 6$

7 The diagram shows triangle PQR, in which $PR = 9$ cm, $PQ = 10$ cm, $PX = 8$ cm and angle $PXR = 90°$.

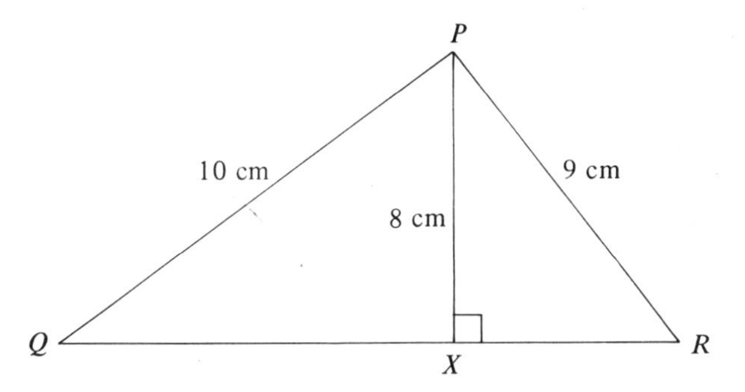

(i) Calculate the length QX in centimetres.
(ii) Calculate the length XR in centimetres.
(iii) Write as a decimal fraction the value of sin angle PQX.
(iv) Use your tables to find angle PQX in degrees.
(v) Calculate the value of cos angle RPX to 3 significant figures.
(vi) Use tables to find angle RPX in degrees.
(vii) Use your answers to (iv) and (vi) to calculate angle QPR.

8 (i) Given that $A = \left(\frac{a+b}{2}\right) \times h$

(*a*) calculate A when $a = 6, b = 8, h = 13$
(*b*) calculate A when $a = 8.3, b = 9.9, h = 9.1$

(ii) Given that $h = \frac{2A}{(a+b)}$

(*a*) calculate h when $A = 96, a = 7, b = 9$
(*b*) calculate h when $A = 0.9, a = 0.5, b = 0.7$

(iii) Given that $a = \left(\frac{2A}{h}\right) - b$

(*a*) calculate a when $A = 16, b = 3$ and $h = 4$
(*b*) calculate a when $A = 8.75, b = 1.25, h = 5$

9 For each part of this question, draw x and y axes on squared paper, scaled from −4 to +4 on both axes. Use 1 cm : 1 unit and then follow the instructions carefully.

(i) (*a*) Draw the graphs of $y = 1$ and $y = -3$
(*b*) Shade the region R where $-3 < y < 1$

(ii) (*a*) Draw the graph of $x = 3$
(*b*) Shade the region S where $0 < x < 3$

(iii) (*a*) Draw the graphs of $y = -x$ and $x + y = 2$
(*b*) Shade the region T where $y > -x$ and $x + y < 2$

(iv) M is the region where $-2 < x < -1$
N is the region where $1 < y < 2$
Shade clearly the region: $M \cap N$.

10 $A = \begin{pmatrix} 3 & 2 \\ 4 & 3 \end{pmatrix}$, $B = \begin{pmatrix} 3 & -2 \\ -4 & 3 \end{pmatrix}$, $C = \begin{pmatrix} -3 & 2 & 4 \\ -4 & 2 & 5 \end{pmatrix}$

(i) Evaluate $A - B$
(ii) Evaluate $B - A$
(iii) Evaluate AC
(iv) Evaluate AB
(v) Evaluate BA
(vi) Explain your answers to (iv) and (v).

11 The journey from A to B shown in the diagram may be written in vector notation as $\vec{AB} = \begin{pmatrix} -5 \\ -2 \end{pmatrix}$

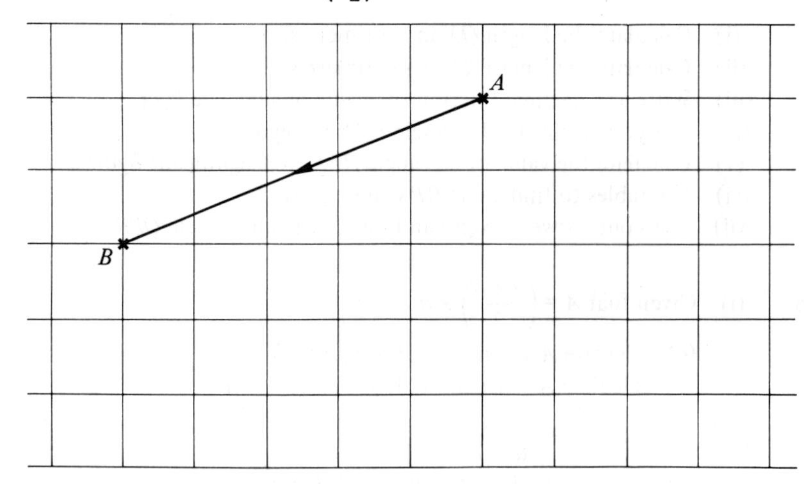

Copy the diagram on squared paper and then:

(i) draw the vector $\vec{BC} = \begin{pmatrix} 3 \\ -2 \end{pmatrix}$

(ii) draw the vector $\vec{AD} = \begin{pmatrix} 3 \\ -2 \end{pmatrix}$

(iii) join CD and name the quadrilateral $ABCD$
(iv) write down the vector $\overrightarrow{DC}$
(v) write down 2 facts about the opposite sides of $ABCD$ which a comparison of their vectors illustrates.

12 Evaluate:

(i) $\frac{1}{3} + \frac{1}{4}$

(ii) $\frac{1}{3} - \frac{1}{4}$

(iii) $(\frac{1}{3} - \frac{1}{4}) \times 24$

(iv) $\frac{1}{3} \times \frac{1}{4}$

(v) $(\frac{1}{3} \times \frac{1}{4}) \div \frac{1}{60}$

13 (Squared paper) Draw x and y axes. Scale both axes from -4 to $+4$, 1 cm : 1 unit.

(i) Plot the trapezium whose vertices are:
(−4,1), (−3,2), (−2,2), (−1,1), and label it T.

(ii) T_1 is the image of T under the translation vector $\begin{pmatrix} 5 \\ 2 \end{pmatrix}$
Draw and label T_1.

(iii) T_2 is the image of T under a rotation of $+90°$ about the origin. Draw and label T_2.

(iv) T_3 is the image of T under a rotation of $180°$ about the origin. Draw and label T_3.

(v) Describe a single transformation which would map T_3 on to T_1.

14 Mr. Worker's gross wages are £64.00 a week.
He gets a $12\frac{1}{2}\%$ increase, but he will, however, have to pay income tax on the increase at the rate of 30%.

(i) Calculate the $12\frac{1}{2}\%$ increase.
(ii) Calculate the tax on the increase.
(iii) Write down the value of the increase after tax has been deducted.
(iv) Calculate the actual increase after tax has been deducted as a percentage of his original wages.
(v) Calculate 70% of $12\frac{1}{2}\%$.

Set 4

1

& = {1,2,3,4,5,6,7,8,9,10,11,12,13}
A, B and C are sets as shown in the diagram.

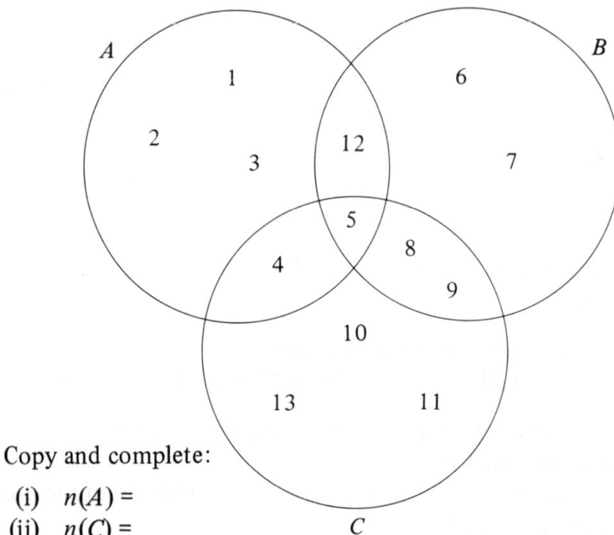

Copy and complete:

(i) $n(A) =$
(ii) $n(C) =$
(iii) $n(A \cup B) =$
(iv) $A \cap B = \{\quad\}$
(v) $A \cap B \cap C = \{\quad\}$
(vi) $n(A') =$

2 (i) If sunrise is at 0715 hours and sunset at 1645 hours, how many minutes are there between the two events?

(ii) A speed of 50 mph is approximately equal to a speed of 80 km/h. What would a speed of 10 mph be approximately equal to in km/h ?

(iii) To change degrees Fahrenheit to degrees Centigrade, the rule is: subtract 32, multiply by 5 and then divide by 9. Use this rule to change 68 °F to degrees Centigrade.

(iv) Write down the rule for changing degrees C to degrees F (it is the reverse process of the rule given in the previous question).

(v) Five miles is equivalent to 8.047 km. Write down 8.047 km in kilometres and metres.

3 Solve the equations:

(i) $x + x + x = 27$
(ii) $2x + 6 = 0$
(iii) $\frac{1}{4}x = 6$
(iv) $4(2x - 1) = 20$
(v) $\frac{1}{2}x^2 = 12.5$

4 The marks scored in a test were as follows:

9, 5, 6, 8, 7, 6, 4, 10, 8, 7, 6, 6, 7, 5, 4, 9, 9, 7, 6, 5

Draw up and complete a frequency table as shown below and then find:

(i) the mode
(ii) the median
(iii) the mean

(*a*) Score	(*b*) Tally marks	(*c*) Frequency	(*a*) x (*c*) $S \times F$
4			
5			
6			
7			
8			
9			
10			
	Totals:		

5 (i) Write as a fraction in its lowest terms, 0.25
(ii) Write $\frac{1}{8}$ as a decimal fraction.
(iii) Add $\frac{1}{2} + \frac{1}{8} + \frac{1}{8}$ and give your answer as a decimal fraction.
(iv) Add 0.003 and 0.017 and give your answer as an ordinary fraction in its lowest terms.
(v) Multiply 0.5 by 0.5 and express the result as a percentage.

6 Factorise:

(i) $3a + 9$
(ii) $14y + 21z$
(iii) $2a^2 + a$
(iv) $x(a + b) + y(a + b)$
(v) $x^2 + 6x + 8$

7

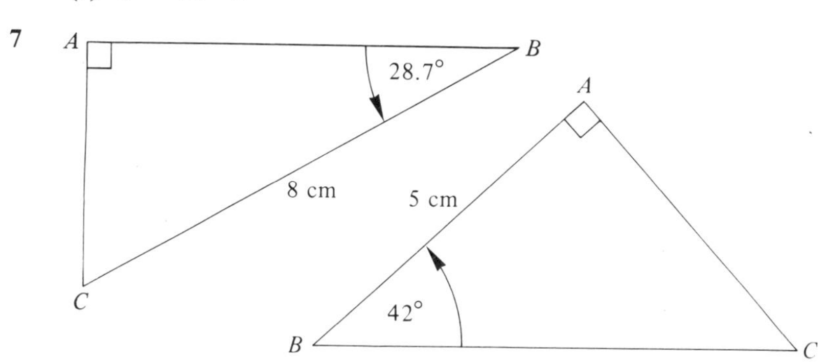

(i) Calculate: (*a*) AC
(*b*) AB

(ii) Calculate: (*a*) AC
(*b*) BC

8 (i) Given that: $2x + a = b$

(*a*) change the subject to x

(*b*) find x when $a = 25$ and $b = 8$

(*c*) find x when $a = -3$ and $b = -12$

(ii) Given that: $xy^2 = z$

(*a*) change the subject to x

(*b*) find x when $y = 3$ and $z = 63$

(*c*) change the subject to y

(*d*) find y when $z = 25$ and $x = 4$

9 For each part of this question you need to draw x and y axes on squared paper scaled from -4 to $+4$ on both axes. Use 1 cm : 1 unit. Follow the instructions carefully.

(i) (*a*) Draw the graphs of $y = x + 2$ and $y = x - 2$

(*b*) Shade the region R, where $y < x + 2$ and $y > x - 2$

(ii) (*a*) Draw the graphs of $x + y = 2$ and $x + y = -2$

(*b*) Shade the region S where $x + y < 2$ and $x + y > -2$

(iii) (*a*) Draw the graphs of $x = 2$ and $y = -3$

(*b*) M is the region where $0 < x < 2$

N is the region where $-3 < y < 0$

Shade the region $M \cap N$

(iv) (*a*) Draw the graphs of $y = x + 2$, $y = x$ and $y = -2$

(*b*) A is the region where $y < x + 2$ and $y > x$

B is the region where $-2 < y < 0$

Shade the region $A \cap B$

10 $P = \begin{pmatrix} 3 & 2 \\ -4 & 0 \end{pmatrix}$, $Q = \begin{pmatrix} -2 & 3 \\ 1 & -4 \end{pmatrix}$, $R = \begin{pmatrix} 1 & 0 & -2 \\ 3 & -1 & 4 \end{pmatrix}$

Evaluate:

(i) $P + Q$

(ii) $Q - P$

(iii) Q^2

(iv) PR

(v) PQR

11 The journey from A to B may be written in vector notation as

$$\vec{AB} = \begin{pmatrix} 3 \\ 3 \end{pmatrix}$$

Copy the diagram on squared paper and then:

(i) draw in the vector $\vec{BC} = \begin{pmatrix} 5 \\ -3 \end{pmatrix}$

(ii) draw in the vector $\vec{CD} = \begin{pmatrix} -5 \\ -3 \end{pmatrix}$

(iii) join DA and complete: $\vec{DA} = \begin{pmatrix} \ \end{pmatrix}$

(iv) name the quadrilateral $ABCD$.

(v) write down 2 facts about the sides of $ABCD$ which are shown by a comparison of their vectors.

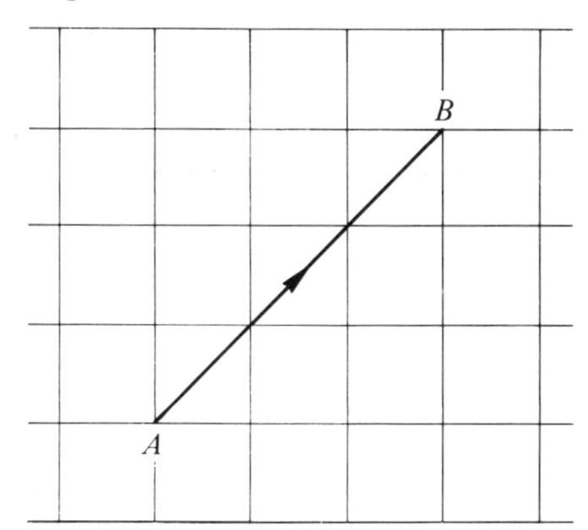

12 Evaluate:

(i) $\frac{2}{3} + \frac{1}{5}$

(ii) $\frac{2}{3} - \frac{1}{5}$

(iii) $(\frac{2}{3} - \frac{1}{5}) \times \frac{3}{7}$

(iv) $(\frac{2}{3})^4$

(v) $(\frac{2}{3} \times \frac{1}{5}) \div (\frac{2}{3} \times \frac{3}{10})$

13

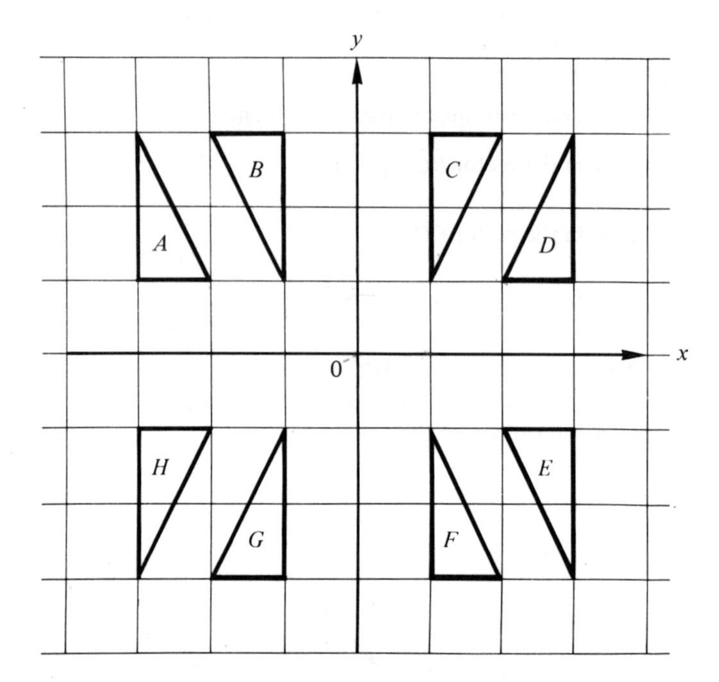

Describe fully the transformations which would map:

(i) *A* onto *E*
(ii) *B* onto *C*
(iii) *C* onto *G*
(iv) *G* onto *D*
(v) *F* onto *B*
(vi) *C* onto *H*.

14 (i) Write 15%:
(*a*) as a fraction with a denominator of 100
(*b*) as a decimal fraction
(*c*) as a fraction in its lowest terms.

(ii) 15% of the pupils in a school cycle to school.
If there are 540 pupils on roll, how many are cyclists?

(iii) A discount of 15% is allowed at a sale. The list price of a radio is £46.00
Calculate:
(*a*) the discount
(*b*) the sale price of the radio.

(iv) An article bought for £12.40 was sold again at a profit of 15%.
Calculate:
(*a*) the profit
(*b*) the selling price.

Set 5

1

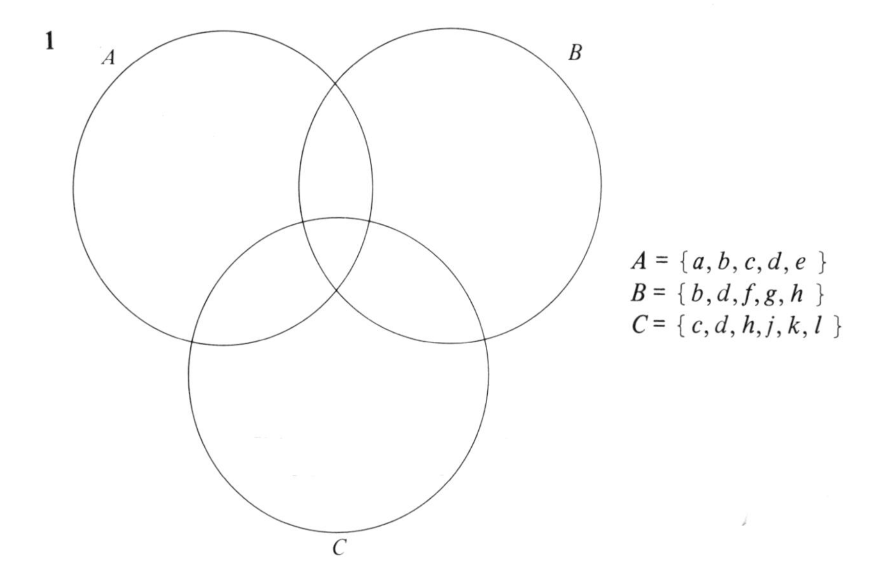

$A = \{a, b, c, d, e\}$
$B = \{b, d, f, g, h\}$
$C = \{c, d, h, j, k, l\}$

(i) Copy the diagram and then write the members of the sets A, B and C in the correct sections of the diagram.

(ii) Copy and complete:

(*a*) $A \cap B = \{\quad\}$
(*b*) $A \cap B \cap C = \{\quad\}$
(*c*) $n(B') =$

2 (i) A petrol tank holds 41 litres. How much would it cost to fill it if petrol costs 35p a litre?

(ii) On a journey of 432 km a car uses 36 litres of petrol. Calculate the petrol consumption of the car in kilometres per litre.

(iii) If a boy runs 100 m in 12 seconds, what is his speed in kilometres per hour.

(iv) A car travels a distance of 400 km at an average speed of 60 km/h. How long in hours and minutes did the journey take?

(v) If 5 miles is approximately 8.05 km, how many kilometres would 60 miles be equivalent to?

3 Solve the equations:

(i) $3y + 2y = 35$
(ii) $2.5x + 1.5x = 18$
(iii) $5(x + 2) = -5$
(iv) $\frac{3}{4}x - \frac{1}{4}x = 13$
(v) $(x + 4)(x - 3) = 0$

4 A group of 11 pupils took a Mathematics test and their marks out of 12 were as follows: 5,8,2,7,6,4,6,9,6,5,8

(i) Arrange the marks in order, smallest first.
(ii) Write down the mode.
(iii) Write down the median.
(iv) Calculate the mean.
(v) An absentee took the test the following day and a new mean was calculated, which was 6.5. How many marks did the absentee score?

5 (i) Approximate 209.8 correct to 3 significant figures.
(ii) Approximate 2.979 correct to 2 decimal places.
(iii) Write £36.48 correct to the nearest 10p.
(iv) Calculate $(0.4)^3$ and give your answer in Standard Form notation.
(v) A number x is approximated to 3 significant figures and is given as 276. Write down the lowest possible value of x before approximation.

6 Factorise:

(i) $6a + 9b$
(ii) $a^3 + a^2$
(iii) $12x + 24y + 60z$
(iv) $\pi A^2 + \pi B^2$
(v) $a^2 - 25$

7

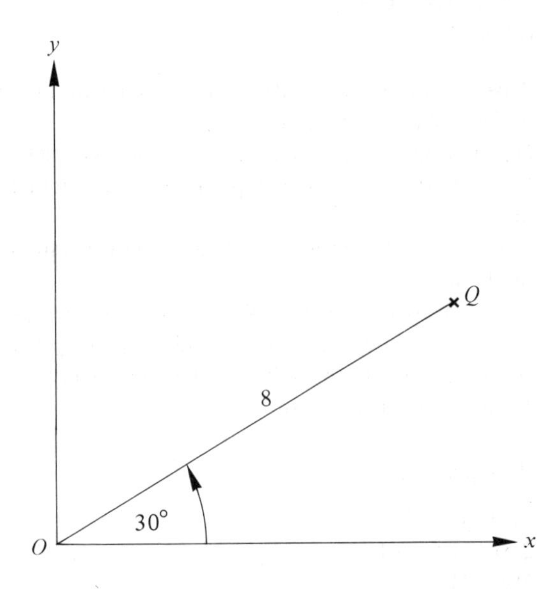

Calculate the x and y coordinates of the point Q.

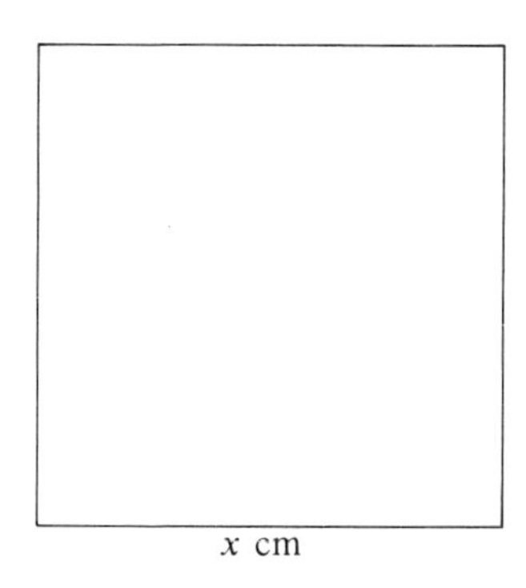

The square in the diagram has a side of length x cm.

(i) Write down a formula in terms of x for:
 (*a*) the perimeter : $P =$
 (*b*) the area : $A =$
 (*c*) the length of a diagonal: $D =$

(ii) If P, the perimeter is 22 cm, calculate:
 (*a*) the area, A
 (*b*) the length of a diagonal, D.

9 You will need to draw x and y axes for each part of this question, scaled from -4 to $+4$, with 1 cm : 1 unit on both axes.

(i) (*a*) Draw the graphs of $y = 2x + 2$, $y = -2$ and $x = 1$
 (*b*) Shade the region S where $y < 2x + 2$, $y > -2$ and $x < 1$
 (*c*) Write down the area of the region S

(ii) (*a*) Draw the graphs of $y = 2$ and $x = -3$
 (*b*) P is the region where $-3 < x < 0$
 Q is the region where $0 < y < 2$
 Shade the region $P \cap Q$

(iii) (*a*) Draw the graphs of $x + y = 3$ and $y = x - 3$
 (*b*) Shade the region T where $x \geqslant 0$, $x + y \leqslant 3$ and $y \geqslant x - 3$

(iv) (*a*) Draw the graphs of $y = 3$, $y = x$ and $y = x + 1$
 (*b*) Shade the region M where $y \leqslant 3$, $y \geqslant x$, $y \leqslant x + 1$ and $y \geqslant 0$
 (*c*) Write down the area of M.

10 The diagram shows two set-squares ABC and XYZ. The small one has been placed on top of the large one.

$AB = 12$ cm, $XY = 8$ cm.
angle ABC = angle $XYZ = 30°$.

Calculate:

(i) AC
(ii) BC
(iii) XZ
(iv) YZ
(v) area of ABC
(vi) area of XYZ
(vii) the area of ABC which is visible when XYZ is placed on top of ABC
(viii) ratio of area XYZ : area ABC

11

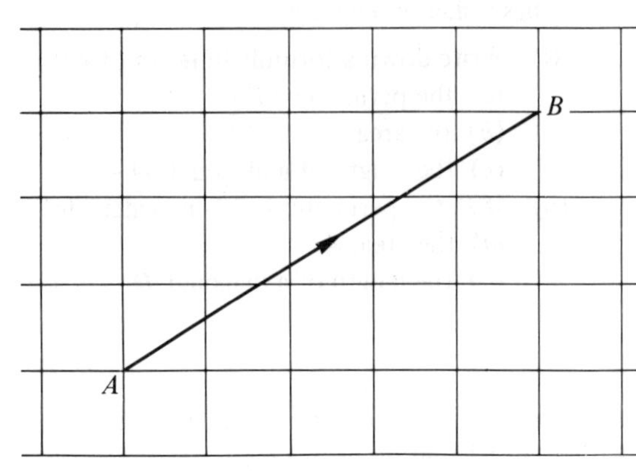

The journey from A to B shown in the diagram may be written in vector notation as:

$\vec{AB} = \begin{pmatrix} 5 \\ 3 \end{pmatrix}$

Copy the diagram on squared paper.

(i) Draw the vector $\vec{BC} = \begin{pmatrix} 5 \\ -3 \end{pmatrix}$

(ii) Plot the point D which would make $ABCD$ a rhombus.

(iii) Write down the vector $\vec{DC}$.

(iv) Write down the vector $\vec{BD}$.

(v) Write down the vector $\vec{AC}$.

(vi) Write down 1 fact about the diagonals of a rhombus which your answers to (iv) and (v) illustrate.

12 Evaluate:

(i) $\frac{3}{4} + \frac{1}{5}$

(ii) $\frac{3}{4} - \frac{1}{5}$

(iii) $\frac{1}{2} \times \frac{2}{3} \times \frac{3}{4} \times \frac{4}{5}$

(iv) $(\frac{3}{4} \times \frac{1}{5}) \div (\frac{3}{8} \times \frac{2}{5})$

(v) $(\frac{3}{4}$ of $76) + (\frac{1}{5}$ of $12\frac{1}{2})$

13 (Squared paper) Draw x and y axes and scale both of them from -4 to $+4$, using 1 cm : 1 unit. Be sure to label all triangles.

(i) Plot the triangle T, whose vertices are: (1,1), (2,1), (1,2)

(ii) Enlarge triangle T, Scale Factor +2, centre of enlargement O, to give T_1.

(iii) Enlarge T, Scale Factor -2, centre of enlargement O, to give T_2.

(iv) Reflect T_2 in the line $x = 0$, to give T_3.

(v) Enlarge T_3, Scale Factor $-\frac{1}{2}$, centre of enlargement O, to give T_4.

(vi) Describe a single transformation which would map T_1 onto T_2.

14 (i) Write as fractions:

(*a*) 10% (*b*) 5% (*c*) $2\frac{1}{2}$% (*d*) $17\frac{1}{2}$%

(ii) Write as percentages:

(*a*) $\frac{1}{4}$ (*b*) $\frac{1}{8}$ (*c*) $\frac{3}{8}$ (*d*) $\frac{5}{8}$

(iii) 55% of the pupils at a mixed school are girls. If there are 760 pupils on roll, how many boys are there in school?

(iv) In a certain year, house values appreciated by 18%. Calculate the increase in value of a house valued at £14 000 at the beginning of the year.

(v) An article was bought for £27.00 in a sale after 10% discount had been deducted. What was the original price of the article?

Set 6

1 (i) In a form of 28 boys, some play soccer, some play rugby and some play both. If 20 play soccer and 15 play rugby, copy and complete the Venn diagram, showing clearly how many boys play both games.

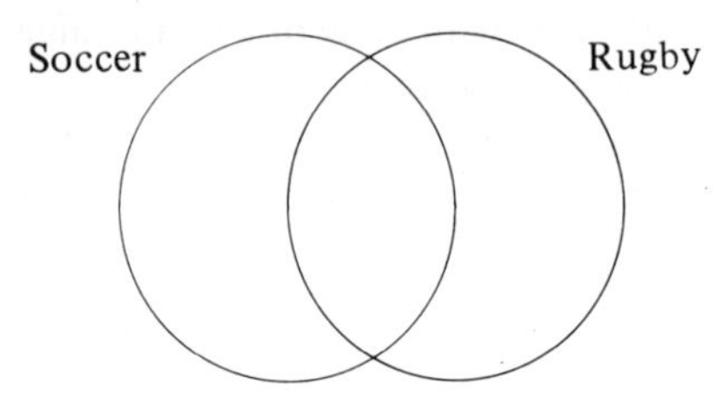

(ii) In a group of girls, 23 girls played netball and 19 played tennis, but 12 girls played both games.
(*a*) Copy and complete the diagram to show these sets.
(*b*) Write down the number of girls in the group.

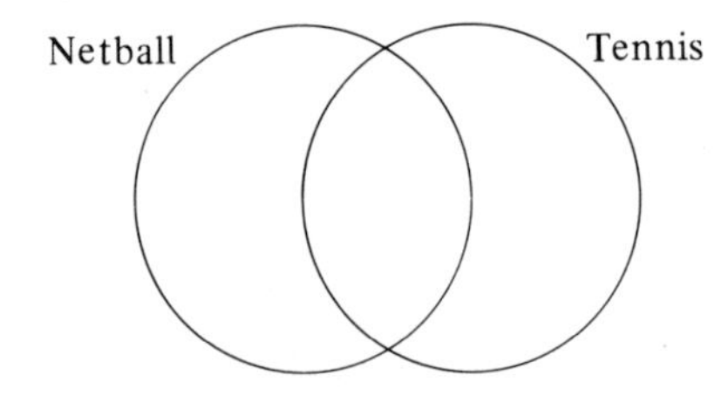

2 (i) The area of a rectangle is 24 cm^2 and its length is 3.2 cm. Calculate the width of the rectangle.
(ii) The area of a square is 75 cm^2. Use tables to find the length of a side of the square.
(iii) The perimeter of a square is 30 m. Calculate the area of the square.
(iv) Calculate the volume of a cube if the edge is 0.3 m, giving your answer in cubic metres.
(v) If the volume of a cube is 0.008 m^3, find the length of the edge of the cube in metres.

3 Solve the equations:

(i) $8a - 2a = 21$
(ii) $0.5x = 25$
(iii) $2b + 3b - b = 21$
(iv) $\frac{1}{2}(x + 4) = 6$
(v) $(a + 4)(2a - 6) = 0$

4 The frequency table shows the total scores which were obtained when 2 dice were rolled in a number of trials.

(*a*) Score	(*b*) Frequency	(*c*) $S \times F$
2	3	
3	6	
4	9	
5	11	
6	14	
7	17	
8	15	
9	9	
10	7	
11	5	
12	4	
Totals		

(i) Make a neat copy of the table and add up column (*b*) to give the number of trials.
(ii) Multiply column (*a*) by column (*b*) and put your answer in column (*c*).
(iii) Add up column (*c*) to give the total of all scores.
(iv) Calculate the mean score.
(v) From these trials, the probability of scoring a total of 7 is $\frac{17}{100}$. What is the theoretical probability of a total score of 7 ?

5 (i) Evaluate $2^5 \times 2^3 \div 2^6$
(ii) Evaluate $x^4 \times x^3 \times x^{-2}$
(iii) Write the number 2 500 000 in Standard Form notation
(iv) Evaluate $(4 \times 10^6) + (4 \times 10^2)$
(v) If the number 0.000 008 2 is written in Standard Form notation as 8.2×10^n, write down the value of n.

6 Factorise:
(i) $by + 4bz$
(ii) $10a + 20b - 25c$
(iii) $2x^2 + 2x$
(iv) $9a^2 - 16b^2$
(v) $a^2 + 8a + 15$

7 The diagram shows triangle ABC, an isosceles triangle. $AB = 12.5$ and $BC = 20$ m.

Calculate:
(i) BD in metres
(ii) AD in metres (correct to 2 significant figures)
(iii) tan angle ABC as a decimal fraction
(iv) angle ABC in degrees
(v) angle BAC in degrees
(vi) area of triangle ABC.

A
12.5 m
B
D
C
20 m

8 Given that $A = 2\pi rh$,

(i) calculate A when $\pi = 3.14$, $r = 5$ and $h = 4$
(ii) calculate A when $\pi = \frac{22}{7}$, $r = 3\frac{1}{2}$ and $h = 11$
(iii) change the subject of the formula to r
(iv) calculate r when $\pi = 3.14$, $h = 5$ and $A = 94.2$
(v) calculate r when $\pi = \frac{22}{7}$, $h = 7$ and $A = 176$
(vi) when would you use this formula?

9 You will need to draw x and y axes for each part of this question, scaled from -4 to $+4$ and using 1 cm : 1 unit.

(i) M is the region where $2 < x < 3$
N is the region where $-3 < y < -1$
Shade the region $M \cap N$

(ii) (*a*) Draw the graphs of $y = x + 2$, $y = x - 2$
$x + y = 2$ and $x + y = -2$
(*b*) Shade the region R where $y \leqslant x + 2$, $y \geqslant x - 2$
$x + y \geqslant -2$ and $x + y \leqslant 2$

(iii) (*a*) Draw the graphs of $y = \frac{1}{2}x$, $y = -\frac{1}{2}x$ and $y = -2$
(*b*) Shade the region S where $y \leqslant \frac{1}{2}x$, $y \leqslant -\frac{1}{2}x$
and $y \geqslant -2$
(*c*) Write down the area of S

(iv) (*a*) Draw the graphs of $y = x - 1$, $y = -x - 1$
$y = 3x + 3$ and $y = -3x + 3$
(*b*) Shade the region K where $y \geqslant x - 1$, $y \geqslant -x - 1$
$y \leqslant 3x + 3$ and $y \leqslant -3x + 3$
(*c*) Why is K a suitable letter to describe the region?

10 $A = \begin{pmatrix} 2 & -3 \\ 4 & 0 \end{pmatrix}$, $B = \begin{pmatrix} 4 & -2 \\ -1 & 3 \end{pmatrix}$, $C = \begin{pmatrix} 1 & 0 \\ 0 & 1 \end{pmatrix}$

(i) Evaluate:

(*a*) $A + B$
(*b*) $A - B$
(*c*) $B - A$
(*d*) AC
(*e*) ABC

(ii) Write down the special name for matrix C.

11 The journey from P to Q shown in the diagram may be described in vector notation as: $\vec{PQ} = \begin{pmatrix} 2 \\ 2 \end{pmatrix}$

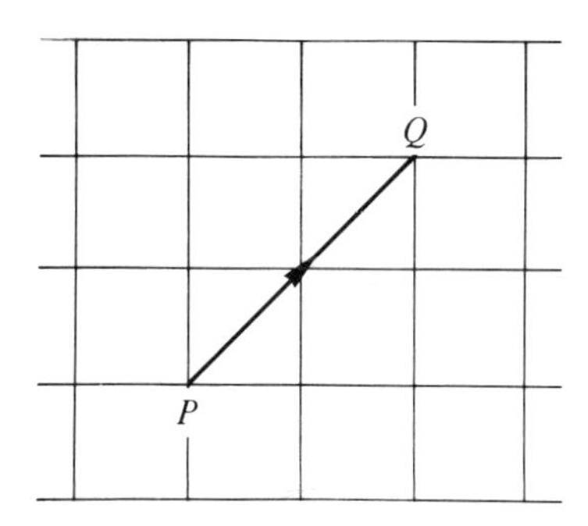

Copy the diagram on squared paper.

(i) Draw in the vector $\vec{QR} = \begin{pmatrix} -2 \\ 2 \end{pmatrix}$

(ii) Draw in the vector $\vec{RS} = \begin{pmatrix} 6 \\ -2 \end{pmatrix}$

(iii) Join PS to give the quadrilateral $PQRS$.
Write down the name of the quadrilateral.

(iv) Write down the vector $\vec{SP}$.

(v) Evaluate $\vec{QR} + \vec{RS}$.

(vi) Write down the letters of the vector which is equivalent to $\vec{QR} + \vec{RS}$.

12 Evaluate:

(i) $\frac{5}{12} + \frac{3}{8}$

(ii) $\frac{5}{12} - \frac{3}{8}$

(iii) $\frac{5}{12} \div \frac{3}{8}$

(iv) $\frac{3}{8} \div \frac{5}{12}$

(v) $\left(\frac{5}{12} \div \frac{3}{8}\right) \times \left(\frac{3}{8} \div \frac{5}{2}\right)$

13 (Squared paper) The following table, partly completed, shows values of x and y for the graph of $y = 2x^2$

x	−3	−2	−1	0	1	2	3
y	18	8	2	0			

$y = 2x^2$

(i) Copy and complete the table.
(ii) Draw x and y axes.
(iii) Scale the x axis from −3 to +3, using 2 cm : 1 unit.
Scale the y axis from 0 to 18, using 1 cm : 2 units.
Plot the values in the table and join up the points to give a smooth curve for the graph of $y = 2x^2$
(iii) Draw the line whose equation is $y = 6$ and from your graph obtain the approximate values of x which would solve the equation: $2x^2 = 6$
(iv) Draw the line whose equation is $y = 14$ and from your graph read off the approximate values of x which would satisfy the equation $2x^2 = 14$

14 A girl decided to deposit £80 in the Safe Savings Bank, which pays interest on savings at the rate of 10% per annum.

(i) Calculate the interest for 1 completed year.

This interest is added to her £80 at the beginning of year 2 and at the end of year 2 she receives 10% interest on the amount she had in at the beginning.

(ii) How much has she altogether at the beginning of year 2 ? (this is called the amount).
(iii) Calculate the interest she gets on this amount at the end of the year 2.
(iv) How much has she altogether at the beginning of year 3 ?
(v) How much interest will she receive at the end of year 3 ?
(vi) What is the total amount of her savings, plus interest at the end of year 3 ?

Set 7

1

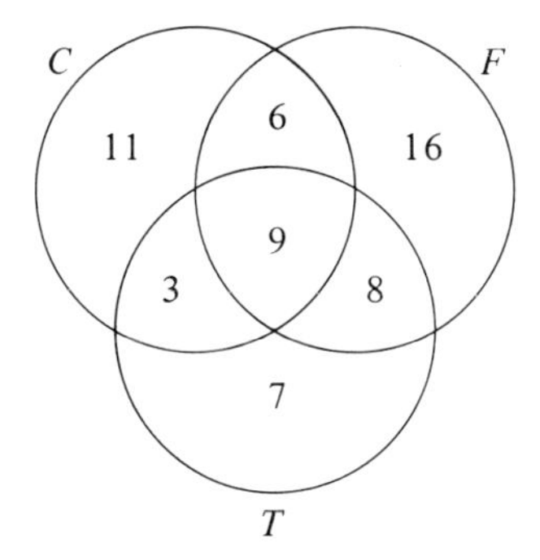

The diagram shows the games played by a group of boys:

C = { the cricket set }
F = { the football set }
T = { the tennis set }

Write down:

(i) the number of boys who played football
(ii) how many boys did not play tennis
(iii) the number of boys who played both cricket and football
(iv) how many boys did not play cricket
(v) the total number of boys in the group
(vi) the percentage of boys who played both tennis and cricket.

2 (i) To change hectares to acres, the rule is multiply by 2.47
Change 12 hectares to acres.

(ii) To change acres to hectares, multiply by 0.4047
Change 1000 acres to hectares.

(iii) The unit of metric land measure is the are and

$$1 \text{ are} = 100 \text{ m}^2$$

A rectangular plot of land measures 80 m by 55 m.
Calculate its area in ares.

(iv) 1 hectare = 100 ares and 1 are = 100 m^2.
How many square metres will there be in 10.5 hectares ?

(v) A plot of land covered 1 000 000 m^2.
How many hectares would this be equivalent to ?

3 Solve the equations:

(i) $3a + 4a = 49$
(ii) $2.5x + 3.5x = 27$
(iii) $3(2x + 1) + x = 38$
(iv) $x^2 + x^2 + x^2 = 48$
(v) $x^2 - 9 = 0$

4 In nine completed innings, a batsman made the following scores:

23, 13, 38, 0, 58, 17, 82, 49, 104

(i) Calculate his batting average (mean score).
(ii) After his next innings, his average was exactly 45. How many runs did he score in that innings?

5 (i) Write down $\frac{1}{20}$ as a decimal fraction.

(ii) Write down $\frac{1}{20}$ as a percentage.

(iii) Write down $\frac{2}{3}$ as a decimal fraction, correct to 2 significant figures.

(iv) Write down $\frac{2}{3}$ as a percentage, giving your answer as a mixed number.

(v) Evaluate $\frac{1}{8} + \frac{1}{8} + \frac{1}{8} + \frac{1}{4}$ and give your answer as a decimal fraction.

6 Factorise:

(i) $4x + 8y + 24z$
(ii) $12a - 18b - 24c$
(iii) $4a^2 + 2a$
(iv) $x^2 - 1$
(v) $x^2 - 3x - 10$

7

(i) Calculate angle ABC

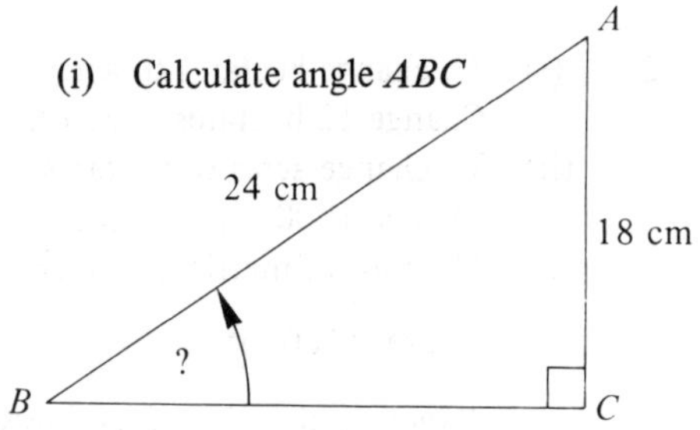

(ii) Calculate angle BAC

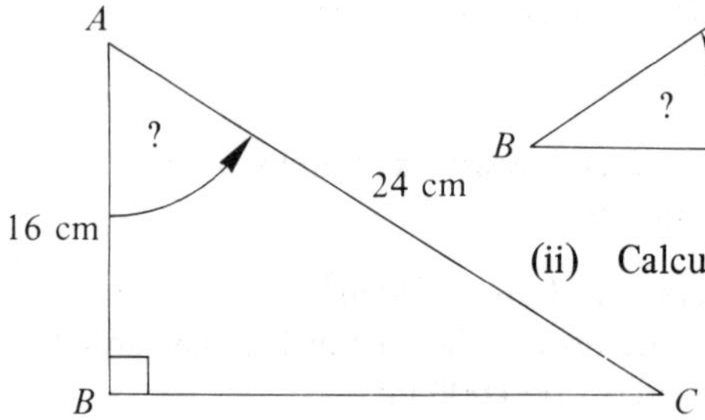

(iii) Calculate angle ABC

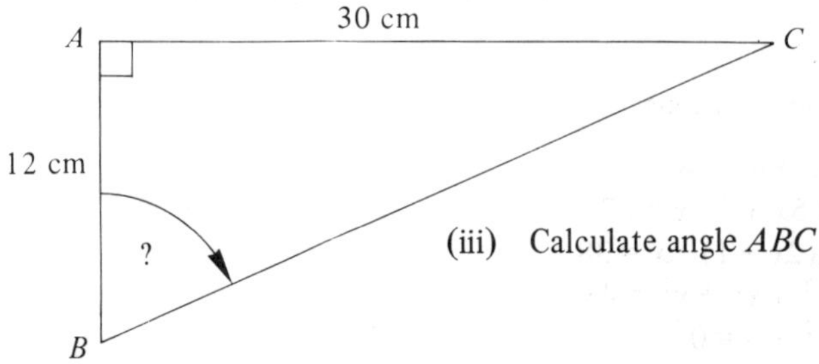

8

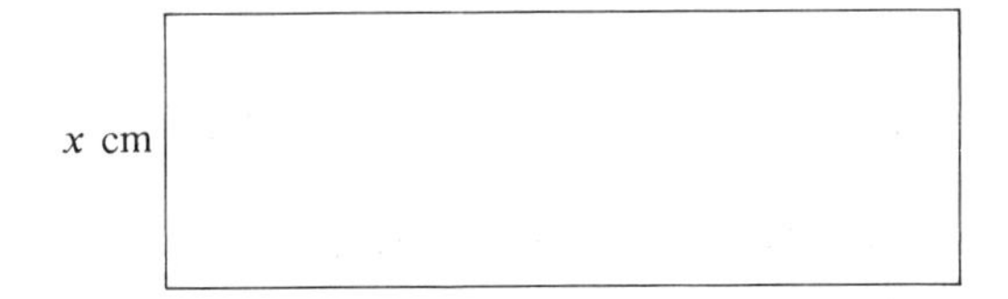

The diagram shows a rectangle whose width is x cm.
The length of the rectangle is 3 times the width.

(i) (*a*) Write down an expression for the length.
(*b*) Write down a formula for P, the perimeter.
(*c*) Write down a formula for A, the area.
(*d*) Write down a formula for D, the length of a diagonal.

(ii) If the perimeter, $P = 20$ cm, calculate:

(*a*) A, the area of the rectangle.
(*b*) D, the length of a diagonal.

9 The diagram shows a circle, centre O, radius 2.3 cm, OX, inscribed in an equilateral triangle of side 8 cm.
X is the mid point of BC
Y is the mid point of AB

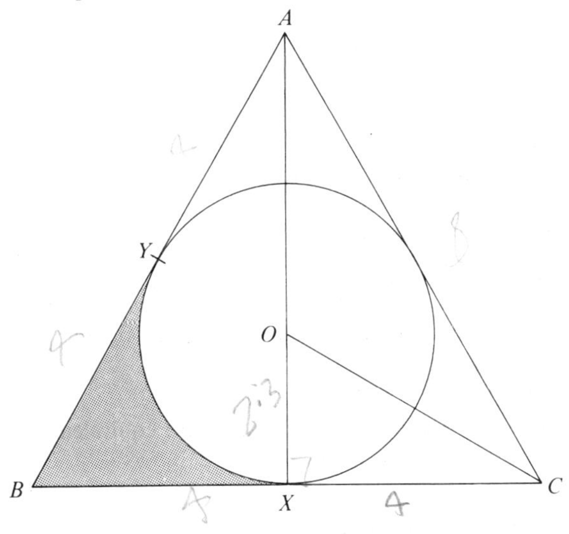

(i) Write down the size of angle OCX.
(ii) Use the sine ratio of angle OCX to calculate OC.
(iii) Using your answer to (ii), write down the length AX.
(iv) Calculate the area of the circle, correct to 3 significant figures ($\pi = 3$).
(v) Calculate the area of the triangle ABC.
(vi) Calculate the shaded area BXY.

10 A school shop sold:

5 pencils, 4 rulers and 2 rubbers on Monday:
3 pencils, 6 rulers and 4 rubbers on Tuesday:
4 pencils and 6 rubbers on Wednesday:
6 pencils on Thursday:
3 pencils and 5 rulers on Friday.

(i) Show these sales as a 5 by 3 matrix: call it A.

Pencils are 9p, rulers 11p and rubbers are 8½p each.

(ii) Show these prices as a column matrix: call it B.
(iii) Evaluate AB and find the total of the sales.

11 The journey from P to Q shown in the diagram may be written in vector notation as: $\overrightarrow{PQ} = \begin{pmatrix} 4 \\ 4 \end{pmatrix}$

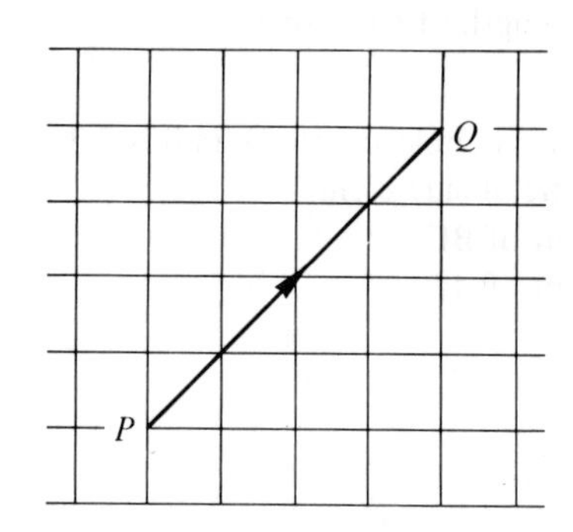

Copy the diagram on squared paper.

(i) Draw in the vectors:
$\overrightarrow{QR} = \begin{pmatrix} 3 \\ 0 \end{pmatrix}$, $\overrightarrow{RS} = \begin{pmatrix} 4 \\ -4 \end{pmatrix}$, $\overrightarrow{ST} = \begin{pmatrix} -4 \\ 2 \end{pmatrix}$, $\overrightarrow{TU} = \begin{pmatrix} -3 \\ 0 \end{pmatrix}$
(ii) Join PU and write down the name of the figure $PQRSTU$.
(iii) Write down the vector $\overrightarrow{UP}$.
(iv) Evaluate: $\overrightarrow{PQ} + \overrightarrow{QR}$
(v) Draw on your diagram from P a vector equivalent to $\overrightarrow{PQ} + \overrightarrow{QR}$ and write down its name.

12 (i) Express in lowest terms: $\frac{62}{93}$
(ii) Evaluate:

(*a*) $\frac{7}{12} + \frac{5}{8}$

(*b*) $\frac{7}{12} - \frac{5}{8}$

(*c*) $\frac{7}{12} \div \frac{5}{8}$

(*d*) $\frac{7}{12}$ of $144 + \frac{5}{8}$ of 72.

13 (Squared paper) Draw x and y axes, scaled from -4 to $+4$, using 1 cm : 1 unit. Be sure to label all transformations clearly.

(i) Plot the kite whose vertices are: (2,0), (4,1), (4,2), (3,2)
Label the kite K.

(ii) Draw in the line whose equation is $y = x$ and reflect K in $y = x$ to give K_1.

(iii) K_2 is the image of K under a rotation of $+90°$ about the origin. Draw in K_2.

(iv) Draw in the line $y = -x$. K_3 is the image of K_1 under a reflection in $y = -x$. Draw in K_3.

(v) K_4 is the image of K under a translation vector $\begin{pmatrix} -4 \\ -4 \end{pmatrix}$
Draw in K_4.

(vi) Describe fully a single transformation which would map K_3 on to K_4.

14 A boy gets the following marks in three tests:

English: 28 out of 40
Science: 17 out of 25
Maths: 57 out of 75

(i) Calculate each mark as a percentage and arrange the results in order, highest first.

(ii) Calculate his mean percentage mark in the tests, giving your answer correct to 1 decimal place.

(iii) After a fourth test in French, his mean percentage mark was exactly 74%. How many marks did he score in the French test if it was marked out of 60?

Set 8

1 (i) Find the value of 1111_8 as a base 10 number.
(ii) Write 65_{10} as a number in base 8.
(iii) Add $111_2 + 111_2 + 1111_2$ and give your answer in base 10.
(iv) Evaluate $100_8 - 11_8$ and give your answer in base 2.
(v) If $123_x = 66_{10}$ find the value of x.

2 (i) After spending amounts of £1.76, £3.08, 59p and £6.82, how much change should you have from a £20.00 note ?
(ii) If 1 k of cheese costs £1.60, how much would 200 g cost ?
(iii) A boy saved 40p a week for 1 year. Find his total savings in a year of 52 weeks.
(iv) A girl saved 5p a day for the first three months of a year which was not a leap year. Find her total savings.
(v) A boy planned to save weekly for a radio costing £39.00. How much should he save per week if his plan was to be spread over 1 year ?

3 Solve the equations:

(i) $5a - 3a = 7$
(ii) $6x + 2x + x = 45$
(iii) $x^2 = 1$
(iv) $3(4a - 2) = 18$
(v) $a^2 - 9 = 0$

4 In cricket, a bowler's average is calculated by dividing the number of runs scored from his bowling by the number of wickets he takes, i.e. $\frac{\text{runs}}{\text{wickets}}$. The table shows statistics for three bowlers.

	Runs	Wickets
I.N. Swinger	1204	40
L.A. Spinner	231	15
M. Pace	234	12

(i) Calculate the bowling averages of the three bowlers.
(ii) Arrange the averages in order, the lowest first.

5 (i) Evaluate $10^4 + 10$
(ii) Approximate 299.7 correct to 3 significant figures.
(iii) If the number 29 000 000 is written in Standard Form notation as $A \times 10^7$ and $1 \leqslant A < 10$, write down the value of A.
(iv) If the number 0.000 009 9 is written in Standard Form notation as 9.9×10^n, write down the value of n.
(v) Simplify $(4.0 \times 10^3) \times (2.0 \times 10^4)$, leaving your answer in Standard Form notation.

6 Factorise:

(i) $13a + 26b + 52c$
(ii) $2x^2 + 4y^2 + 10z^2$
(iii) $m(a + b) + n(a + b)$
(iv) $\frac{1}{2}a + \frac{1}{2}b + 2c$
(v) $3x^2 - 3y^2$ (factorise fully)

7

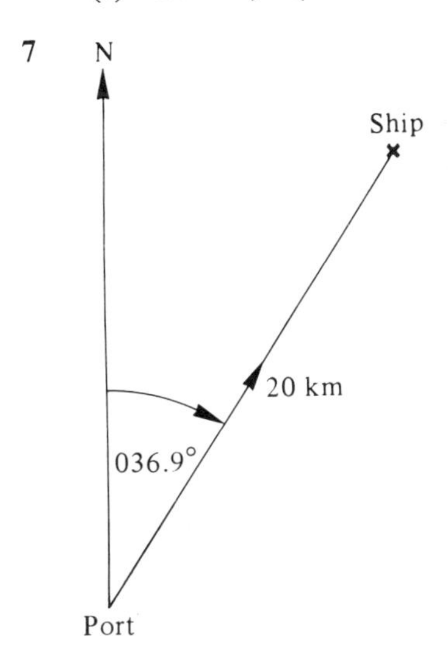

The diagram shows the position of a ship after travelling for 20 km on a bearing of 036.9°.

Calculate:

(i) the bearing of the port from the ship.
(ii) the northerly component of the journey.
(iii) The easterly component of the journey.

8 (i) Given that $C = \pi D$

(a) calculate C when $\pi = 3.14$ and $D = 0.5$
(b) change the subject to D
(c) calculate D when $\pi = \frac{22}{7}$ and $C = 66$

(ii) Given that $v = u + ft$

(a) change the subject to u
(b) calculate u when $v = 25, f = 6$ and $t = -2$

9

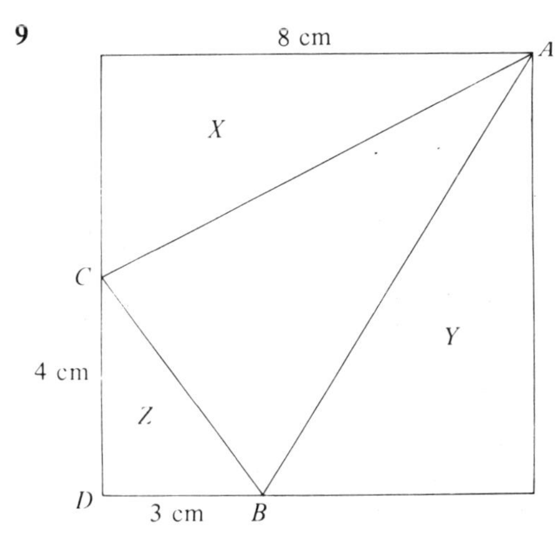

The diagram shows triangle ABC inside a square of side 8 cm.
$CD = 4$ cm and $DB = 3$ cm.

(i) By calculating the areas of the triangles marked X, Y and Z, find the area of triangle ABC.

(ii) Calculate:

(a) BC
(b) AB
(c) AC

10 (i) Evaluate:

(*a*) $\begin{pmatrix} 6 & 1 \\ 0 & 2 \end{pmatrix} \begin{pmatrix} 3 \\ 4 \end{pmatrix}$

(*b*) $\begin{pmatrix} 3 & 2 & 4 \\ 1 & 2 & 0 \end{pmatrix} \begin{pmatrix} 6 \\ 2 \\ 1 \end{pmatrix}$

(*c*) $\begin{pmatrix} 3 & 2 & 1 & 6 \end{pmatrix} \begin{pmatrix} 2 & 1 \\ 3 & 2 \\ 4 & 5 \\ 6 & 0 \end{pmatrix}$

(*d*) $\begin{pmatrix} 6 & 8 & 4 \\ 3 & 2 & 9 \end{pmatrix} \begin{pmatrix} 7 \\ 3 \\ 6 \end{pmatrix}$

(ii) A is a 3 by 4 matrix and B is a 4 by 3 matrix.
What will be the order of the matrix AB ?

11 The journey from A to B may be written in vector notation as:

$\vec{AB} = \begin{pmatrix} -2 \\ -4 \end{pmatrix}$.

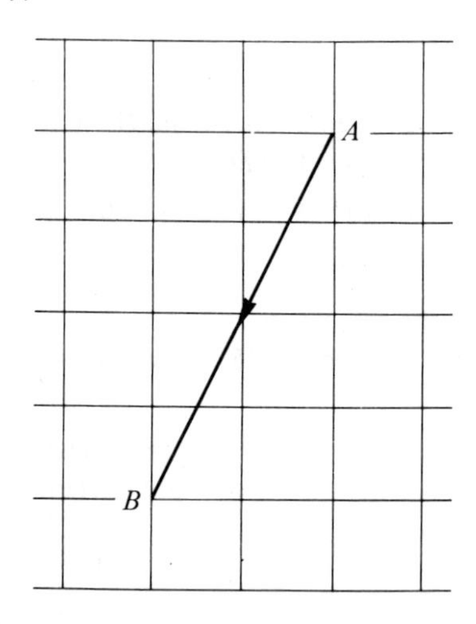

Copy the diagram on squared paper and then follow the instructions:

(i) Draw in the vectors: $\vec{BC} = \begin{pmatrix} -4 \\ 0 \end{pmatrix}$, $\vec{CD} = \begin{pmatrix} -2 \\ 4 \end{pmatrix}$, $\vec{DE} = \begin{pmatrix} 4 \\ 3 \end{pmatrix}$

(ii) Join EA and write down the vector $\vec{EA}$.

(iii) Write down the name of the polygon $ABCDE$.

(iv) Evaluate: $\vec{BC} + \vec{CD} + \vec{DE}$.

(v) Draw from B, the vector which is equivalent to $\vec{BC} + \vec{CD} + \vec{DE}$ and write down its name.

12 This question concerns the set of fractions:

$$\{\tfrac{2}{3}, \tfrac{7}{12}, \tfrac{5}{6}, \tfrac{5}{8}, \tfrac{3}{4}\}.$$

(i) Write down an equivalent fraction for each member of the set with a denominator of 24.
For example: $\frac{1}{3} = \frac{8}{24}$

(ii) Arrange the set in order of size, smallest first.
(iii) Subtract the smallest member of the set from the largest.

(iv) Evaluate $\frac{2}{3} + \frac{7}{12} + \frac{5}{6} + \frac{5}{8} + \frac{3}{4}$

(v) Write $\frac{3}{4}$ and $\frac{5}{8}$ as percentages, subtract them and give your answer as a percentage.

13 The partly completed table shows values of x and y for the graph of: $y = (x + 1)^2$

x	-4	-3	-2	-1	0	1	2
y	9	4	1	0	1		

(i) Copy and complete the table.
(ii) Draw x and y axes.
Scale the x axis from -4 to +2 using 2 cm : 1 unit
Scale the y axis from 0 to 9, using 1 cm : 1 unit
Plot the values of x and y from the table and join up the points to give you the smooth curve graph of $y = (x + 1)^2$
(iii) Draw on your graph the straight line whose equation is $y = 6$ and from your graph obtain the approximate values of x which would solve the equation $(x + 1)^2 = 6$
(iv) The graph of $y = (x + 1)^2$ is symmetrical.
Draw in the line of symmetry and write down its equation.

14 (i) A shopkeeper adds $33\frac{1}{3}\%$ to the cost price of the goods which he sells when he calculates the selling price.
An article costs him £17.25

Calculate:

(*a*) the profit
(*b*) the selling price of the article.

(ii) A man is given a wage increase of 12% but has to pay income tax on the increase at the rate of $33\frac{1}{3}\%$.

(*a*) What is the net amount of his increase after tax has been deducted (as a percentage) ?
(*b*) If he earned £75 a week before the increase, what will he earn now after tax has been deducted?

Set 9

1 The Venn diagram shows the numbers of pupils studying foreign languages at a school:

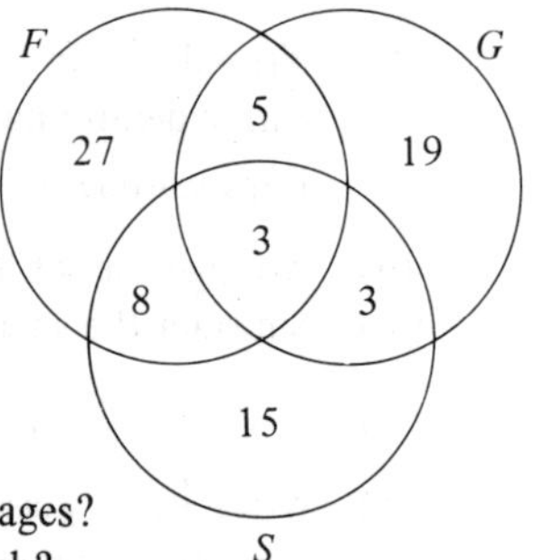

F = {French}
G = {German}
S = {Spanish}

(i) How many pupils study German?
(ii) How many pupils study foreign languages?
(iii) How many pupils do not study Spanish?
(iv) How many pupils study both French and Spanish?
(v) How many pupils study both French and German but not Spanish?
(vi) What percentage of the pupils study German?

2 (i) The edge of a cube is 9 mm. Calculate its volume in cubic centimetres.
(ii) How many 2 cm cubes would fit exactly into a 10 cm cube?
(iii) A rectangle is twice as long as it is wide.
If the length is 8.5 cm, calculate:

(*a*) the area
(*b*) the perimeter

(iv) A cube has an edge of 19 mm.
Calculate the total length of the edges of the cube in centimetres.

3 Solve the equations:

(i) $x + 2x + 3x + 4x = 25$
(ii) $3y + 9 = 0$
(iii) $\frac{1}{4}x = 8$
(iv) $2(2a - 4) = 12$
(v) $x^2 + x^2 + x^2 + x^2 = 64$

4 The following scores were recorded in a test out of 10 marks:

2, 8, 7, 6, 4, 8, 6, 7, 3, 5, 7, 9, 7, 5, 10

(i) Arrange the marks in order, smallest first.
(ii) Write down the mode.
(iii) Write down the median.
(iv) Calculate the mean, giving your answer correct to 3 significant figures.
(v) What percentage of the pupils scored more than 50%?

5 (i) Write down $\frac{1}{3}$ as a decimal fraction, giving your answer correct to 2 significant figures.

(ii) Write down $\frac{1}{3}$ as a percentage, giving your answer as a mixed number.

(iii) Write down 0.125 as a fraction in its lowest terms.

(iv) Write down 0.125 as a percentage.

(v) Evaluate $(0.9)^3$ and approximate your answer correct to 2 decimal places.

6 Factorise:

(i) $14r + 21s - 35t$

(ii) $d^2 + d$

(iii) $4 - 8a - 12b + 4c$

(iv) $x(a + 4) + y(a + 4)$

(v) $x^2 - x - 6$

7 (i) A is 6 km due North of B and 3 km due West of C.

(*a*) Draw a diagram showing these facts.

(*b*) Calculate the bearing of C from B.

(*c*) Write down the bearing of B from C.

(ii) An aircraft flies 200 km bearing North-East.
Calculate the northerly and easterly components of the position of the aircraft.

8 The diagram shows the rhombus $ABCD$, whose diagonals intersect at O.

$AC = 2x$ and $BD = 4x$ cm.

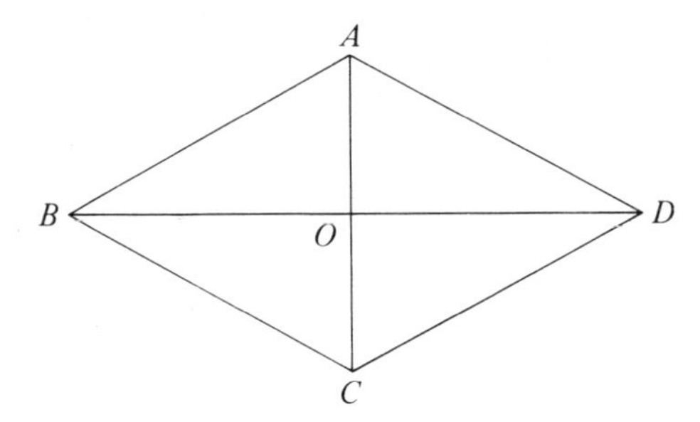

(i) Write down the terms of x, an expression for:

(*a*) the area of the triangle AOB

(*b*) the area of the rhombus $ABCD$

(*c*) the length AB.

(ii) If the area of the rhombus is 100 cm^2,

Calculate:

(*a*) the length of AB

(*b*) the perimeter of the rhombus.

9 You will need to draw x and y axes for each part of this question. Scale both axes from −4 to +4, using 1 cm : 1 unit.

(i) (*a*) Draw the graphs of $y = 2x - 2$, $y = -2x - 2$ and $y = 3$
(*b*) Shade the region R where $y > 0$, $y < 3$, $y > 2x - 2$ and $y > -2x - 2$

(ii) (*a*) Draw the graphs of $x = -3$, $y = \frac{1}{4}x - 1$ and $y = -\frac{1}{4}x + 1$
(*b*) Shade the region S where $x > -3$, $y < -\frac{1}{4}x + 1$ and $y > \frac{1}{4}x - 1$

(iii) (*a*) Draw a circle, centre O, the origin, radius 3 cm.
(*b*) Draw the graphs of $y = 2x$ and $y = -2x$
(*c*) Shade the region T where points are $<$ 3 cm from O. where $y > 2x$ and $y > -2x$

(iv) (*a*) Draw the graphs of $x = -3$, $x = 3$, $y = -3$, $y = 3$, $y = -x$ and $y = x$
(*b*) Shade the region M where $y > x$, $x > 0$ and $y < 3$
(*c*) If (x, y) are integers, write down the coordinates of a point which is inside the region M.

10 $A = \begin{pmatrix} 3 & -2 \\ -2 & 3 \end{pmatrix}$, $B = \begin{pmatrix} 4 & -1 \\ 0 & -3 \end{pmatrix}$, $C = \begin{pmatrix} 2 & -2 \\ -4 & 3 \end{pmatrix}$

Evaluate:

(i) $A - B$
(ii) $B - C$
(iii) $A + B + C$
(iv) AB
(v) ABC

11 The journey from A to B shown in the diagram may be written in vector form as:

$\vec{AB} = \begin{pmatrix} 2 \\ -3 \end{pmatrix}$

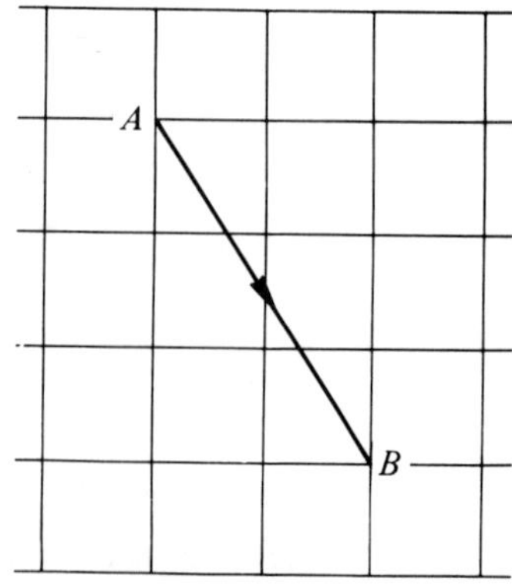

Copy the diagram on squared paper.

(i) Draw in the vectors: $\vec{BC} = \begin{pmatrix} 4 \\ 0 \end{pmatrix}$, $\vec{CD} = \begin{pmatrix} 2 \\ 3 \end{pmatrix}$, $\vec{DE} = \begin{pmatrix} -2 \\ 3 \end{pmatrix}$, $\vec{EF} = \begin{pmatrix} -4 \\ 0 \end{pmatrix}$
(ii) Join FA to give the polygon $ABCDEF$ and write down its name.
(iii) Write down the vector $\vec{AF}$.
(iv) Evaluate $\vec{BC} + \vec{CD} + \vec{DE} + \vec{EF}$.
(v) Draw from B the vector which is equivalent to $\vec{BC} + \vec{CD} + \vec{DE} + \vec{EF}$ and write down its name.

12 This question concerns the set of fractions:

$$\{\tfrac{11}{15}, \tfrac{19}{30}, \tfrac{7}{10}, \tfrac{3}{5}\}$$

(i) Write down an equivalent fraction for each member of the set with a denominator of 30.

e.g. $\frac{2}{5} = \frac{12}{30}$

(ii) Arrange the set in order of size, smallest first.
(iii) Add up the set.
(iv) Calculate the mean of the members of the set.
(v) Express the smallest member of the set as a percentage.

13

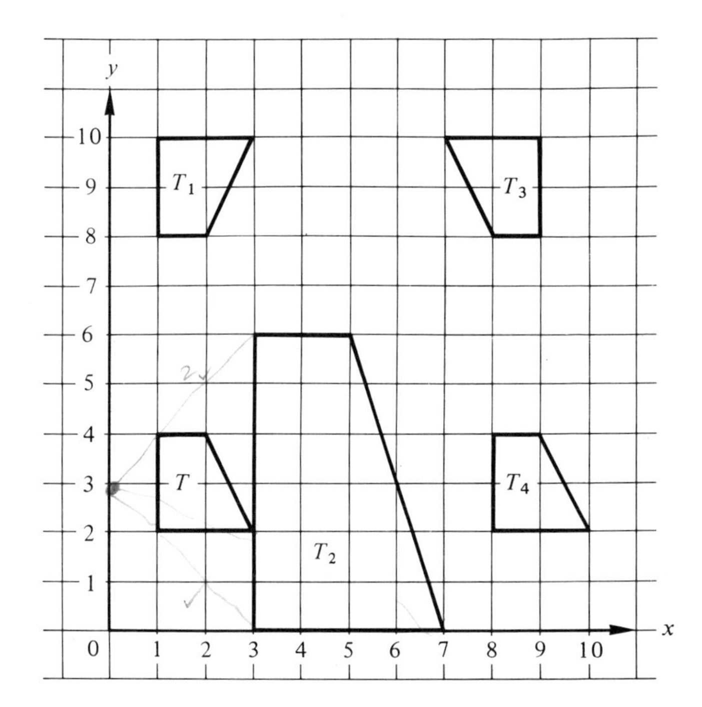

(i) Describe fully the transformations which map:

(*a*) T onto T_1
(*b*) T onto T_2
(*c*) T_1 onto T_3
(*d*) T onto T_4
(*e*) T_4 onto T_3

(ii) Write down the ratio of the areas: $\dfrac{T}{T_2}$

14 A man bought a family car for £2800 and during the first year it depreciated by 15%. He travelled a total of 9600 km and his average petrol consumption was 1 litre for every 8 km. Petrol costs 25p a litre. He also spent £138 on insurance and servicing.

Calculate:

(i) the depreciation in the first year
(ii) litres of petrol consumed
(iii) cost of petrol consumed
(iv) the total cost of depreciation, petrol, insurance and servicing
(v) the cost per week of the family car in a year of 52 weeks.

Set 10

1

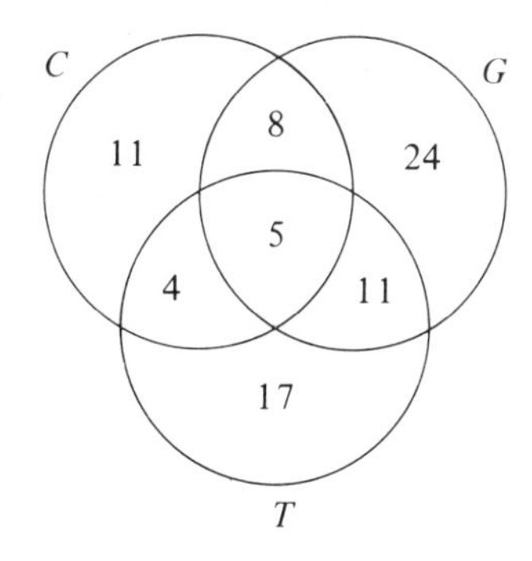

The Venn diagram shows the number of pupils who are members of various school clubs:

$C =$ {the chess club}
$G =$ {the gym club}
$T =$ {the table tennis club}

(i) How many pupils took part in club activities?
(ii) How many pupils were in the table tennis club?
(iii) How many pupils did not play chess?
(iv) How many pupils were members of both the gym club and the table tennis club?
(v) How many members had the most popular club?
(vi) What percentage of the pupils were members of the most popular club?

2 (i) The cost price of an article was £27.50 but it could be purchased by paying 12 monthly instalments of £2.60 How much is saved by paying the cash price?
(ii) When it is noon in Greenwich it is 1530 hours in Teheran. What time is it in Greenwich when it is noon in Teheran?
(iii) It takes 6 hours 55 minutes to fly from London to Cairo. What time (Greenwich Mean Time) will a plane leaving Heathrow at 0945 hours land in Cairo?
(iv) A jet left Heathrow at 1845 hours and arrived at San Francisco the following day at 0615 hours (Greenwich Mean Time). How long did the journey take?
(v) Actual flying time from London to Sydney is 24 hours 50 minutes, but Sydney Time is 10 hours ahead of Greenwich Mean Time. Calculate the day and time in Sydney when a jet leaving Heathrow at 1130 hours (Greenwich Mean Time) on a Saturday should arrive.

3 Solve the equations:

(i) $4x = 22$
(ii) $1.5x + 2.5x + 3x = 49$
(iii) $\frac{3}{4}x = 18$
(iv) $x(x - 5) = 0$
(v) $(3a - 9)(2a + 7) = 0$

4 The frequency table shows the statistics recorded in a survey of the number of children in a family.

(*a*) Children in family	(*b*) Frequency	(*c*) $C \times F$
1	7	
2	9	
3	7	
4	4	
5	2	
6	1	
Totals:		

(i) Copy and complete the table.
(ii) Draw a bar chart to illustrate the statistics using the data in columns (*a*) and (*b*).
(iii) Calculate the mean.

5 (i) Write the number 7 480 000 in Standard Form notation, correct to 2 significant figures.
(ii) Write the number 0.000 000 87 in Standard Form notation.
(iii) Simplify $(2.0 \times 10^6) \times (4.5 \times 10^3)$ leaving your answer in Standard Form notation.
(iv) Simplify $(4.0 \times 10^7) \times (5.0 \times 10^5)$ leaving your answer in Standard Form notation.
(v) Evaluate fully: $(2.0 \times 10^7) \times (3.0 \times 10^{-5})$

6 Factorise:

(i) $18a - 27b - 36c + 81$
(ii) $4x^2 + 12x$
(iii) $a(x + 5) + b(x + 5)$
(iv) $a^2 - 8a + 7$
(v) $25x^2 - 25$

7 The diagram shows a quadrilateral *ABCD* which is divided into two triangles by the diagonal *AC*. Triangle *ABC* is equilateral and triangle *ADC* is isosceles and right-angled at *D*.

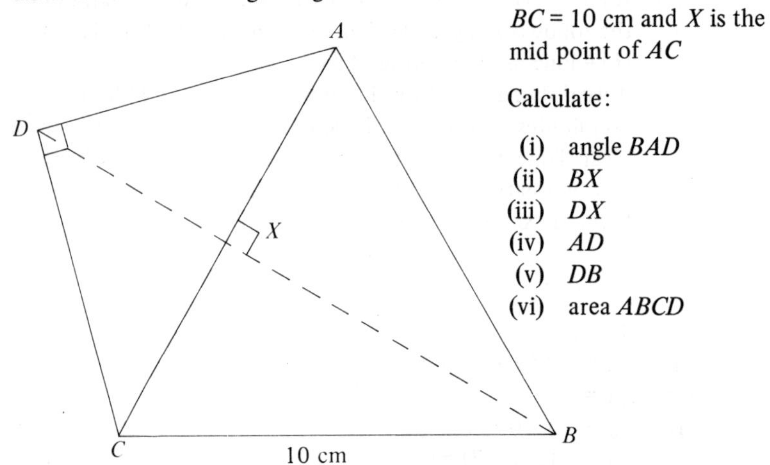

BC = 10 cm and X is the mid point of AC

Calculate:

(i) angle *BAD*
(ii) *BX*
(iii) *DX*
(iv) *AD*
(v) *DB*
(vi) area *ABCD*

8 Given that $A = \pi r^2$,

(i) calculate A when $\pi = 3.14$ and $r = \sqrt{10}$
(ii) calculate A when $\pi = \frac{22}{7}$ and $r = 3\frac{1}{2}$
(iii) change the subject of the formula to r
(iv) calculate r when $\pi = 3.14$ and $A = 28.26$
(v) calculate r when $\pi = \frac{22}{7}$ and $A = 132$

9

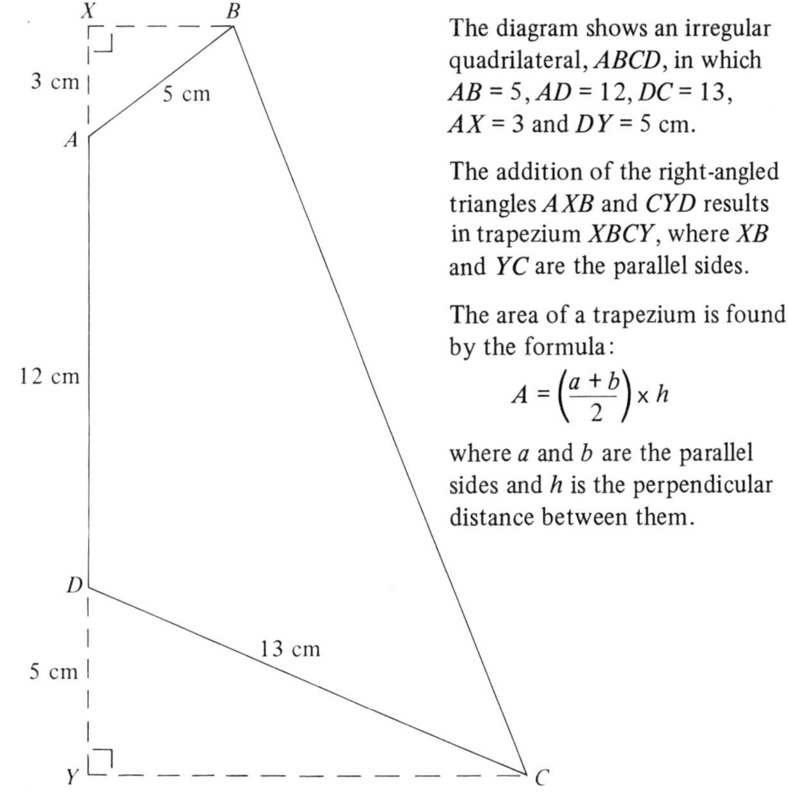

The diagram shows an irregular quadrilateral, $ABCD$, in which $AB = 5$, $AD = 12$, $DC = 13$, $AX = 3$ and $DY = 5$ cm.

The addition of the right-angled triangles AXB and CYD results in trapezium $XBCY$, where XB and YC are the parallel sides.

The area of a trapezium is found by the formula:

$$A = \left(\frac{a+b}{2}\right) \times h$$

where a and b are the parallel sides and h is the perpendicular distance between them.

(i) Calculate XB and YC.
(ii) Use the formula to calculate the area $XBCY$.
(iii) Calculate the area of triangle AXB.
(iv) Calculate the area of triangle CYD.
(v) Calculate the area $ABCD$.

10 $A = \begin{pmatrix} 3 & -1 \\ -2 & 0 \end{pmatrix}$, $B = \begin{pmatrix} -2 & 4 \\ 1 & -3 \end{pmatrix}$, $C = \begin{pmatrix} 4 & 0 \\ -2 & -3 \end{pmatrix}$

Evaluate:

(i) $A + B + C$
(ii) $A - C$
(iii) AB
(iv) ABC

11 (Squared paper) Draw x and y axes. Scale both axes from 0 to 10, 1 cm : 1 unit;

(i) Plot and join up the following points to give the polygon $ABCDEFGH$:

A: (5,10), B: (3,7), C: (0,5), D: (3,3), E: (5,0), F: (7,3), G: (10,5), H: (7,7).

(ii) Name the polygon.

(iii) Write down the vectors: $\vec{AB}, \vec{BC}, \vec{DE}, \vec{FG}, \vec{HA}$.

(iv) $\vec{DC} = \vec{GH} = \begin{pmatrix} -3 \\ 2 \end{pmatrix}$. In a similar way, copy and complete:

(*a*) $\vec{BA} = \ldots = \begin{pmatrix} \\ \end{pmatrix}$

(*b*) $\vec{GF} = \ldots = \begin{pmatrix} \\ \end{pmatrix}$

(v) (*a*) Evaluate: $AB + BC$.

(*b*) Draw on the diagram from A the single vector which is equivalent to $\vec{AB} + \vec{BC}$. Write down its name.

12 Evaluate:

(i) $\frac{5}{6} + \frac{5}{9}$

(ii) $\frac{5}{9} - \frac{5}{6}$

(iii) $\frac{5}{9} \div \frac{5}{6}$

(iv) $\frac{5}{6} \div \frac{5}{9}$

(v) $(\frac{5}{6})^2 + (\frac{5}{9} \times \frac{1}{4})$

(vi) $(\frac{5}{6}$ of $54) - (\frac{5}{9}$ of $81)$

13 The partly completed table shows values of x and y for the graph of: $y = (x - 1)^2$

x	−2	−1	0	1	2	3	4
y	9	4	1	0	1		

(i) Copy and complete the table of values.

(ii) Draw x and y axes.
Scale the x axis from − 2 to + 4, using 2 cm : 1 unit.
Scale the y axis from 0 to 9, using 1 cm : 1 unit.

(iii) Plot the values of x and y from the table of values and join up the points to give you the smooth curve graph of $y = (x - 1)^2$

(iv) Estimate from your graph the value of y when:

(*a*) $x = 2.4$
(*b*) $x = -0.5$

(v) Use the graph to solve the equation: $(x - 1)^2 = 6.2$

14 Two dice are made in the shape of a tetrahedron with 4 faces and have the scores 1,2,3,4 on their respective faces.

2nd die

	+	1	2	3	4
	1				
1st	2				
die	3				
	4				

(i) Copy and complete the table to show the possible combinations and total scores when the dice are thrown together, the scores being taken from the horizontal faces.

(ii) Write down the probability of:

(*a*) a total score of 2
(*b*) a total score of 5
(*c*) a total score of 7
(*d*) a total score greater than 4
(*e*) a total score less than 5
(*f*) throwing any double
(*g*) a total score greater than 0 and less than 9.

Section 2

Set 1

1 The diagram shows the net of a solid which measures 6 cm by 3.5 cm by 2 cm.

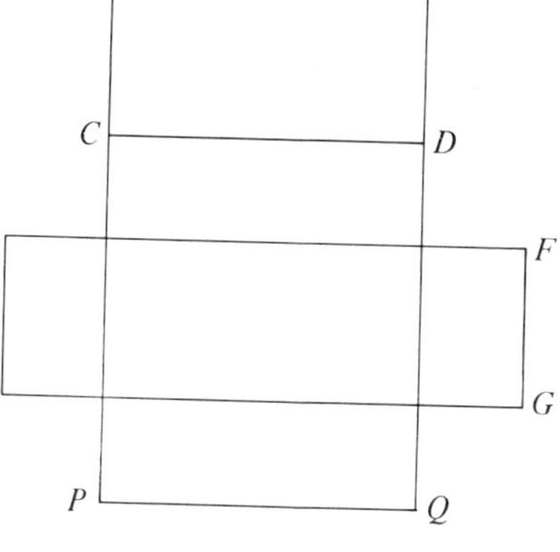

(i) Write down:

(*a*) the name of the solid
(*b*) the number of edges
(*c*) the number of vertices.

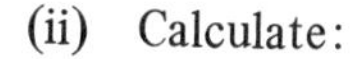

(ii) Calculate:

(*a*) the total surface area of the solid
(*b*) the volume of the solid.

(iii) If the net is made into the solid:

(*a*) which points will meet at the vertex where B is one of the points?
(*b*) calculate the length of the diagonal FH.

2 The Venn diagram shows the number of boys in the first year and their games activities:

C = {boys who play cricket}
T = {boys who play tennis}
A = {boys who do athletics}

Write down:

(i) the total number of boys in the first year group
(ii) the number of boys who play cricket and tennis
(iii) the number of boys who do athletics
(iv) the number of boys who do not play tennis
(v) the percentage of the number of boys in the group who play tennis and do athletics.

3 (i) (*a*) Evaluate $(0.0016)^2$
(*b*) Write your answer to (*a*) in Standard Form notation.

(ii) (*a*) Evaluate 2.39×10^5 fully in decimal notation.
(*b*) Write your answer to (*a*) correct to 2 significant figures.

(iii) The sides of a rectangle are 8.0×10^{-3} m and 7.0×10^{-2} m.

(*a*) Write down the area of the rectangle in Standard Form notation.
(*b*) Write your answer to (*a*) fully in decimal notation.
(*c*) Write your answer to (*b*) in square centimetres.

4 Solve the equations:

(i) $2a + 10 = 2$
(ii) $4x - 7 = 3$
(iii) $x^2 + 3x = 0$
(iv) $2x^2 = 72$
(v) $x^2 - 6x - 7 = 0$

5 A car was purchased for £3200 and depreciation during the first year was 25%, during the second year 15% and during the third year $12\frac{1}{2}$%.

Calculate:

(i) the value of the car at the end of
(*a*) 1 year (*b*) 2 years (*c*) 3 years
(ii) the total depreciation over the three years as a percentage of the purchase price, to the nearest whole per cent.

6 A survey was taken of the number of passengers in cars passing a check point. The results are shown in the table.

Number of passengers:	1	2	3	4	5	6
Frequency :	15	32	21	15	10	7

(i) Write down the number of cars in the survey.
(ii) Calculate the total number of passengers in the cars.
(iii) Write down the mode.
(iv) Write down the median.
(v) Calculate the mean for the distribution.
(vi) If the statistics were shown on a pie chart, how many degrees would be needed for the sector angle for cars with 5 passengers?
(vii) What percentage of the cars in the survey carried less than 4 passengers?

7 $A = \begin{pmatrix} 3 & -5 \\ -1 & 2 \end{pmatrix}$, $B = \begin{pmatrix} 3 & 6 \\ 2 & 4 \end{pmatrix}$

(i) Evaluate:

(*a*) $2A + B$
(*b*) $B - A$
(*c*) AB

(ii) (*a*) Write down the matrix which is singular.
(*b*) Calculate the inverse, X, of the non-singular matrix.
(*c*) Evaluate XA.

8 The diagram shows a hollow triangular prism, the cross section of which is formed by the equilateral triangles ABC and XYZ.
$AB = 10$, $XY = 5$ and $AE = 40$ cm.

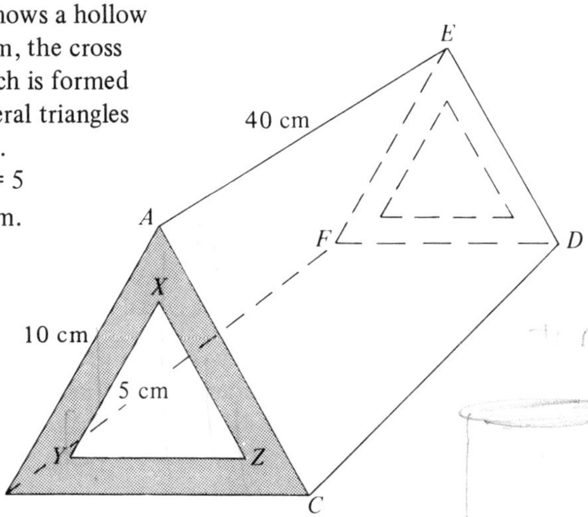

Calculate:

(i) area of triangle ABC
(ii) area of triangle XYZ
(iii) the shaded area of the cross-section
(iv) the volume of the prism
(v) the ratio $\dfrac{\text{Area } ABC}{\text{Area } XYZ}$
(vi) the distance BD in centimetres.

9 (i) Solve the equation $x^2 + 2x - 15 = 0$ by the method of factorising.
(ii) Solve the equation $2x^2 + 3x - 4 = 0$ by using the formula:

$$x = \frac{-b \pm \sqrt{b^2 - 4ac}}{2a}$$

10 An aircraft is airborne at point A on a runway and climbs at 20° to the horizontal, AB. Point B is 1000 m from A and point X is perpendicularly above B.

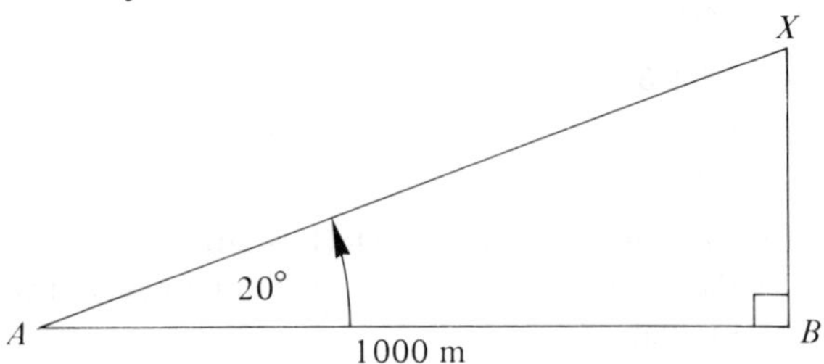

Calculate:

(i) the height of the plane above B in metres
(ii) the distance AX, correct to 3 significant figures
(iii) the average speed of the take-off, if it takes the plane 8 seconds to reach point X; give your answer

(*a*) in metres per second (*b*) in kilometres per hour.

11 The diagram shows 3 lines, AB, CD and EF.

(i) Write down the equations of the lines AB, CD and EF.
(ii) Write down 3 inequalities which define the shaded region S.
(iii) Find the area of the shaded region S.

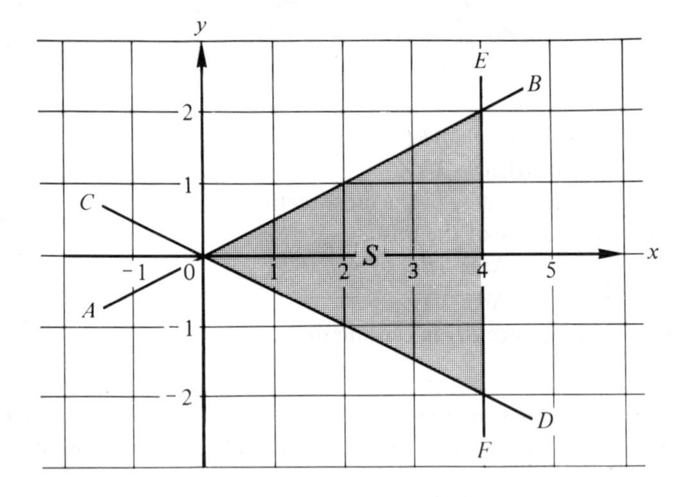

12 Copy the diagram on squared paper and then follow the instructions below.

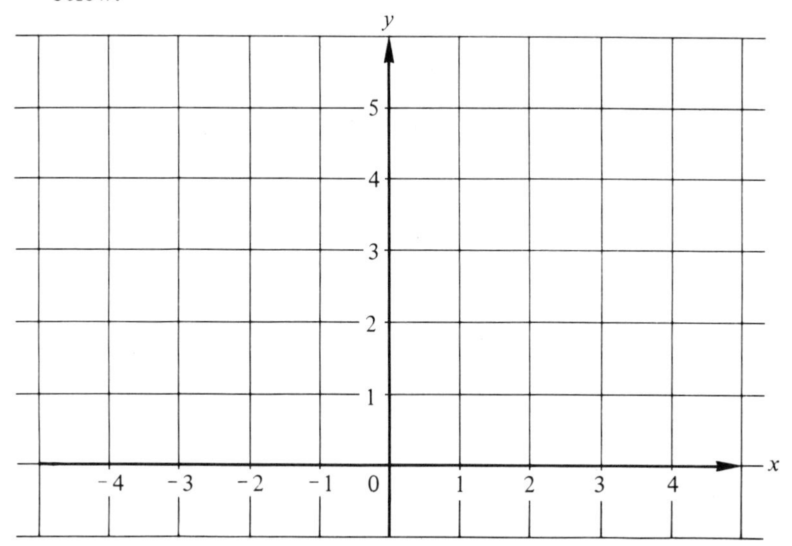

(i) Copy and complete the following table for the values of y for the graph of $y = \frac{1}{4}x^2$.

x	−4	−3	−2	−1	0	1	2	3	4
y	4				0			$2\frac{1}{4}$	

(ii) Plot these values and draw the graph of $y = \frac{1}{4}x^2$

(iii) On the same diagram draw the graph of $y = \frac{1}{4}x + 1$

(iv) Write down the values of x at the points of intersection of the graphs $y = \frac{1}{4}x^2$ and $y = \frac{1}{4}x + 1$

(v) Write down the single equation which satisfies these values of x.

13 In the diagram, angle $DAC = 40°$, angle $CDB = 80°$, angle $DBC = 30°$ and $CD = 5$ m.

Calculate:

(i) angle ACD

(ii) AD

(iii) DB

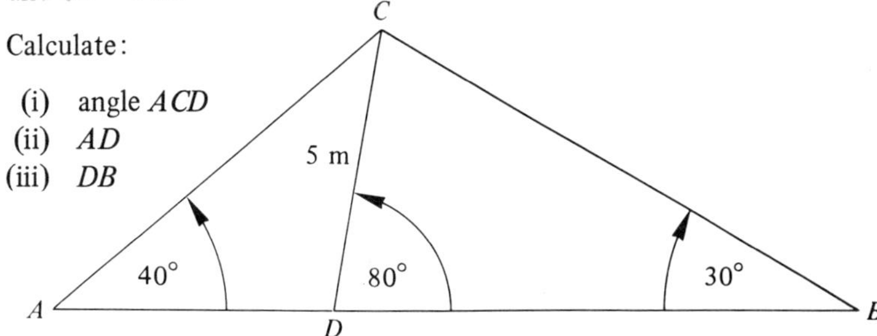

(iv) the shortest distance from C to AB

(v) the height of a vertical post PC, erected at C so that the angle of elevation from D to the top of the post is 63.4°

14 On squared paper draw x and y axes.
Scale the x axis from -6 to $+4$ and the y axis from -7 to $+4$, using 1 cm : 1 unit.
ABCD is a V-Bomber kite: $A(2,3)$, $B(2,2)$, $C(3,2)$, $D(1,1)$

(i) Plot the points and letter the kite *ABCD*.

(ii) Write down below the grid, a 2 by 4 matrix to show the journeys from the origin of A, B, C and D. Call it K.

(iii) Multiply K by the matrix $\begin{pmatrix} -2 & 0 \\ 0 & -2 \end{pmatrix}$ on the left.

(iv) Plot and letter on the diagram the image $A'\,B'\,C'\,D'$ of *ABCD* under the transformation $\begin{pmatrix} -2 & 0 \\ 0 & -2 \end{pmatrix}$

(v) Describe in words the transformation by the matrix $\begin{pmatrix} -2 & 0 \\ 0 & -2 \end{pmatrix}$

15 (i) Three coins are tossed. Make an array of the possible outcomes and find the probability of obtaining:

(*a*) only 1 head
(*b*) three heads.

(ii) A 2 digit number is made by combining any 2 of the digits 1,3,5,7,9 at random.

(*a*) Make an array of the possible combinations.
(*b*) Find the probability that the number is divisible by 3.

Set 2

1 Factorise:

(i) $4x - 16$
(ii) $a^2 + ab + ac$
(iii) $3x^2 - 12$
(iv) $p^2 + 4p - 5$
(v) $1 - p^2$

2 The diagram shows the net of a regular solid which has 4 faces, each one being an equilateral triangle.
$DF = 20$ cm and X is the mid point of DE

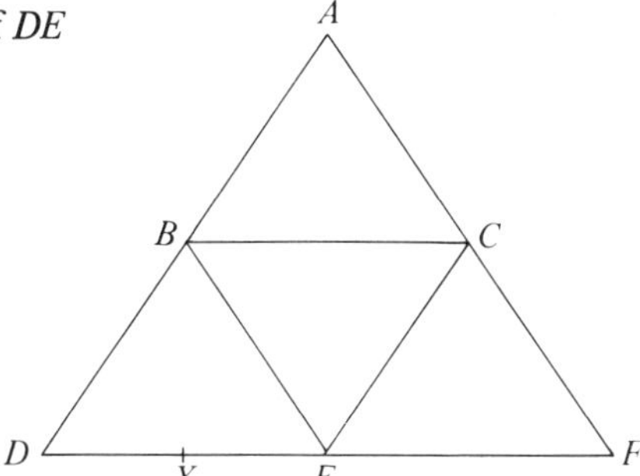

(i) Write down:

(a) the name of the quadrilateral $ABEC$
(b) the name of the quadrilateral $BCFD$
(c) the name of the solid you could make from the net.

(ii) Calculate:

(a) the size of angle ABE
(b) the cosine ratio of angle CEF
(c) the distance BX
(d) the area of triangle BDE
(e) the area of triangle ADF.

3 (i) $P = \begin{pmatrix} 2 & -1 \\ 3 & -2 \end{pmatrix}$, $Q = \begin{pmatrix} -1 & 5 \\ 2 & -3 \end{pmatrix}$, $R = \begin{pmatrix} 0 & 3 \\ -2 & 1 \end{pmatrix}$

Evaluate:

(a) $P + Q$
(b) $P - Q$
(c) $P(Q + R)$

(ii) If $A = \begin{pmatrix} 4 & 2 \\ 3 & 2 \end{pmatrix}$, find A^{-1}

4 Solve the equations:

(i) $x + x + x = 39$
(ii) $4a - 2 = 8$
(iii) $y^2 + y^2 + y^2 = 108$
(iv) $x^2 - 15 = 66$
(v) $x^2 - 2x - 15 = 0$

5 (i) Express the number 101_8 as a binary number.
(ii) Write 231_4 as an octal number.
(iii) Evaluate $321_8 + 467_8 - 257_8$ and leave your answer in base 8.
(iv) Evaluate $10\,010_2 \times 11_2$ and give your answer in base 10.
(v) If $234_x = 105_8$, calculate the value of x.

6 The width of a rectangle is x cm and the length is y cm. x and y are integers and must satisfy the following inequalities:
$y > x, y < 3x, x + y > 15$ and $x + y < 20$.

(i) Draw x and y axes on squared paper. Scale both axes from 0 to 20, using 2 cm : 5 units.
(ii) Draw and label the straight lines whose equations are:
$y = x, y = 3x, x + y = 15$ and $x + y = 20$.
(iii) By reference to the inequalities given above, identify the region inside which the points (x,y) must be found. Outline this region in ball point or colour.
(iv) Mark with neat crosses the points whose coordinates satisfy the inequalities.
(v) Write down the number of rectangles which satisfy the inequalities.

7 (i) Write in Standard Form notation:

(*a*) 5 400 000 000 (*b*) 0.000 017

(ii) Write fully in decimal notation:

(*a*) 8.0×10^{-6} (*b*) 1.7×10^4

(iii) Calculate: $9\,000\,000 \div 600$ by writing both numbers in Standard Form notation and then using the laws of indices. Leave your answer in Standard Form notation.
(iv) The sides of a rectangle are 0.006 km and 0.004 km.

(*a*) Write these dimensions in Standard Form notation.
(*b*) Find the area of the rectangle and write your answer:
(i) in Standard Form notation in square kilometres;
(ii) in decimal notation in square metres.

8 $n(\mathscr{E}) = 100$.

X, Y and Z are three sets and the numbers of elements are shown in their appropriate regions in the Venn diagram.

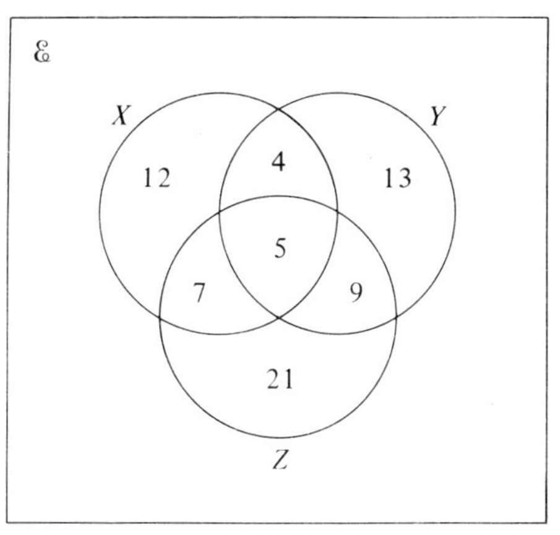

Write down:

(i) $n(X \cap Y)$
(ii) $n(X \cap Z)$
(iii) $n(X \cup Y)$
(iv) $n(X \cap Y \cap Z)$
(v) $n(X \cup Y) \cap Z$
(vi) $n(X \cup Y \cup Z)'$

9 In the diagram, TA is a tangent to the circle, centre O, radius 5 cm. Angle $ATO = 22°$

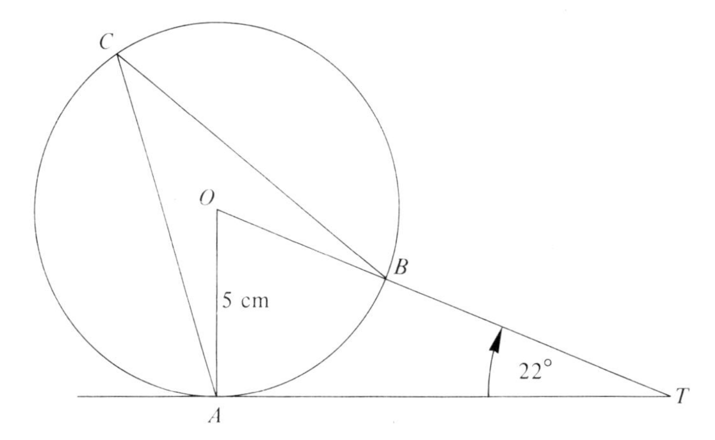

(i) Calculate:

(*a*) angle AOT
(*b*) angle OAB
(*c*) angle BAT
(*d*) angle ACB

(ii) (*a*) Write down from the tables the value of tan angle AOT.
(*b*) Calculate the length of the tangent TA.

10 (Squared paper) Draw x and y axes. Scale the x axis from 0 to 5 and the y axis from 0 to 3, using 1 cm : 1 unit.

(i) Plot triangle ABC: $A(2,2)$, $B(0,0)$ and $C(2,0)$.
(ii) Write down the journeys from the origin to A, B and C as a 2 by 3 matrix, call it K.
(iii) Multiply K by $\begin{pmatrix} 1 & 1 \\ 0 & 1 \end{pmatrix}$ on the left to give $A'B'C'$, the image of ABC under the transformation matrix $\begin{pmatrix} 1 & 1 \\ 0 & 1 \end{pmatrix}$
(iv) Plot $A'B'C'$
(v) Describe the transformation fully.

11 The diagram shows PA, a vertical pole erected at A, the corner of triangle ABC, which is in a horizontal plane.
$PA = 8$ m and $PC = 16$ m and $AB = 6$ m.

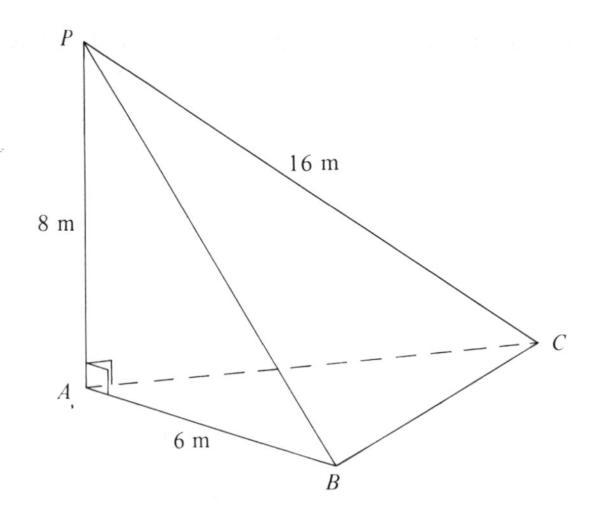

(i) Write down sin angle PCA as a decimal fraction.
(ii) Use tables to find angle PCA in degrees.
(iii) Calculate PB.
(iv) Write down tan angle ABP as a decimal fraction.
(v) Use tables to find angle ABP in degrees.
(vi) If angle $CAB = 90°$, calculate the length of BC.

12 (i) Evaluate $\begin{pmatrix} 6 & 8 & 4 \\ 3 & 2 & 9 \end{pmatrix}\begin{pmatrix} 7 \\ 3 \\ 6 \end{pmatrix}$

(ii) If $A = \begin{pmatrix} 2 & -2 \\ -3 & 4 \end{pmatrix}$, find A^2

(iii) Evaluate $\begin{pmatrix} 2 & -\frac{1}{2} \\ 4 & -2 \end{pmatrix}\begin{pmatrix} 4 & 2 \\ -\frac{1}{2} & 4 \end{pmatrix}$

(iv) If $\begin{pmatrix} 2 & -2 \\ 0 & -3 \end{pmatrix}\begin{pmatrix} a \\ 2 \end{pmatrix} = \begin{pmatrix} 4 \\ b \end{pmatrix}$ find a and b.

(v) If $\begin{pmatrix} 4 & 2 \\ -1 & 0 \end{pmatrix}\begin{pmatrix} 3 & x \\ x & 1 \end{pmatrix} = \begin{pmatrix} 8 & z \\ y & 2 \end{pmatrix}$ find x, y and z.

13 The diagram, not drawn to scale, shows the side of a wooden shed, 8.3 m by 4.2 m, from which spaces for 2 identical windows 1.4 m by 1.3 m and a door 0.90 m by 2.8 m have been cut out.

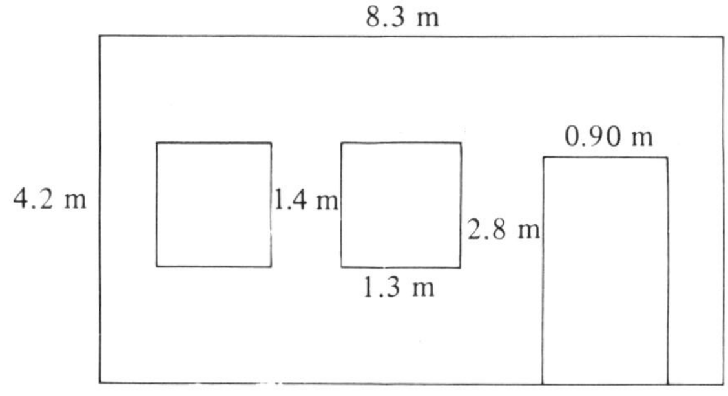

Calculate:

(i) the total area of the wooden side before the spaces are cut out
(ii) the total area of the two windows
(iii) the area of the door
(iv) the total area cut out
(v) the area remaining.

14 (Squared paper) 2 planes leave an airport at the same time. Plane A is on a course of 090° and travels at an average speed of 800 km/h. Plane B is on a course of 315° and travels at an average speed of 600 km/h.

(i) Use a scale of 1 cm : 100 km and make an accurate drawing to show the position of the planes after 1 hour.
(ii) From your drawing estimate the distance the planes are apart after:

(*a*) 15 minutes (*b*) 30 minutes (*c*) 45 minutes (*d*) 1 hour.

(iii) Measure the bearing of plane B from plane A after 1 hour.

15 The diagram shows triangle ABC. D, E and F are the mid points of AB, BC and AC respectively.

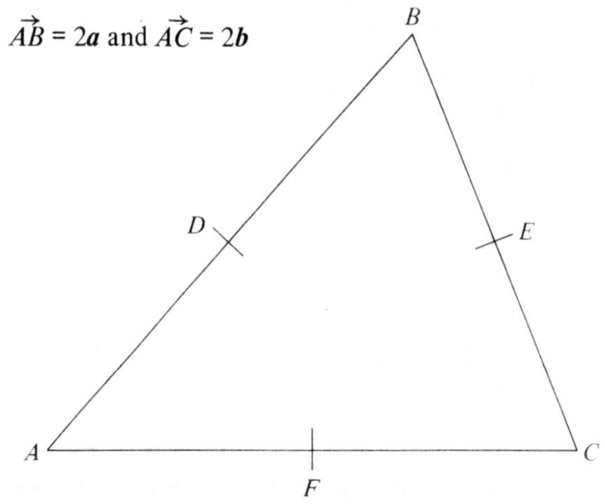

(i) Express $\vec{AD}$ in terms of $\mathbf{a}$.
(ii) Express $\vec{AF}$ in terms of $\mathbf{b}$.
(iii) Express $\vec{BC}$ in terms of $\mathbf{a}$ and $\mathbf{b}$.
(iv) Express $\vec{DE}$ in terms of $\mathbf{a}$ or $\mathbf{b}$.
(v) Express $\vec{FE}$ in terms of $\mathbf{a}$ or $\mathbf{b}$.
(vi) From your results what two facts can you state about:

(*a*) AC and DE ?
(*b*) AB and FE ?

(vii) Write down the name of the quadrilateral $ADEF$.

Set 3

1 (i) Find $12\frac{1}{2}\%$ of £16.08

(ii) Find the number of which 24 is 60%.

(iii) After an 8% price increase an article costs £27.00
Calculate the original price of the article.

(iv) Express the fraction $\frac{5}{8}$ as a percentage.

(v) An article bought for £6.48 is sold at a profit of $33\frac{1}{3}\%$.
Calculate the selling price.

2 The diagram shows rectangle $ABCD$ in which $AB = 6$ cm and $AD = 4$ cm.
The arc DX is the quadrant of a circle and XY is the diameter of the semi-circle in rectangle $XBCY$.

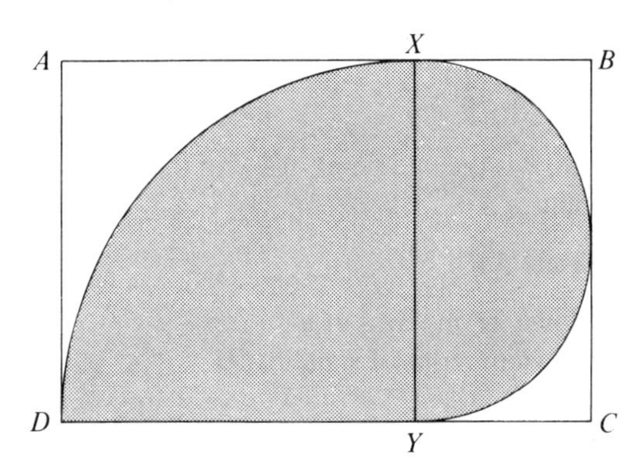

(i) Write down:

(*a*) the radius of the quadrant DXY

(*b*) the radius of the semi-circle in rectangle $XBCY$.

(ii) Calculate and give your answers in terms of π:

(*a*) the length of the arc DX

(*b*) the length of the arc XY

(*c*) the curved length DXY

(*d*) the shaded area

(*e*) the unshaded area.

3 A competition consists of finding hidden discs.
There are white, green, red and orange discs. 3 teams, X, Y and Z compete.
Team X finds 3 white, 4 green, 10 red and 3 orange discs.
Team Y finds 4 white, 6 green, 9 red and 4 orange discs.
Team Z finds 6 white, 9 green, 4 red and 2 orange discs.

(i) Write this information as a 3 by 4 matrix, call it P.
White discs score 12 points each, green score 6, red score 3, but orange discs score 2 penalty points (i.e. -2 points each)
(ii) Write the scoring system as a column matrix, call it Q.
(iii) Evaluate PQ and thus find the total score of each team.

4 (i) Solve the simultaneous equations:

$$3x + 2y = 9$$
$$x + 2y = 7$$

(ii) Use the formula $x = \dfrac{-b \pm \sqrt{b^2 - 4ac}}{2a}$ to solve the quadratic equation: $2x^2 + 7x - 4 = 0$

5 In the diagram, P is the mid point of YZ in the triangle XYZ and M is the mid point of XP.

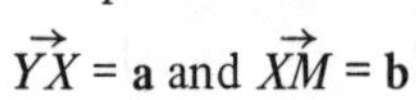

$$\overrightarrow{YX} = \mathbf{a} \text{ and } \overrightarrow{XM} = \mathbf{b}$$

(i) Write down $\overrightarrow{XP}$ in terms of $\mathbf{b}$.
(ii) Write down in terms of $\mathbf{a}$ and $\mathbf{b}$:

(*a*) $\overrightarrow{YP}$
(*b*) $\overrightarrow{YZ}$
(*c*) $\overrightarrow{XZ}$

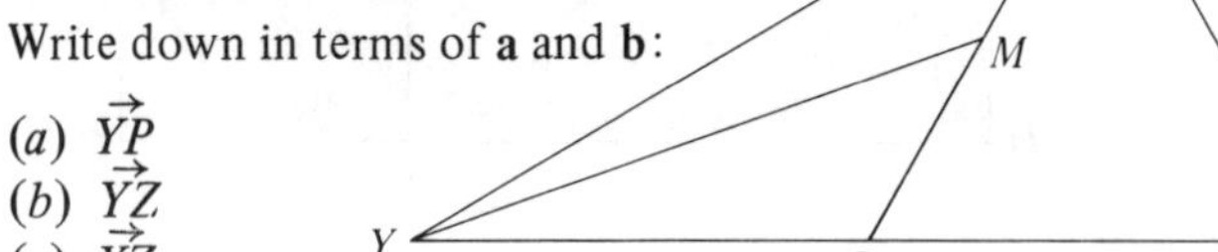

6 The diagram shows the Earth and the position on its surface of Greenwich and Mauritius. Latitude is the angle from 0° to 90°, measured North or South of the Equator. 1° of latitude is equal to 60 nautical miles on the Earth's surface.

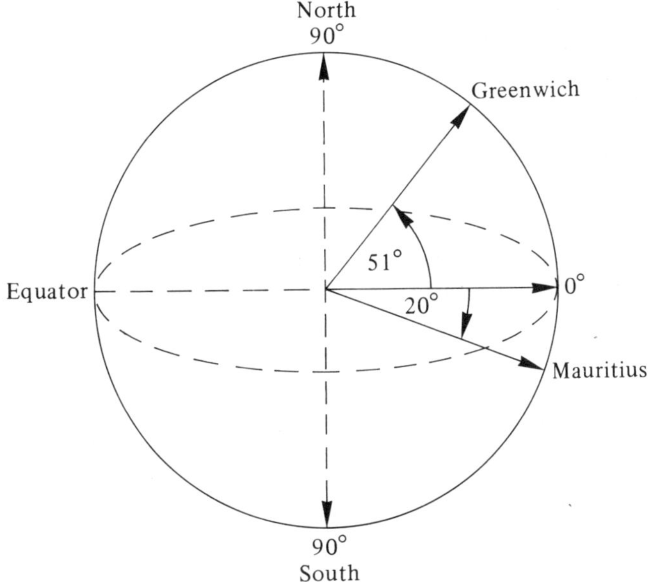

(i) Write down:

(*a*) the latitude of Greenwich (angle and direction N or S)
(*b*) the latitude of Mauritius.

(ii) Calculate:

(*a*) the distance from the Equator to Greenwich in nautical miles
(*b*) the distance from the Equator to Mauritius in nautical miles
(*c*) the latitude of a point which is 2700 nautical miles north in the Equator
(*d*) the latitude of a point which is 2220 nautical miles south of the Equator
(*e*) the circumference of the Earth in nautical miles.

7 (i) Simplify:

(*a*) $\dfrac{x^2 - 7x - 18}{2x + 4}$

(*b*) $\dfrac{x}{1 + x} + \dfrac{1 - x}{x}$

(ii) Solve the equation: $\dfrac{x}{2} + \dfrac{2}{x} = \dfrac{5}{2}$

8 $x * y$ denotes the operation x^2y^2.

(i) Evaluate:

(*a*) $2 * 4$
(*b*) $4 * 2$
(*c*) $(1 * 2) * 3$
(*d*) $1 * (2 * 3)$

(ii) Find k if $k * 3 = 225$.
(iii) If $k * k = 81$, evaluate k.

9 The diagram shows a circle, centre O, radius 6 cm.
AT and AY are tangents. Angle $TAY = 42°$
and POY is a straight line.

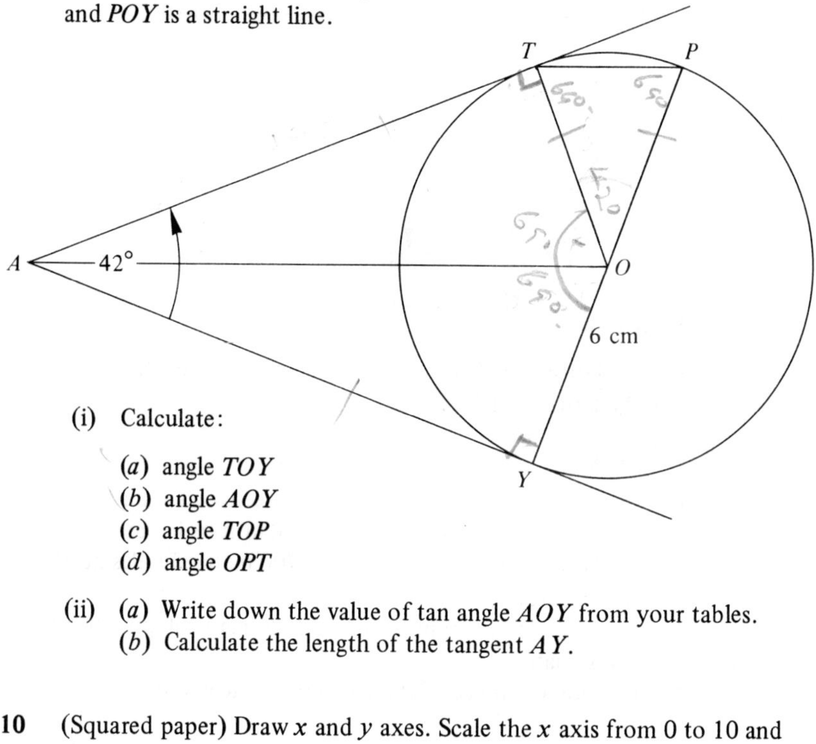

(i) Calculate:

(*a*) angle TOY
(*b*) angle AOY
(*c*) angle TOP
(*d*) angle OPT

(ii) (*a*) Write down the value of tan angle AOY from your tables.
(*b*) Calculate the length of the tangent AY.

10 (Squared paper) Draw x and y axes. Scale the x axis from 0 to 10 and the y axis from 0 to 3, using 1 cm : 1 unit.

(i) Plot the parallelogram $ABCD$: $A(2,3)$, $B(0,0)$, $C(2,0)$, $D(4,3)$.
(ii) Write down the coordinates of $ABCD$ as a 2 by 4 matrix, K.
(iii) Multiply K, on the left, by the transformation matrix
$T\begin{pmatrix}1 & 2\\ 0 & 1\end{pmatrix}$, to give the coordinates of $A'B'C'D'$, the image of $ABCD$ under the transformation.
(iv) Plot $A'B'C'D'$.
(v) Describe the transformation T fully.
(vi) Write down two invariants under the transformation.

11 Use the cosine formula: $\cos A = \dfrac{b^2 + c^2 - a^2}{2bc}$

to calculate the size of angle BAC.

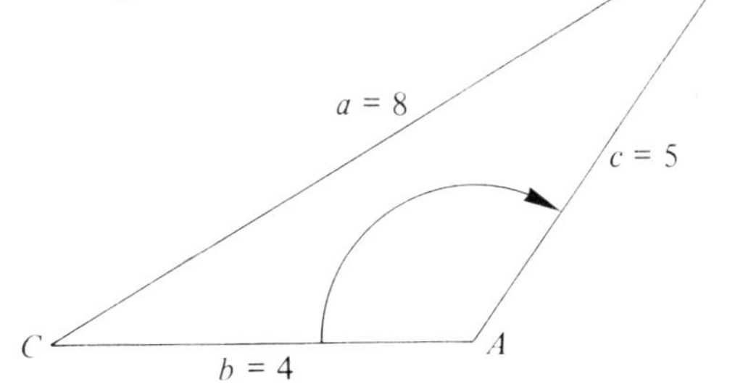

12 (i) Evaluate

$$\begin{pmatrix} -2 & 3 & 4 \\ 1 & -2 & 3 \end{pmatrix}\begin{pmatrix} 1 & 4 & 0 \\ 2 & -2 & -3 \\ 3 & 1 & 2 \end{pmatrix}$$

(ii) $A = \begin{pmatrix} 2 & 4 \\ 1 & 3 \end{pmatrix}$, $B = \begin{pmatrix} 3 & 1 \\ 2 & 0 \end{pmatrix}$, $C = \begin{pmatrix} 2 & -2 \\ -3 & 1 \end{pmatrix}$

Solve the equations (in which X is a 2 by 2 matrix):

(*a*) $A + X = C$ (*b*) $B - X = A$ (*c*) $2C + X = A$

(iii) If $\begin{pmatrix} 3 & -2 \\ 2 & 0 \end{pmatrix}\begin{pmatrix} -4 \\ b \end{pmatrix} = \begin{pmatrix} 0 \\ c \end{pmatrix}$, find b and c.

13 (Squared paper) A plane is flying on a bearing of 074° from base and has travelled 720 km when it is hijacked and has to change course to 225°. It flies on this bearing for 980 km before being ordered to land.

(i) Using a scale of 1 cm : 100 km, draw an accurate plot of the journey and the plane's position when it lands.

(ii) From your diagram, estimate and write down:

(*a*) the distance of the plane from base

(*b*) the bearing of the plane from base.

14 (Squared paper)

x	0	1	2	3	4	5	6
y	1	2			16		

$y = 2^x$

The table shows values of x from 0 to 6 for the graph of $y = 2^x$

(i) Copy and complete the table.
(ii) Draw x and y axes. Scale the x axis from 0 to 6, using 2 cm : 1 unit and the y axis from 0 to 70 using 1 cm : 5 units. Plot the values and join up the points to give a smooth curve.
(iii) From your graph write down approximate values for:

(*a*) $2^{2.3}$ (*b*) $2^{3.6}$ (*c*) $2^{4.7}$

(iv) From the graph write down the following numbers in the form 2^x:

(*a*) 10 (*b*) 22 (*c*) 45

15 In the trapezium $ABCD$, $\vec{AB} = 3\mathbf{a}$ and $\vec{AD} = \mathbf{b}$

The ratio $\frac{AX}{XC} = 3$ and $AB = 3DC$

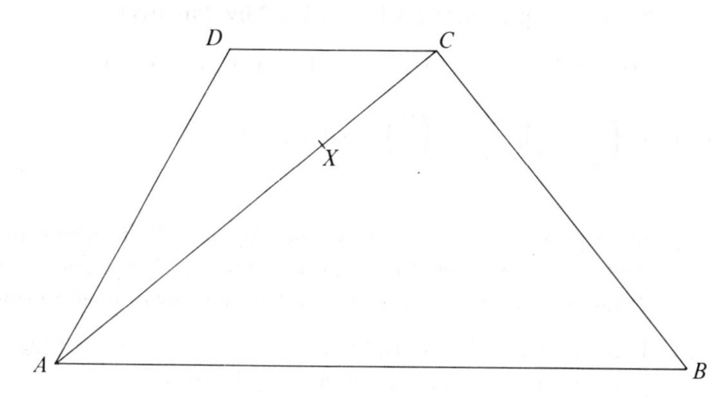

(i) Express $\vec{AC}$ in terms of **a** and **b**.
(ii) Express $\vec{AX}$ in terms of **a** and **b**.
(iii) Express $\vec{XC}$ in terms of **a** and **b**.
(iv) Express $\vec{BD}$ in terms of **a** and **b**.
(v) Express $\vec{DX}$ in terms of **a** and **b**.
(vi) Express $\vec{XB}$ in terms of **a** and **b**.
(vii) From your answers to (v) and (vi) what can you deduce about:

(*a*) the points D, X and B ?

(*b*) the ratio $\frac{DX}{XB}$?

Set 4

1 Factorise:

(i) $2a + 4b + 6c$
(ii) $x^2 - x$
(iii) $a^2 - b^2$
(iv) $a^2 + 15a + 26$
(v) $5k^2 - 3k - 2$

2 The diagram shows a regular hexagon $ABCDEF$, centre O.
$ABXY$ is a square with an area of 100 cm^2.

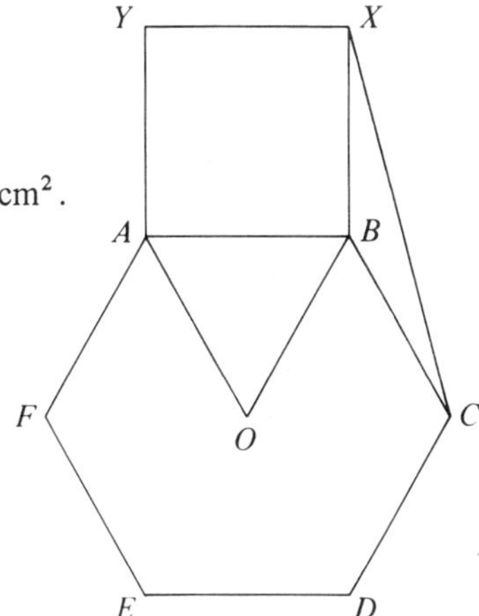

Calculate:

(i) angle ABC
(ii) angle XBC
(iii) angle BCX
(iv) area of triangle AOB
(v) area of hexagon $ABCDEF$.

3 $\mathscr{E}$ = {3,4,5,6,7,8,9,10,11,12}

List the members of the following sets:

(i) A {multiplies of 4}
(ii) B {prime numbers}
(iii) C {factors of 108}
(iv) D $\{x: 5x \leqslant 30\}$
(v) E $\{y: y^2 - 10 \geqslant 30\}$
(vi) $A \cap C$

4 The diagram shows 4 towns P, Q, R and S.
From P to Q there are 2 routes, from Q to R, 3 routes, from P to S three routes and from S to R, 4 routes.

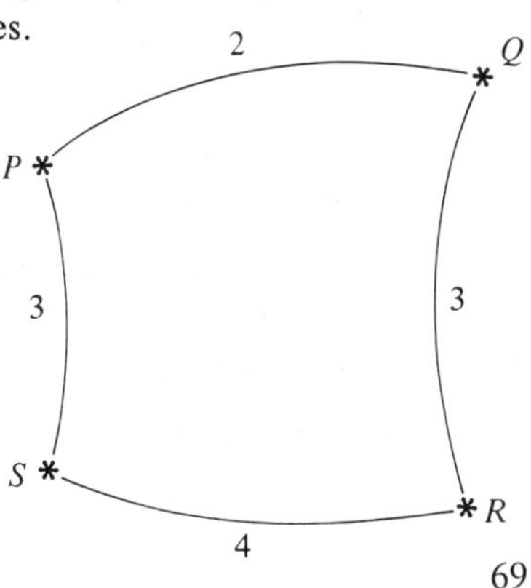

(i) Write down the number of different routes from P to R, passing through Q.
(ii) Write down the number of different routes from P to R passing through S.
(iii) If a route is chosen at random from P to R, what is the probability that it passes through Q?

5 A sphere has a radius of 5 cm. Take $\pi = 3.14$ and calculate:

(i) its volume (use $V = \frac{4}{3}\pi r^3$)
(ii) its surface area (use $A = 4\pi r^2$)

6 (i) Simplify:

(*a*) $\dfrac{x^2 - y^2}{6x^2 + 5xy - y^2}$

(*b*) $\dfrac{1}{x + 1} - \dfrac{1}{x - 1}$

(ii) Solve the equations:

(*a*) $x^2 - 8x + 15 = 0$
(*b*) $x^2 - x = 6$.

7 In the diagram, angle PXY = angle $XYP = 60°$.
$XP = 9$ cm and $PZ = 10$ cm.

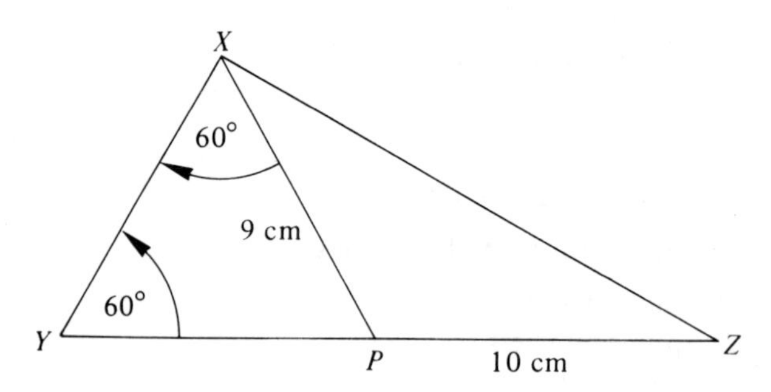

(i) Write down the length of XY.
(ii) Find angle XPZ.
(iii) Calculate XZ.

8 (Squared paper) Draw x and y axes.
Scale the x axis from -6 to $+4$, using 1 cm : 1 unit.
Scale the y axis from -6 to $+5$, using 1 cm : 1 unit.

(i) Plot and join up the points: $A(-1,4)$, $B(3,0)$, $C(-1,-5)$, $D(-5,0)$ to give the quadrilateral $ABCD$.
(ii) Write down:

(*a*) the name of $ABCD$
(*b*) the equations of the lines AB and DA
(*c*) the equation of the line of symmetry of $ABCD$.

(iii) Calculate the area of $ABCD$.

9 (Squared paper) Draw x and y axes.
Scale the x axis from 0 to 4, the y axis from 0 to 8, using 1 cm : 1 unit on both axes.

(i) Plot the triangle: $A(0,2)$, $B(0,0)$, $C(3,1)$.
(ii) Write down the coordinates of ABC as a 2 by 3 matrix, call it M.
(iii) Multiply M, on the left, by T the transformation matrix $\begin{pmatrix} 1 & 0 \\ 2 & 1 \end{pmatrix}$ to give the coordinates of $A'\,B'\,C'$, the image of ABC under the transformation T.
(iv) Plot $A'B'C'$.
(v) Describe the transformation T fully.
(vi) Write down the areas of ABC and $A'B'C'$.

10 The diagram shows triangle PQR in which:
angle PSR = angle $PXS = 90°$ and angle $PRS = 64.1°$
$PR = QS = 10$ cm and
RSQ is a straight line.

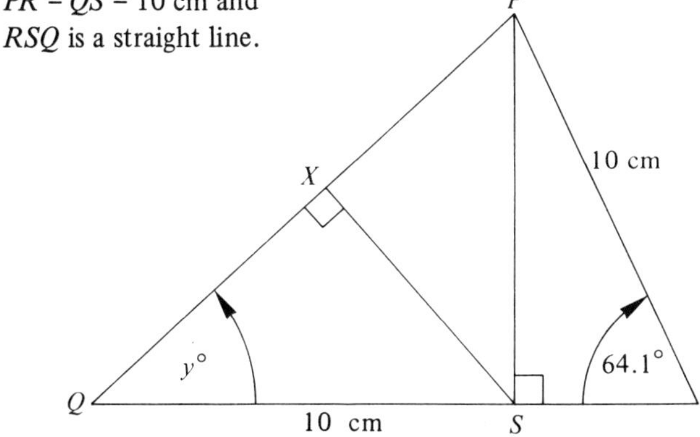

Calculate:

(i) PS
(ii) SR
(iii) angle y
(iv) angle PSX
(v) QX.

11 (i) If $\begin{pmatrix} a & b \end{pmatrix}\begin{pmatrix} 0 & 2 \\ 3 & 0 \end{pmatrix} = \begin{pmatrix} 6 & 20 \end{pmatrix}$ find a and b.

(ii) If $A = \begin{pmatrix} 7 & 3 \\ 5 & 2 \end{pmatrix}$ evaluate A^{-1}.

(iii) Solve the simultaneous equations by a matrix method:

$$7x + 3y = 11$$
$$5x + 2y = 8$$

12 The cumulative frequency graph shows the distribution of marks gained by 400 candidates in an examination.

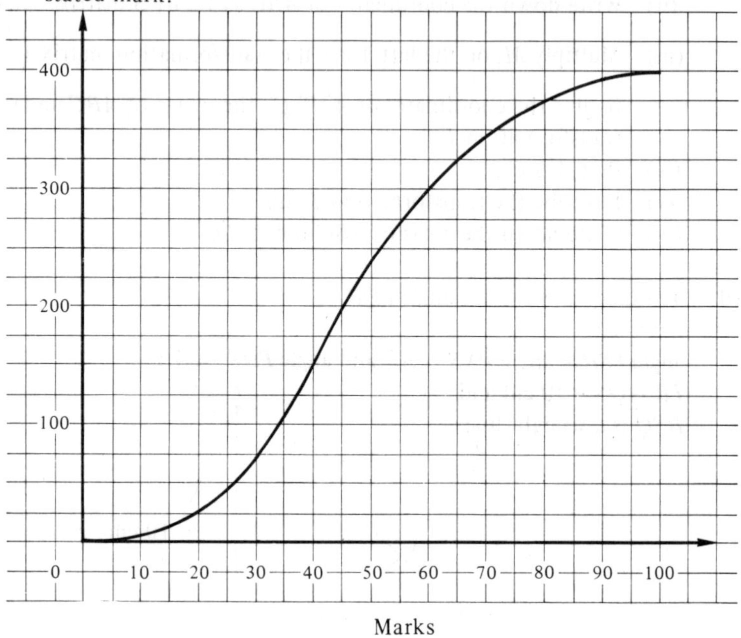

From the graph, estimate:

(i) the median mark
(ii) the number of candidates who scored more than 65 marks
(iii) the number of candidates who scored less than 40 marks
(iv) the pass mark, if 50% of the candidates passed
(v) the upper quartile
(vi) the lower quartile
(vii) the interquartile range.

13 A model of a Jumbo jet is constructed to a scale of 1 : 200

(i) The wing span of the jet is 50 m, what will the wing span measure on the model in centimetres?
(ii) The length of the model is 35 cm, how long is the actual plane in metres?
(iii) If the width of the cabin on the plane is 8 m, what will be the model cabin width (in centimetres)?
(iv) The vertical height of the tailplane is 20 m, calculate the height of the tailplane on the model, in centimetres.
(v) If the wing area of the model is 125 cm^2, calculate the wing area of the jet in square metres.

14 In the parallelogram $ABCD$, $\overrightarrow{AB} = \mathbf{a}$ and $\overrightarrow{AD} = \mathbf{b}$

X is a point on DC such that $\dfrac{DX}{DC} = \dfrac{1}{3}$ and

Y is a point on BC such that $\dfrac{BY}{BC} = \dfrac{1}{3}$

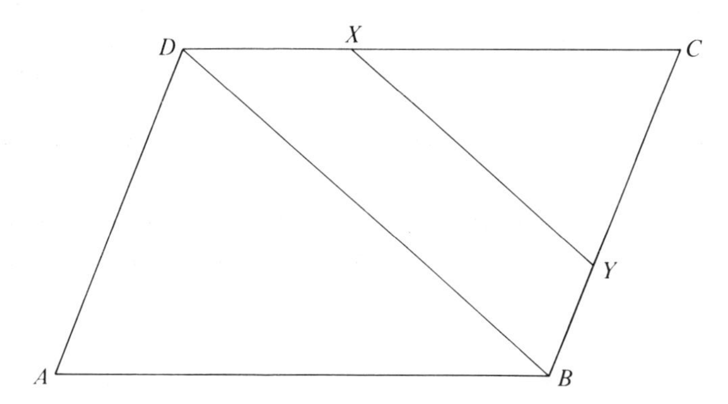

(i) Express $\overrightarrow{DC}$ in terms of $\mathbf{a}$.
(ii) Express $\overrightarrow{XC}$ in terms of $\mathbf{a}$.
(iii) Express $\overrightarrow{BC}$ in terms of $\mathbf{b}$.
(iv) Express $\overrightarrow{CY}$ in terms of $\mathbf{b}$.
(v) Express $\overrightarrow{DB}$ in terms of $\mathbf{a}$ and $\mathbf{b}$.
(vi) Express $\overrightarrow{XY}$ in terms of $\mathbf{a}$ and $\mathbf{b}$.
(vii) What is the ratio $\dfrac{DB}{XY}$?

Set 5

1 (i) Find $12\frac{1}{2}\%$ of £19.04

(ii) Express the fraction $\frac{141}{300}$ as a percentage.

(iii) 30p is $12\frac{1}{2}\%$ of a boy's pocket money. How much does the boy get altogether?

(iv) An article costing £1.50 is sold for £2.25. Calculate the profit as a percentage of the cost price.

(v) Calculate the compound interest on £800 for 2 years at 10% per annum.

2 The diagram shows a rhombus which is the cross section of a prism. $AB = 13$ cm, the diagonal $BD = 10$ cm and the prism is 0.5 m long.

Calculate:

(i) diagonal AC

(ii) tan angle DBC as a decimal fraction

(iii) angle ABC

(iv) area $ABCD$

(v) volume of the prism in cubic centimetres.

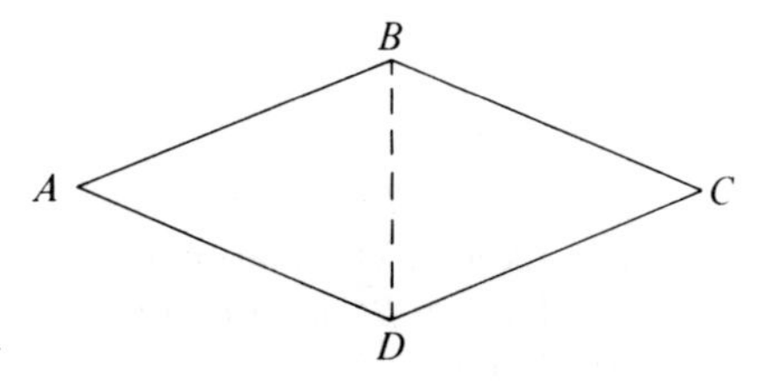

3 $A = \begin{pmatrix} 4 & -1 \\ -2 & 1 \end{pmatrix}$, $B = \begin{pmatrix} -2 & 3 \\ 1 & 4 \end{pmatrix}$

Evaluate:

(i) $B + A$ (ii) $B - A$ (iii) $3(A + B)$

(iv) A^{-1} (v) $A^2 \times A^{-1}$

4 (i) Solve the simultaneous equations:

$$4a - 3b = 15$$
$$a - 3b = 6$$

(ii) Use the formula: $x = \dfrac{-b \pm \sqrt{b^2 - 4ac}}{2a}$

to solve the equation: $3x^2 - 5x - 2 = 0$

5 In the diagram, LM is a tangent to a circle, radius 10 cm, centre O. OL cuts the circumference at X and OL = 20 cm.

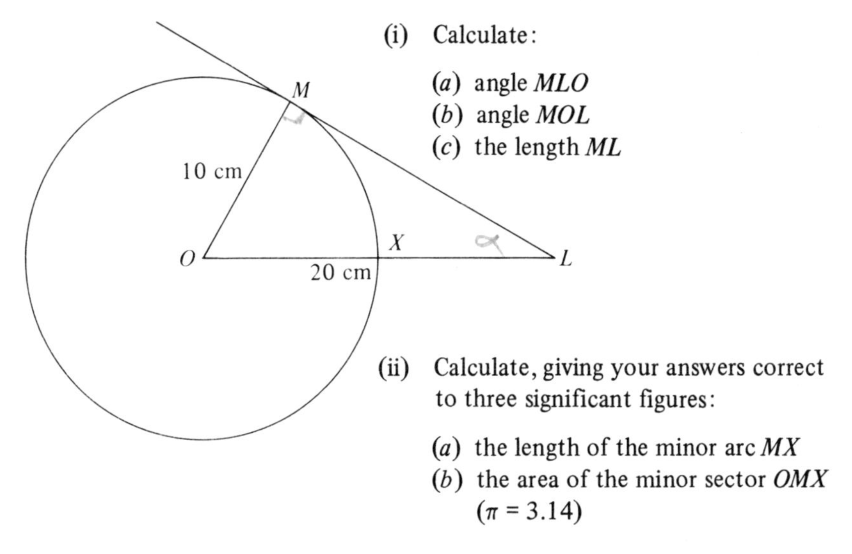

(i) Calculate:

(*a*) angle MLO
(*b*) angle MOL
(*c*) the length ML

(ii) Calculate, giving your answers correct to three significant figures:

(*a*) the length of the minor arc MX
(*b*) the area of the minor sector OMX
($\pi = 3.14$)

6 The diagram shows a solid wooden block from which a solid wooden cylinder is made, diameter 20 cm, length 15 cm.

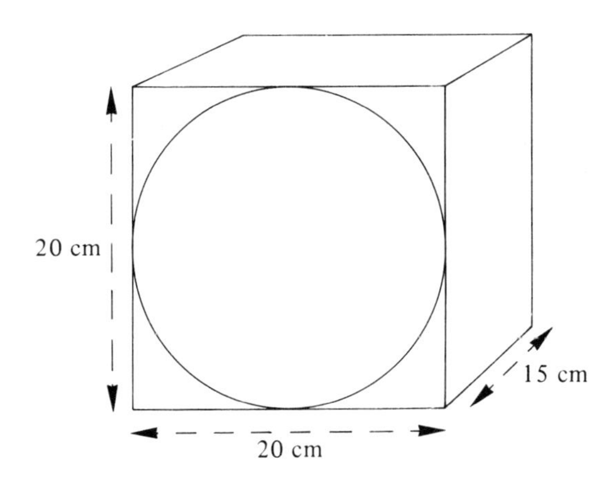

Calculate:

(i) the volume of the block
(ii) the volume of the cylinder
(iii) the volume of wood wasted
(iv) the volume of wood wasted as a percentage of the volume of the block.
($\pi = 3.14$)

7 A car leaves Aberdeen for Dover, which is 896 km away, to catch a ferry which sails at 1800 hours every Tuesday. The driver estimates that they will average 56 km/h and will make three stops of 1 hour each. The car travels approximately 16 km for every litre of petrol used and petrol is 25p a litre.

Calculate:

(i) how long the journey would take without stops at an average speed of 56 km/h

(ii) what time they must leave Aberdeen on Monday if they have to arrive 1 hour before the ferry sails and they make the planned stops.

(iii) the petrol consumption for the journey in litres

(iv) the cost of the petrol used

(v) the time the journey would take travelling by plane at an average speed of 600 km/h, with no stops (give your answer correct to the nearest minute).

8 (i) $a = \frac{1}{8}$ and $b = \frac{1}{2}$, calculate $\frac{a}{b}$ as a percentage.

(ii) Write down a fraction between $\frac{1}{9}$ and $\frac{1}{10}$, giving your answer in the form $\frac{c}{d}$ where c and d are integers.

(iii) If $e * f$ denotes $\frac{e}{f}$, evaluate: $\frac{3}{4} * \frac{3}{16}$

(iv) Evaluate: $\sqrt{10\,000_2}$ and give your answer as a binary number.

(v) $f(x) = x^2 - 2x - 4$, evaluate $f(-3)$

9 The diagram shows a circle, centre O, radius 4 cm.
TA and TM are tangents. Angle $ATM = 32°$

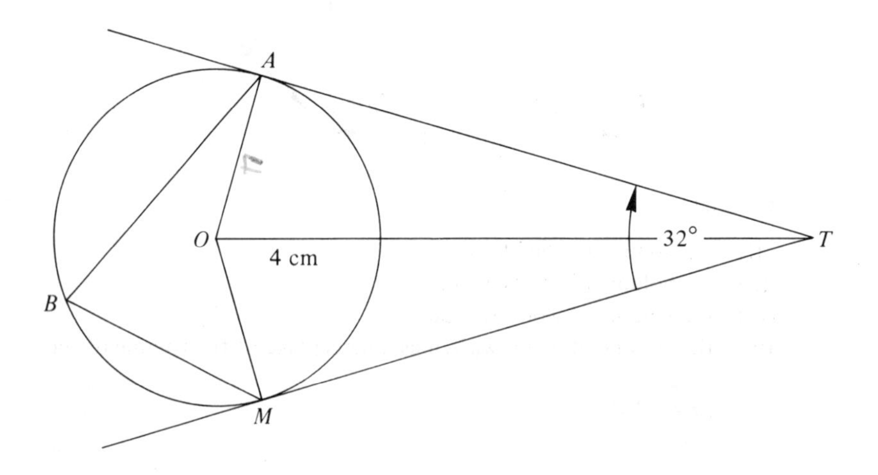

(i) Write down the size of

(*a*) angle TOA (*b*) angle AOM (*c*) angle ABM.

(ii) Write down the value of tan angle TOA from your tables.
(iii) Calculate the length of the tangent TA.
(iv) Find the area of triangle TAO.
(v) Find the area of the quadrilateral $ATMO$.

10 (Squared paper) Draw x and y axes. Scale the x axis from 0 to 6 and the y axis from -5 to $+5$, using 1 cm : 1 unit for both axes.

(i) Plot the kite $ABCD$, $A(1,0)$, $B(5,2)$, $C(3,2)$, $D(3,4)$.
(ii) Write down the coordinates of $ABCD$ as a 2 by 4 matrix, P.
(iii) Multiply P, on the left, by the transformation matrix $T\begin{pmatrix}1 & 0\\ 0 & -1\end{pmatrix}$ to give the coordinates of $A'B'C'D'$, the image of $ABCD$ under the transformation T.
(iv) Plot $A'B'C'D'$.
(v) Describe the transformation T.

11 In the diagram, BCD is an isosceles triangle.
$CD = 5$ cm, $BD = 8$ cm, $AD = 10$ cm.
Angle $BDA = 60°$.

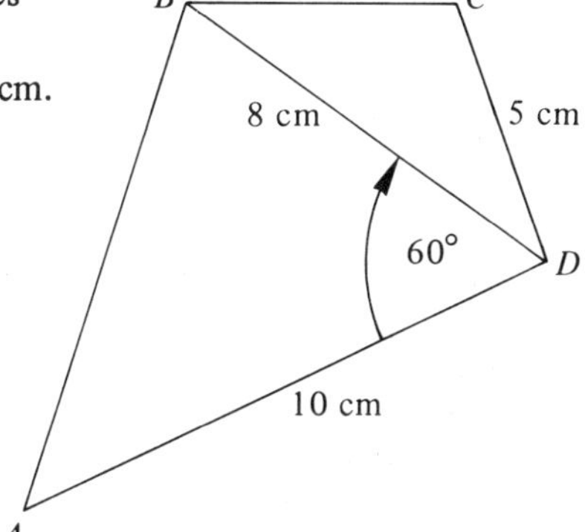

Calculate:

(i) angle CBD
(ii) area of triangle BCD
(iii) angle ADC
(iv) the length AB
(use the cosine formula)

12 (i) A sum of money is shared among three men A, B and C in the ratio 2 : 5 : 11 respectively. If B's share is £90, calculate:

(*a*) A's share (*b*) C's share (*c*) the sum of money.

(ii) On a map drawn to a scale of 1 : 30 000, a straight road is 8 cm long. Calculate the length of the road in kilometres.
(iii) The sides of 2 cubical containers are in the ratio 2 : 5.
If the smaller one has a capacity of 64 litres, what is the capacity of the larger one?

13 The travel graph shows the journeys of Mr Raleigh and Mr Copter from Aytown to Beetown and the journey of Mr Driver from Beetown to Aytown.

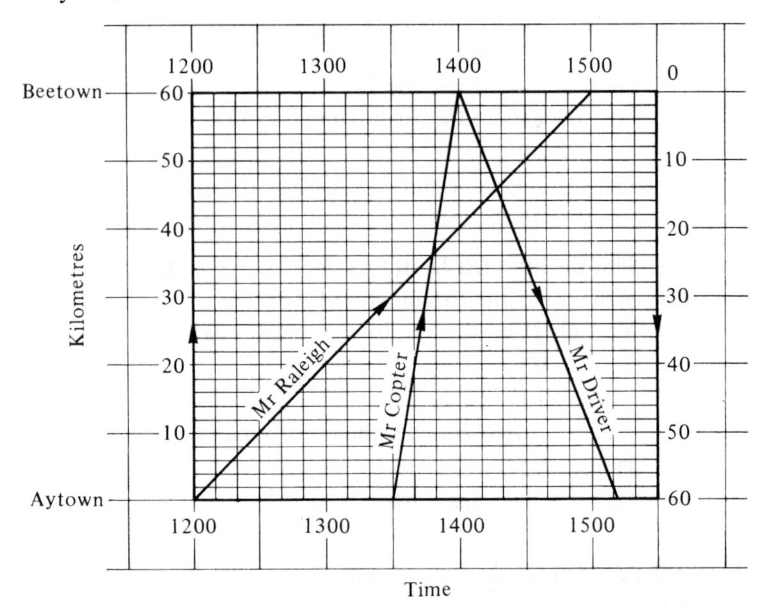

(i) At what time did Mr Driver arrive at Aytown?
(ii) What was his average speed in kilometres per hour?
(iii) At what time did Mr Copter leave Aytown?
(iv) What was his average speed for the journey in kilometres per hour?
(v) What was Mr Raleigh's average speed?
(vi) At what time did Mr Copter overtake Mr Raleigh?
(vii) How far were they from Beetown when this happened?
(viii) At what time did Mr Raleigh and Mr Driver pass each other?
(ix) If Mr Raleigh's average speed had been 24 km/h, at what time would he have arrived?
(x) If 5 miles = 8 km, what is Mr Copter's average speed in miles per hour?

14 In the parallelogram $ABCD$, $\vec{AB} = \mathbf{a}$ and $\vec{AD} = \mathbf{b}$

X is a point on BC such that $\dfrac{BX}{BC} = \dfrac{3}{4}$

Y is a point on CD such that $\dfrac{CY}{CD} = \dfrac{1}{4}$

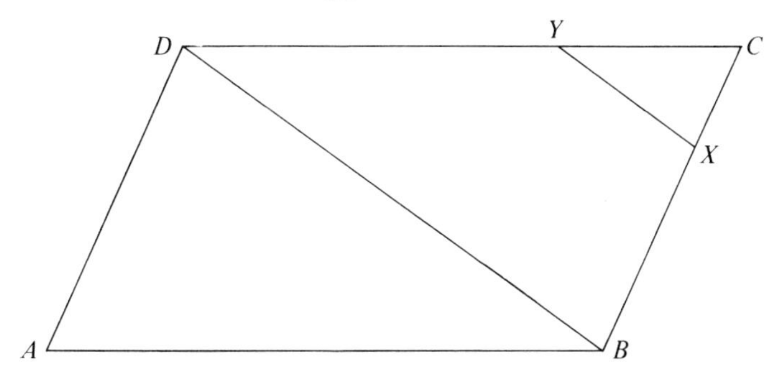

(i) Express $\vec{BC}$ in terms of $\mathbf{b}$.
(ii) Express $\vec{XC}$ in terms of $\mathbf{b}$.
(iii) Express $\vec{CD}$ in terms of $\mathbf{a}$.
(iv) Express $\vec{CY}$ in terms of $\mathbf{a}$.
(v) Express $\vec{BD}$ in terms of $\mathbf{a}$ and $\mathbf{b}$.
(vi) Express $\vec{XY}$ in terms of $\mathbf{a}$ and $\mathbf{b}$.
(vii) Write down the ratio $\dfrac{XY}{BD}$

15 (Squared paper) The area of a rectangle is 24 cm^2. Its length is x cm and its breadth is y cm. x and y are integers.
The partly completed table shows some possible values of x and y.

x	1	2	3	4	6	8	12	24
y	24	12	8					

(i) Copy and complete the table.
(ii) Draw x and y axes. Scale both axes from 0 to 24, 1 cm : 2 units. Plot the values in the table and join them up to give a smooth curve.
(iii) One equation of this curve is $xy = 24$, write down another equation for the curve.
(iv) It is also stated that $y < x$ and $x + y \leqslant 14$. Draw a ring round the points on your curve which satisfy these inequalities.
(v) Write down the values of x and y which satisfy the inequalities.

Set 6

1 Factorise:

(i) $5a + 15b + 10$
(ii) $9a^2 - 25b^2$
(iii) $3x^2 - 48$
(iv) $y^2 + 8y + 15$
(v) $9a^2 - 3ab - 2b^2$

2 The diagram shows the net of a solid. $AB = 8$, $AD = 3$, $EH = 5$ cm.

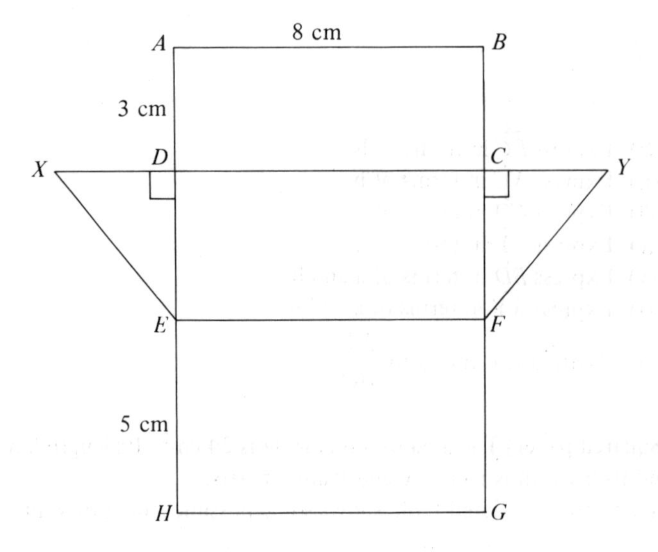

(i) Write down:

(*a*) the name of the solid which could be made from the net
(*b*) the length EX.

(ii) Calculate:

(*a*) the length DE
(*b*) the area of triangle DEX
(*c*) the total surface area of the solid
(*d*) the volume of the solid
(*e*) angle CFY.

3 On squared paper draw x and y axes.
Scale the x axis from -1 to $+5$ and the y axis from -5 to $+5$.
Use 1 cm : 1 unit on both axes.

(i) Draw and label the graphs of $x = 2$, $x + y = 4$ and $y = x - 4$.
(ii) Shade the region R where $x \geqslant 2$, $x + y \leqslant 4$ and $y \geqslant x - 4$.

4 Solve the equations:

(i) $12x - 5x = 56$
(ii) $\frac{1}{4}x + \frac{1}{4}x = 12$
(iii) $2(a + 4) + 3(a - 2) = 12$
(iv) $x^2 - 8x - 20 = 0$
(v) $2x^2 - 5x - 3 = 0$

5 In the diagram, XY and XZ are tangents to the circle, centre O. If angle $YXZ = 50°$, calculate: (i) angle ZOX
(ii) angle OZA
(iii) angle ZAX.

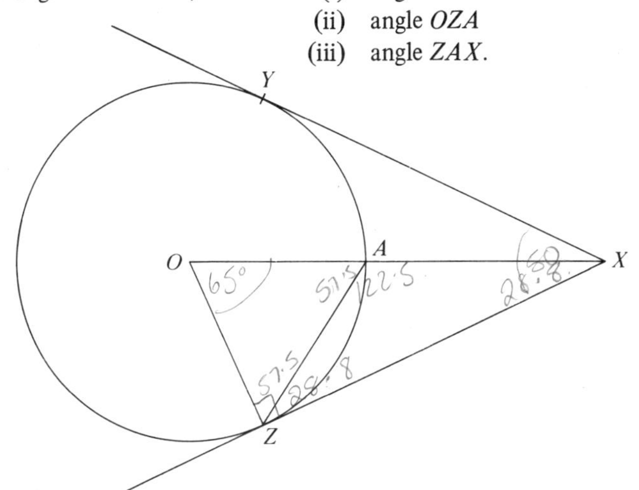

6 The diagram shows a large tent which is composed of a cylindrical base and a conical top. The base has a diameter of 20 m. XY, the vertical side is 2 m and the height of the cone, AB is 7.5 m.

Calculate:

(i) the area of ground covered by the base
(ii) the volume of the cylindrical base
(iii) the volume of the conical top
(iv) the total volume of the tent
(v) the length of the sloping side AX
(vi) the area of canvas needed to make:

(*a*) the conical top
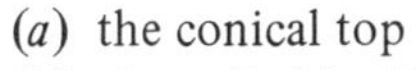
(*b*) the cylindrical base.

($\pi = 3.14$)

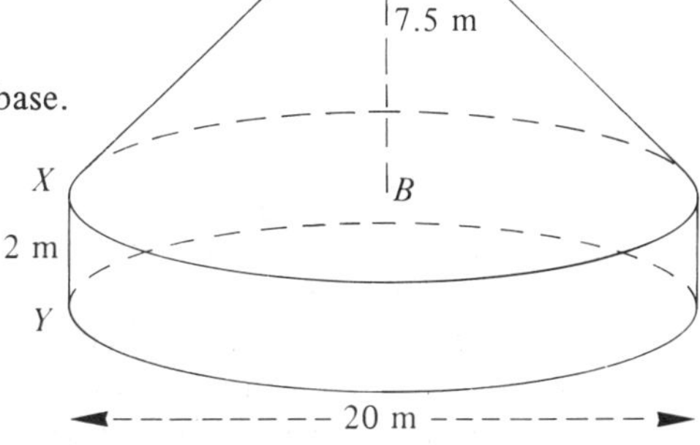

7 If $a = 10^8$, $b = 10^5$ and $c = 10^{-2}$, evaluate the following and leave your answer in index form:

(i) ab (ii) abc (iii) $\frac{ac}{b}$

(iv) $\frac{ab}{c}$ (v) $a^{\frac{1}{2}}$ (vi) b^2c^2

8 (i) £2.70 is shared among Tom, Dick and Harry in the ratio 2 : 3 : 4 Find Harry's share.

(ii) A sum of money was shared among Jill, Joan and Judy in the ratio 2 : 3 : 5. If Joan's share was £9.00, find the sum of money which was shared.

(iii) The sides of a cube are in the ratio 2 : 5. Write down:

(*a*) the ratio of their surface areas
(*b*) the ratio of their volumes.

9 The diagram shows the circumscribed circle, centre O, radius 5 cm, of the regular polygon $ABCDE$. X is the mid point of BC. Angle ABC is an interior angle of the polygon.

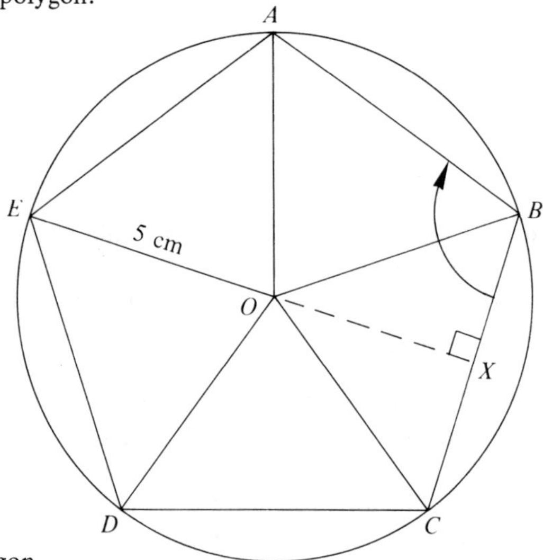

(i) Name the polygon.
(ii) Calculate the size of:

(*a*) angle BOC
(*b*) angle OCX
(*c*) angle ABC

(iii) Write down the sum of the interior angles in right angles.
(iv) Use the sine ratio to calculate BX.
(v) Use the cosine ratio to calculate OX.
(vi) Calculate the perimeter of the polygon.
(vii) Calculate the area of the polygon, correct to 3 significant figures.

10 (Squared paper) Draw x and y axes.
Scale the x axis from -6 to $+6$ and the y axis from -2 to $+6$, using 1 cm : 1 unit on both axes.

(i) Plot the kite $ABCD$: $A(3,5)$, $B(5,3)$, $C(3,-1)$, $D(1,3)$.
(ii) Write down the coordinates of the kite as a 2 by 4 matrix, P.
(iii) Multiply P, on the left, by the transformation matrix T, $\begin{pmatrix} -1 & 0 \\ 0 & 1 \end{pmatrix}$, to give $A'B'C'D'$, the image of $ABCD$ under the transformation T.
(iv) Plot $A'B'C'D'$.
(v) Describe the transformation T.

11 In the diagram, $AC = 8$ cm, $BC = 17$ cm and angle $BAC = 90°$. CX is the bisector of angle ACB.

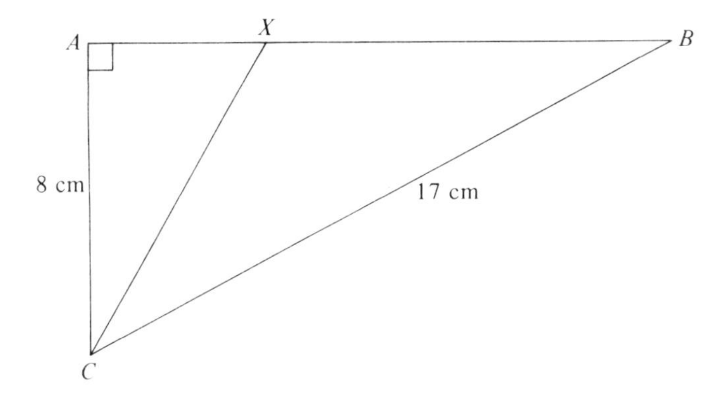

Calculate:

(i) AB
(ii) angle ACB
(iii) angle ACX
(iv) AX.

12 (i) Evaluate $x^2 - x - \frac{1}{x}$ when $x = -4$
(ii) Find the smallest integer x, such that $x + 8 \leqslant 4x - 7$
(iii) Factorise $x^2 + 8x + 16$
(iv) If $x^2 + 8x + 16 \leqslant 36$, write down the possible values of x if x is a positive integer.
(v) If $\frac{(x+1)}{(x-4)} = (x + 1)$ find 2 possible values of x.

13 (Squared paper) The partly completed table shows values of x and y for the graph of the equation, $y = x^2 - 2x - 2$.

x	−2	−1	0	1	2	3	4
y	6			−3		1	

(i) Copy and complete the table.

(ii) Draw x and y axes. Scaled: x axis from − 2 to + 4, 2 cm : 1 unit;
y axis from − 3 to + 6, 1 cm : 1 unit.
Plot the points and join them up to give a smooth curve.

(iii) From your graph write down the values of x which would satisfy the equation: $x^2 - 2x - 2 = 0$

(iv) Draw on the same diagram, the straight line whose equation is $y = x - 2$

(v) Write down the values of x at the points of intersection of $y = x^2 - 2x - 2$ and $y = x - 2$

(vi) Write down a single equation which these values would satisfy. Simplify the equation.

14 In the trapezium $ABCD$, $AB = 2DC$, $\overrightarrow{AB} = 2\mathbf{a}$ and $\overrightarrow{AD} = \mathbf{b}$

X is a point on AC such that $\dfrac{AX}{AC} = \dfrac{1}{2}$

Y is a point on BD such that $\dfrac{BY}{BD} = \dfrac{1}{2}$

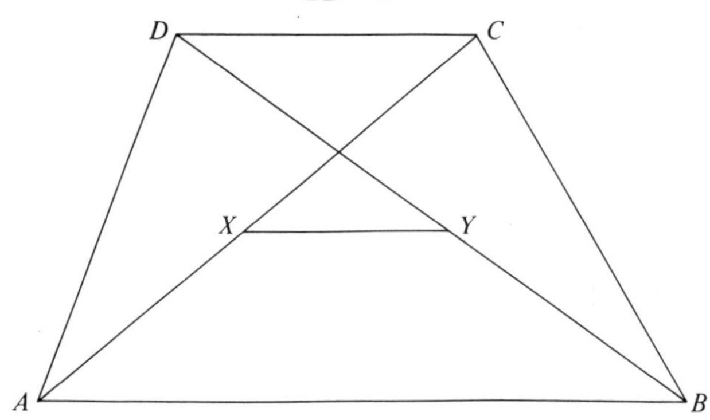

(i) Express in terms of **a** and **b**:

(*a*) $\overrightarrow{BD}$ (*b*) $\overrightarrow{AC}$ (*c*) $\overrightarrow{AX}$ (*d*) $\overrightarrow{BY}$ (*e*) $\overrightarrow{XY}$

(ii) Deduce two relationships about XY and CD.

15 The diagrams show a cylindrical cheese from which a uniform slice has been accurately removed. On the right is a bird's eye view of the cheese, showing the sector angle of 36° which has been removed. O is the centre of the circle. The diameter of the cheese is 20 cm and its height is 25 cm.

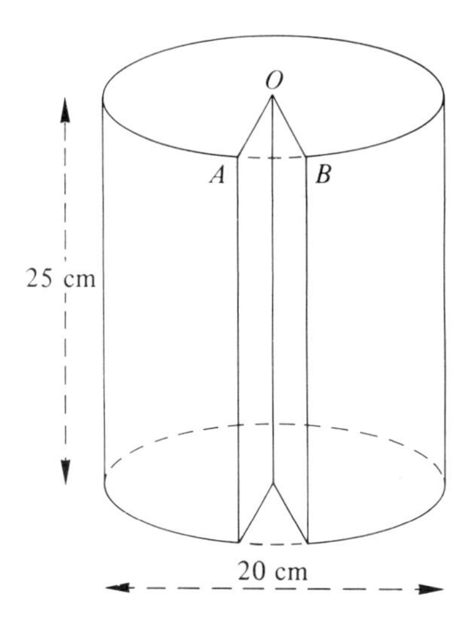

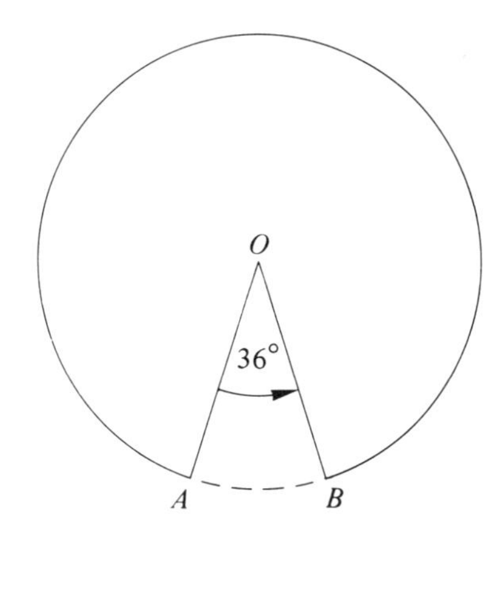

(i) Calculate in terms of π.

(*a*) the area of the minor sector *OAB*
(*b*) the length of the circumference left after the slice has been removed
(*c*) the volume of the whole cheese
(*d*) the volume of the slice cut out.

(ii) A whole cheese costs £15.50, calculate the value of the slice removed.

(iii) A similar cheese has a height of 50 cm and a diameter of 40 cm. Write down:

(*a*) the ratio of the volumes of the two cheeses
(*b*) the value of the larger cheese.

Set 7

1 (i) Find $33\frac{1}{3}\%$ of £16.56

(ii) Express a score of 16 out of 25 as a percentage.

(iii) Find the number of which 24 is $12\frac{1}{2}\%$.

(iv) A man sold his car for £1280 and in doing so lost 20% of the price he paid for it. Find the price he paid for it.

(v) Calculate the compound interest on £400 for 2 years at 5% per annum.

2 The diagram shows a wooden cube of side 12 cm from the corner of which a triangular pyramid is to be cut off.

$PQ = PR = PS = 6$ cm.

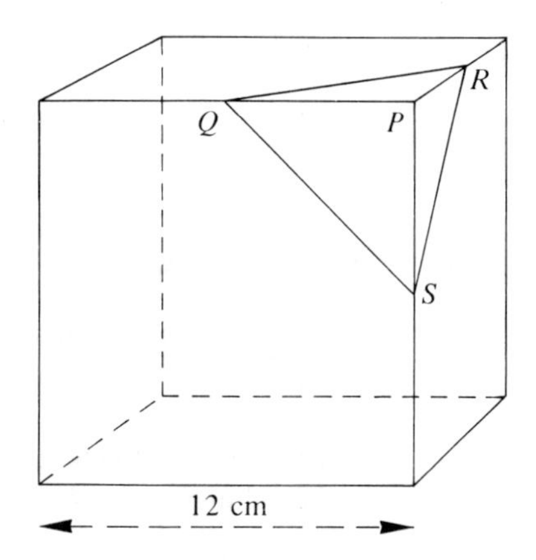

Calculate:

(i) the volume of the cube

(ii) the volume of the pyramid to be removed

(iii) the ratio: $\dfrac{\text{volume of pyramid}}{\text{volume of cube}}$ (in its lowest terms).

3 (i) $P = \begin{pmatrix} 3 & -3 \\ -4 & 4 \end{pmatrix}$

(*a*) Calculate P^2.

(*b*) Why is P a singular matrix?

(ii) If $3\begin{pmatrix} 2 & 1 \\ -2 & 3 \end{pmatrix} + \begin{pmatrix} 2 & 1 \\ 0 & -2 \end{pmatrix}\begin{pmatrix} 3 & -1 \\ 2 & 2 \end{pmatrix} = \begin{pmatrix} a & b \\ c & d \end{pmatrix}$,

find a,b,c and d.

4 The diagram shows a rectangle $(x^2 + 12)$ cm long and $(x^2 + 2)$ cm wide. Its perimeter is 92 cm.

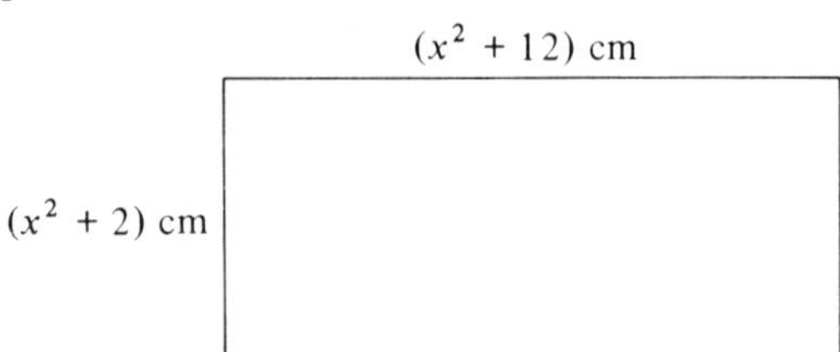

(i) Write down an equation for the perimeter of the rectangle.
(ii) Solve the equation and find:

(*a*) the length of the rectangle
(*b*) the width of the rectangle
(*c*) the area of the rectangle.

5 Triangle ABC is in the semi-circle, diameter AB, centre O. OX is an arc of a circle, centre A, radius AO.

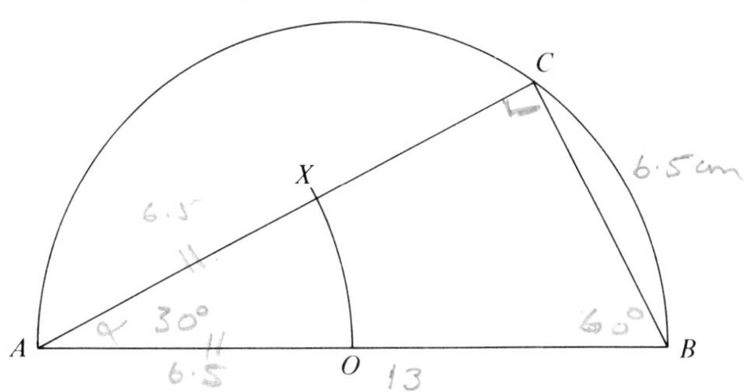

If AB = 13 cm and BC = 6.5 cm,

Calculate:

(i) angle BAC
(ii) angle ABC
(iii) the length XC
(iv) the length of the arc OX
(π = 3.14)

6 The diagram shows the metal weight at the end of a plumb line, which is called a plummet. It consists of a hemisphere attached to a cone. The radius of the hemisphere is 2 cm and the height of the cone is 3 cm.

Calculate:

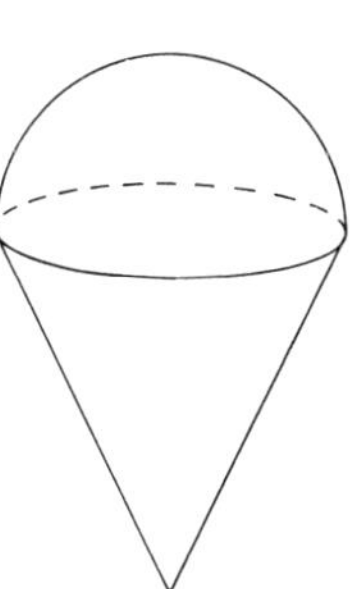

(i) the volume of the hemisphere
(ii) the volume of the cone
(iii) the volume of the plummet
(iv) the weight of the plummet if the metal weighs 11 grams per cm^3.
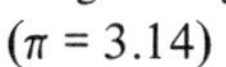
(π = 3.14)

7 The diagram shows a lawn which is in the shape of a square with two semi-circular ends. The side of the square is $2x$ metres.

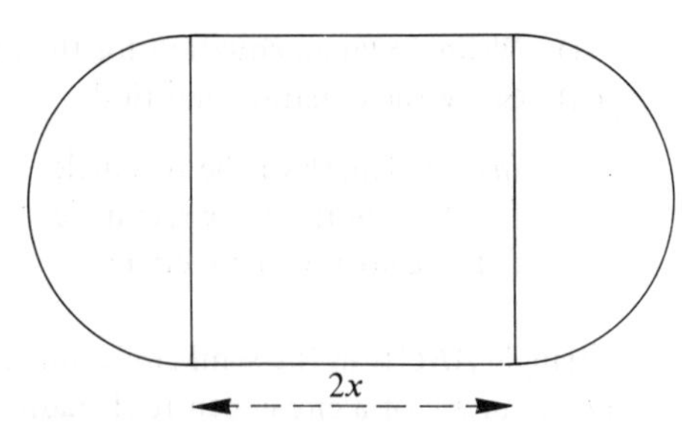

(i) Write down the area of the square in terms of x.
(ii) Write down the total area of the semi-circular ends in terms of π and x.
(iii) Write down a formula for A, the total area of the lawn.
(iv) Factorise your formula.
(v) Use your formula to find A if $x = 10$ m and $\pi = 3.14$

8 (i) Three shareholders, A, B and C hold shares to the value of £20 000, £30 000 and £40 000 respectively.
The profits are shared in the ratio of their holdings.
If the profits in a certain year amounted to £18 000, calculate A's, B's and C's share of the profits.
(ii) If these shares of the profits were shown in a pie chart, calculate the sector angles for A, B and C.

9 The diagram shows 4 lines AB, CD, EF and GH which intersect at the points P, Q, R, S.

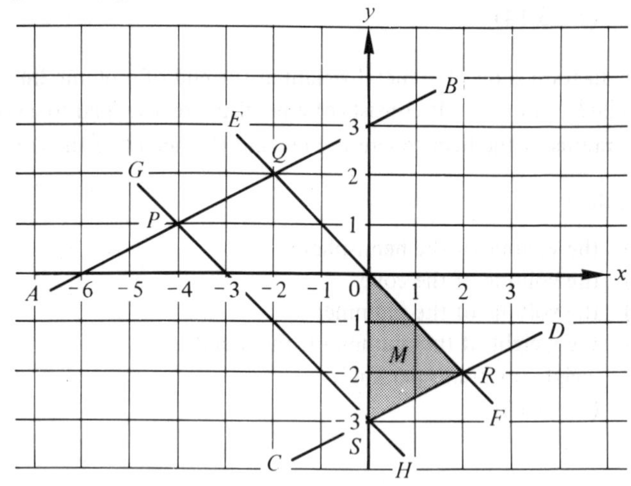

(i) Write down the coordinates of:

(*a*) P (*b*) S

(ii) Write down the equations of the lines:

(*a*) AB (*b*) CD (*c*) EF (*d*) GH

(iii) Write down 3 inequalities which would define the shaded region M.

(iv) Write down the equation of a line which passes through the origin 0 and divides the parallelogram $PQRS$ into 2 parallelograms which are equal in area.

(v) Write down the equation of a line which passes through P and bisects the area $PQRS$.

10 (Squared paper) Draw x and y axes.
Scale the x axis from -4 to $+4$ and the y axis from 0 to 4, using 1 cm : 1 unit on both axes.

(i) Plot the quadrilateral $ABCD$: $A(4,4)$, $B(3,1)$, $C(3,3)$, $D(1,3)$.

(ii) Write down the coordinates of $ABCD$ as a 2 by 4 matrix, Q.

(iii) Multiply Q on the left by T, the transformation matrix $\begin{pmatrix} 0 & -1 \\ 1 & 0 \end{pmatrix}$ to give $A'B'C'D'$ the image of $ABCD$ under the transformation T.

(iv) Plot $A'B'C'D'$.

(v) Describe the transformation under T.

11 Triangle ABC is an isosceles triangle in which:
$AB = BC = 10$ cm and angle $ACB = 23.6°$

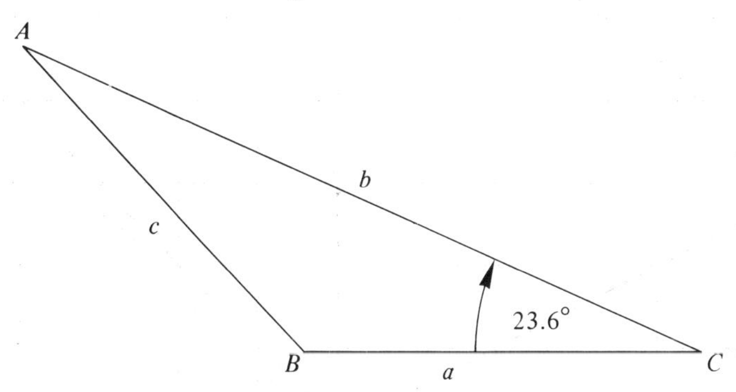

(i) Calculate angle ABC.

(ii) Use the sine formula to calculate the length of AC.

12 The diagram, drawn accurately to a scale of 1 cm : 10 km, shows the positions of 4 towns A,B,C and D.
From A to B is 100 km, from B to C is 50 km and from C to D is 70 km. The dotted lines represent North lines. Copy the diagram on squared paper.

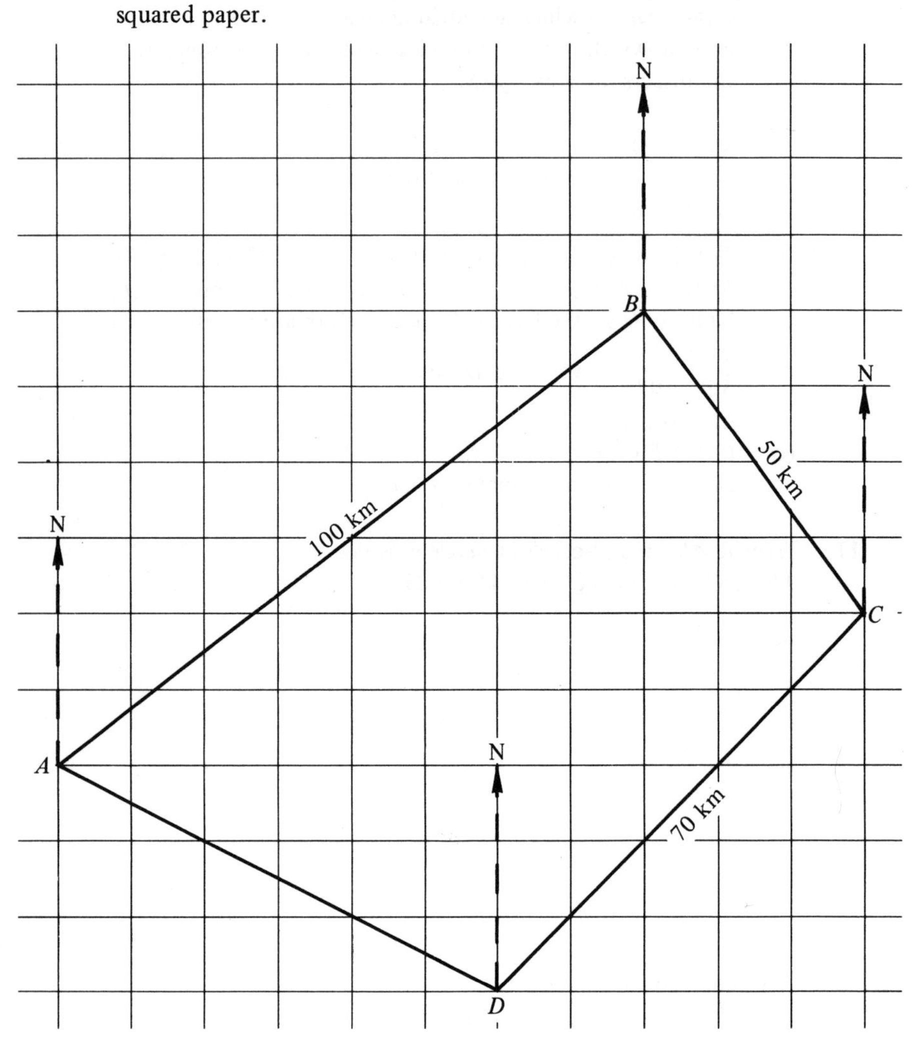

(i) By accurate measuring write down the following bearings:

(*a*) B from A (*b*) C from B (*c*) D from C
(*d*) D from A (*e*) A from C

(ii) Measure and write down the direct distances in kilometres of:

(*a*) A from D (*b*) B from D (*c*) A from C

(iii) A motorist travelled from A to D, via B and C at an average speed of 60 km/h. How long did the journey take him? (Answer in hours and minutes.)

(iv) A motorist left C at 1215 hours to go to A via B. If he arrived at A at 1600 hours, what was his average speed?

13 (i) Solve the simultaneous equations:

$4x - y = 3$
$2x + 3y = 19$

(ii) Use the formula: $x = \dfrac{-b \pm \sqrt{b^2 - 4ac}}{2a}$ to solve the quadratic equation: $3x^2 - 5x - 2 = 0$.

14 In the parallelogram $ABCD$, $\vec{AB} = 3\mathbf{a}$ and $\vec{AD} = 3\mathbf{b}$

X is a point on AD such that $\dfrac{AX}{AD} = \dfrac{1}{3}$

Y is a point on BC such that $\dfrac{BY}{BC} = \dfrac{2}{3}$

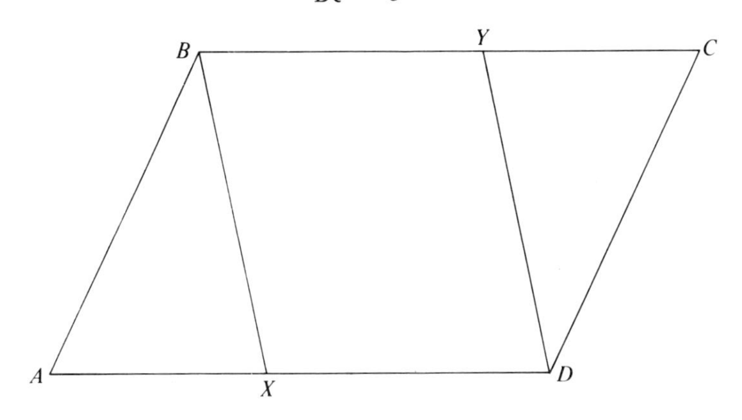

(i) Express in terms of $\mathbf{b}$:

(a) $\vec{AX}$ (b) $\vec{XD}$ (c) $\vec{CY}$ (d) $\vec{BY}$

(ii) Express in terms of $\mathbf{a}$ and $\mathbf{b}$:

(a) $\vec{BX}$ (b) $\vec{YD}$

(iii) From the above answers, what facts can you deduce about:

(a) BX and YD (b) BY and XD

(iv) What name do we give to the quadrilateral $XDYB$?

15 The diagram shows a cube whose side is x cm.

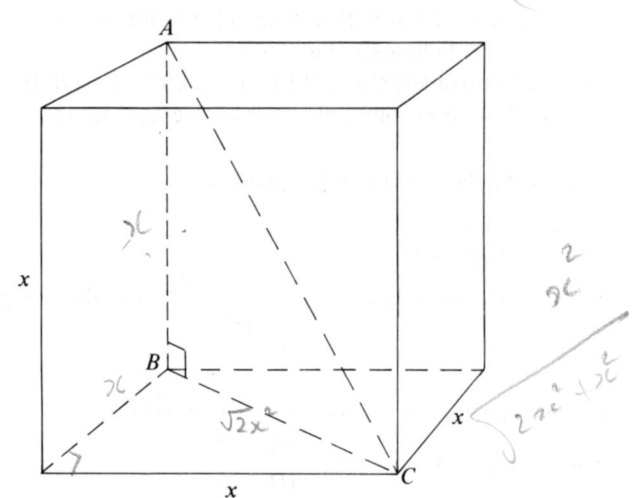

(i) Write down expressions in terms of x for:

(*a*) the total length of the edges
(*b*) the area of one face
(*c*) the total surface area
(*d*) the volume of the cube
(*e*) the length of BC, a face diagonal
(*f*) the length of AC, a space diagonal.

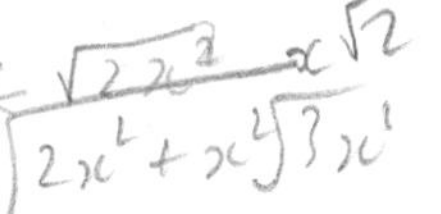

(ii) If the total surface area is 486 cm^2 find:

(*a*) the value of x
(*b*) the volume of the cube in cubic centimetres.

Set 8

1 Factorise:

(i) $6x + 12y + 6$
(ii) $x^3 + x^2$
(iii) $x^2 - 144$
(iv) $a^2 - 12a + 11$
(v) $6x^2 + 24x + 24$

2 *EFGH* is a cardboard square, inside which is drawn the net of a square-based pyramid. $EF = 5.8$ cm and $DC = 3$ cm. This sides of the pyramid are equilateral triangles and X is the mid point of AB.

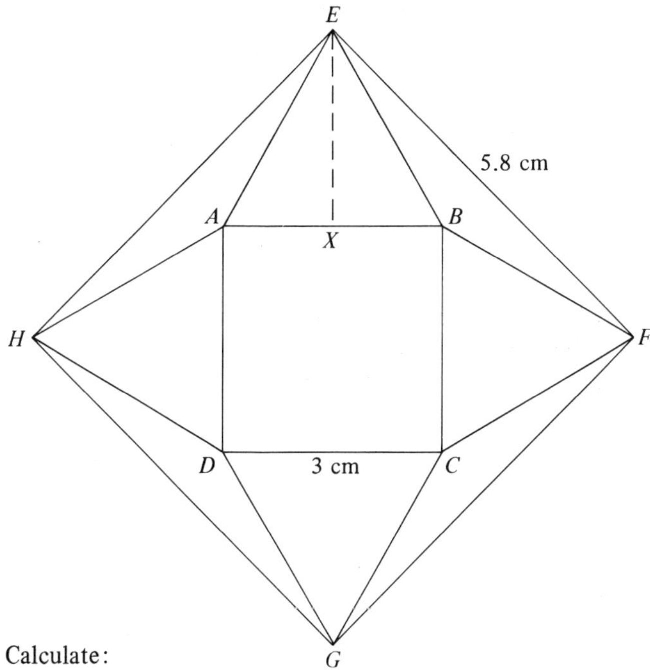

Calculate:

(i) angle *EBF*
(ii) the distance *EX*
(iii) the area of triangle *AEB*
(iv) the total surface area of the pyramid
(v) the area of card wasted when the pyramid is cut out
(vi) the area of triangle *EBF*
(vii) the perpendicular height of the pyramid made from the net (the points *E,F*, *G*, *H* meet directly above the centre of the square *ABCD*).

3 (i) In a General Election there was a 78% poll in a constituency of 38 000 voters. Of the votes cast, 49% went to Mr Talkwell, 36% to Mr Fairwords and the rest to Mrs Yapgood.

Calculate:

(*a*) the total votes cast
(*b*) the number of votes received by Mrs Yapgood.

(ii) During the first year a car depreciated by 25% of its original value and during the second year by 20% of its value at the end of the first year.
Express the total depreciation during the first two years as a percentage of its original value.

4 (i) Solve the simultaneous equations:

$$3x + 2y = 8$$
$$5x - 2y = -8$$

(ii) Solve the quadratic equation: $3x^2 - 5x - 3$ by using the formula:
$$x = \frac{-b \pm \sqrt{b^2 - 4ac}}{2a}$$

5 (i) $\vec{AB} = \mathbf{x}, \vec{BC} = \mathbf{y}, \vec{DC} = \mathbf{z}$.

(*a*) Write $\vec{DB}$ in terms of $\mathbf{y}$ and $\mathbf{z}$.
(*b*) Write $\vec{DA}$ in terms of $\mathbf{x}$, $\mathbf{y}$ and $\mathbf{z}$.
(*c*) Write $\vec{AD}$ in terms of $\mathbf{x}$, $\mathbf{y}$ and $\mathbf{z}$.

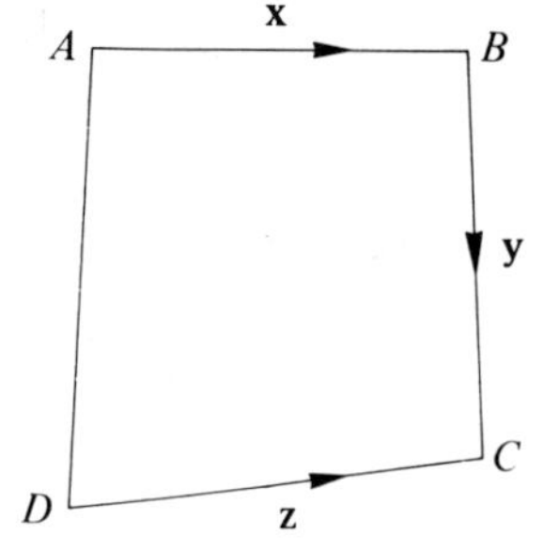

(ii) $\vec{AC} = \mathbf{p}, \vec{BC} = \mathbf{q}, AY = \frac{1}{2}YC$
XY is parallel to BC.

Calculate $\vec{AX}$ in terms of $\mathbf{p}$ and $\mathbf{q}$.

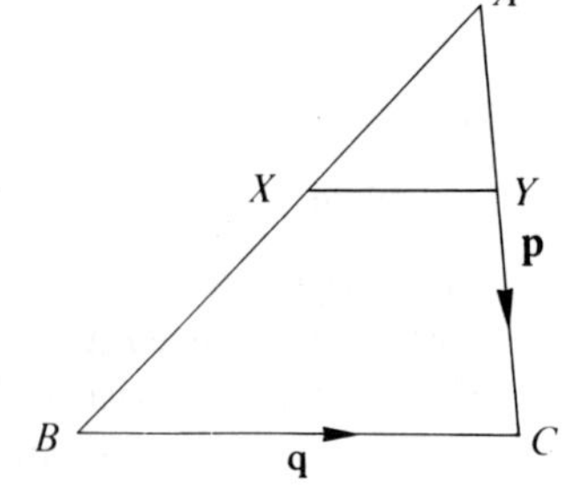

6 The diagram shows a sphere which fits exactly inside a cylinder, radius r cm.

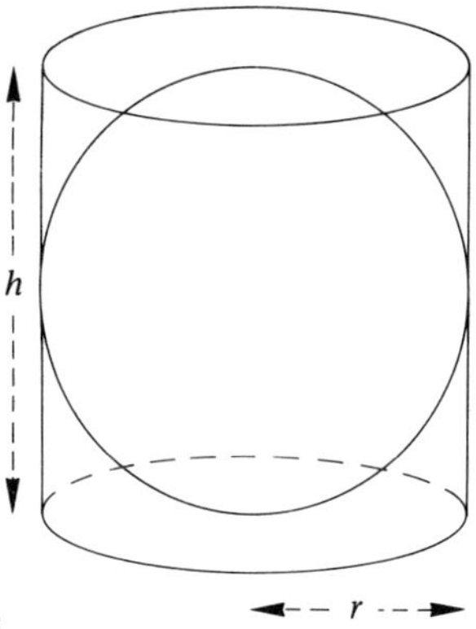

Formulae

	Sphere	Cylinder
Volume	$\frac{4}{3}\pi r^3$	$\pi r^2 h$
Area of curved surface	$4\pi r^2$	$2\pi rh$

(i) Write down h, the height of the cylinder in terms of r.
(ii) Substitute this expression for h in the cylinder formula for volume.
(iii) Simplify the volume formula.
(iv) What fraction of the volume of the cylinder is the volume of the sphere?
(v) Carry out a similar operation on the area of curved surface formula for the cylinder and state the relationship between the area of the curved surface of a cylinder and a sphere which fits inside the cylinder exactly.

7 Given that $C = 2\pi r$:

(i) find C if $\pi = 3.14$ and $r = 2.5$
(ii) find C if $\pi = 3.14$ and $r = 7.5$
(iii) change the subject of the formula to r
(iv) find r if $\pi = 3.14$ and $C = 62.8$
(v) find r if $\pi = \frac{22}{7}$ and $C = 88$

8 (i) $x * y$ denotes the operation $2x + 5y$.

Evaluate:

(*a*) $3 * 6$
(*b*) $-2 * 5$
(*c*) $(2 * 3) * 4$

(ii) $p * q$ denotes the operation q^p

Evaluate:

(*a*) $2 * 8$
(*b*) $-2 * 8$
(*c*) $(1 * 3) * 4$

9 The diagram shows a bird's eye view of a 1p coin placed on top of and exactly in the centre of a 2p coin. The shaded region is the area of the 2p coin which is visible. The diameter of the 1p coin is 2 cm and the diameter of the 2p coin is 2.6 cm.

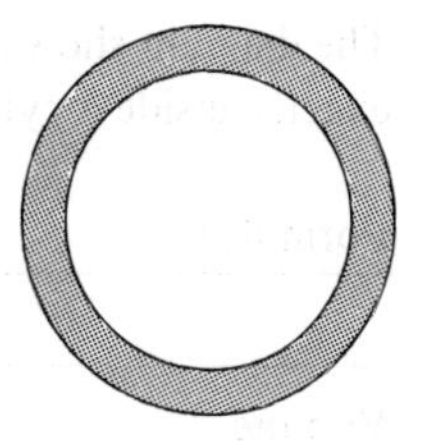

(i) Write down the radius of the 2p coin, call it R (capital R).
(ii) Write down a formula for the area of the 2p coin, using π and R.
(iii) Write down the radius of the 1p coin, call it r (small r).
(iv) Write down a formula for the area of the 1p coin, using π and r.
(v) Use your answers to (ii) and (iv) to write down a formula for the shaded area.
(vi) Factorise your formula from (v):

(*a*) by taking out the common factor
(*b*) by using the 'difference of 2 squares' factors.

(vii) Use your formula to calculate the shaded area, correct to 3 significant figures.
($\pi = 3.14$)

10 (Squared paper) Draw x and y axes. Scale the x axis from 0 to 10 and the y axis from 0 to 4, using 1 cm : 1 unit on both axes.

(i) Plot the triangle ABC: $A(1,2)$, $B(0,0)$, $C(2,0)$.
(ii) Write down the coordinates of ABC as a 2 by 3 matrix, M.
(iii) Multiply M, on the left, by T_1 the transformation matrix $\begin{pmatrix} 1 & 2 \\ 0 & 1 \end{pmatrix}$ to give $A'B'C'$, the image of triangle ABC under the transformation T_1.
(iv) Plot $A'B'C'$.
(v) Find the coordinates of $A''B''C''$, the image of $A'B'C'$ under T_2, the transformation matrix $\begin{pmatrix} 2 & 0 \\ 0 & 2 \end{pmatrix}$
(vi) Plot $A''B''C''$.
(vii) Write down the matrix which would result in the single transformation from ABC to $A''B''C''$.

11 *ABCD* is a quadrilateral in which $AB = 12$ cm, $BC = 5$ cm, $CD = 8$ cm, $DA = 7$ cm and angle *ABC* is a right angle.

(i) Calculate the length of *AC*.

(ii) Use the cosine formula to calculate angle *ADC*.

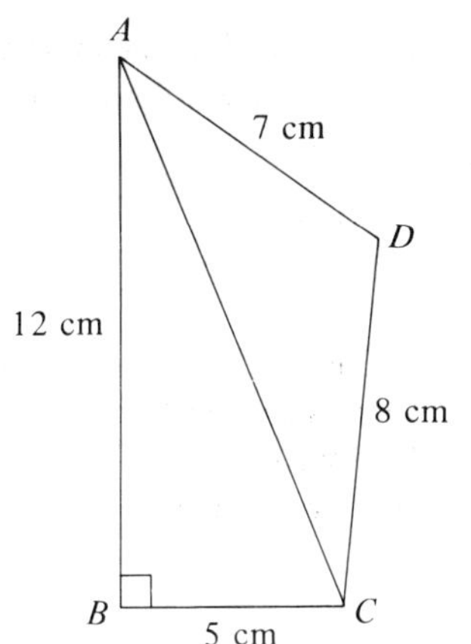

12 *ABCD* is a parallelogram. *E* and *F* are the mid points of *AD* and *DC* respectively. *BEH* and *BFG* are straight lines; *E* is the mid point of *BH* and *F* is the mid point of *BG*.

$\overrightarrow{BC} = \mathbf{a}$ and $\overrightarrow{BA} = \mathbf{b}$

(i) Write down in terms of **a** and **b**

(*a*) $\overrightarrow{BF}$ (*b*) $\overrightarrow{BG}$ (*c*) $\overrightarrow{BE}$ (*d*) $\overrightarrow{BH}$ (*e*) $\overrightarrow{GH}$ (*f*) $\overrightarrow{CA}$

(ii) From your answers to (*e*) and (*f*) write down two facts about *CA* and *GH*.

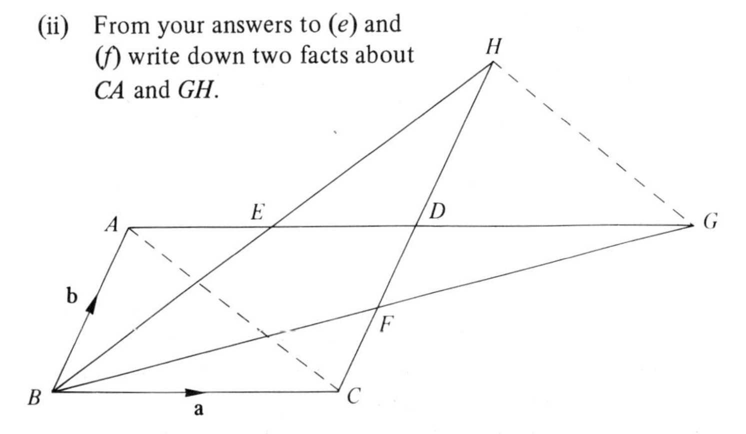

13 (Squared paper) Draw *x* and *y* axes. Scale the *x* axis from −3 to +5 and the *y* axis from −4 to +6, using 1 cm : 1 unit on both axes.

(i) Draw and label the graphs whose equations are:

(*a*) $x + y = 3$
(*b*) $y = x - 4$
(*c*) $y = -3x$

(ii) By shading the unwanted regions, show clearly the region R which satisfies the inequalities:

(*a*) $x + y \leqslant 3$
(*b*) $y \geqslant x - 4$
(*c*) $y \geqslant -3x$

14 ABC is a right-angled triangle.
Angle $ACB = 90°$
$AB = x$ cm, $BC = 15$ cm and
$AC = (x - 9)$ cm

(i) Use the theorem of Pythagoras to form an equation in x and write it down fully.
(ii) Solve the equation.
(iii) Write down the lengths of AB and AC.

15 (Squared paper)

x	-4	-3	-2	-1	0	1	2	3	4
y	6	$2\frac{1}{2}$	0		-2	$-1\frac{1}{2}$			6

$y = \frac{1}{2}x^2 - 2$

The table, which is partly completed, shows values for the graph of the equation: $y = \frac{1}{2}x^2 - 2$

(i) Copy and complete the table.
(ii) Draw x and y axes. Scale the x axis from -4 to $+4$ and the y axis from -3 to $+6$, using 1 cm : 1 unit on both axes.
Plot the values from the table and join up the points to give a smooth curve for the graph of $y = \frac{1}{2}x^2 - 2$
(iii) From the graph, write down the values of x which satisfy the equation: $\frac{1}{2}x^2 - 2 = 0$
(iv) On the same axes, draw the straight line whose equation is $y = -\frac{1}{2}x + 1$
(v) Write down the values of x at the points where the graphs of $y = \frac{1}{2}x^2 - 2$ and $y = -\frac{1}{2}x + 1$ intersect.
(vi) Write down and simplify the single equation which these values satisfy.

Set 9

1 (i) Find 15% of £25.00
(ii) Express $2\frac{1}{3}$ as a percentage.
(iii) Duty on imported watches is 15%. Calculate the duty-free cost of a watch which is worth £23.00 after duty has been paid.
(iv) A discount of 10% was offered at a sale. Find the sale price of an article which was originally £16.50
(v) In a mixed school, 60% of the pupils are girls. 40% of the girls cycle to school, 50% of them travel by bus and the rest walk. 50% of the boys cycle to school, 25% travel by bus and the rest walk.
Find the percentage of the pupils who walk to school.

2 The diagram shows the cross section of a hollow length of piping. The inside radius of the cross section is r cm and the outside radius is R cm. l is the length of piping.

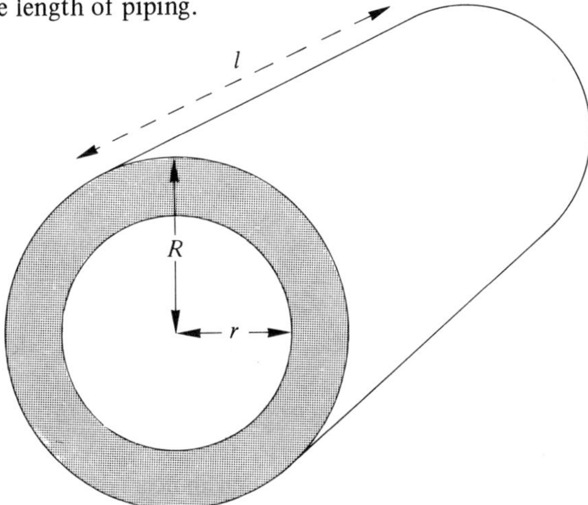

(i) Write down a formula for A, the area of the cross section in terms of π, R and r.
(ii) Factorise your formula by first taking out the common factor π and then by using the 'difference of 2 squares' factors.
(iii) Evaluate A if $R = 6$, $r = 4$ and $\pi = 3.14$
(iv) If $l = 15$ cm, calculate the volume of the length of piping.

3 The two longer sides of a right-angled triangle are given as 8 cm and 6 cm, correct to the nearest centimetre. Calculate the minimum and maximum value of the third side of the triangle.

4 Solve the equations:

(i) $7x - 4x + 9x = 108$
(ii) $5x^2 = 20$
(iii) $4(a - 1) + 2(a + 3) = 17$
(iv) $(2a + 5)(a - 6) = 0$
(v) $x^2 - 11x + 18 = 0$

5 (i) A coin is tossed and comes down tails. It is tossed a second time. What is the probability that it comes down tails again?

(ii) Two dice are rolled and the total score is x. If x is greater than 4, and the probability of scoring x is $\frac{1}{18}$, find x.

(iii) A bag contains a number of black and a number of white beans. If a bean is chosen at random, the probability that it is black is 0.7. There are 35 black beans in the bag. Calculate the number of white beans.

6 *ABCDE* is a pyramid on a rectangular base 18 cm by 16 cm. $AB = AC = AD = AE = 17$ cm. X is the mid point of BC in triangle ABC and AO is the height of the pyramid.

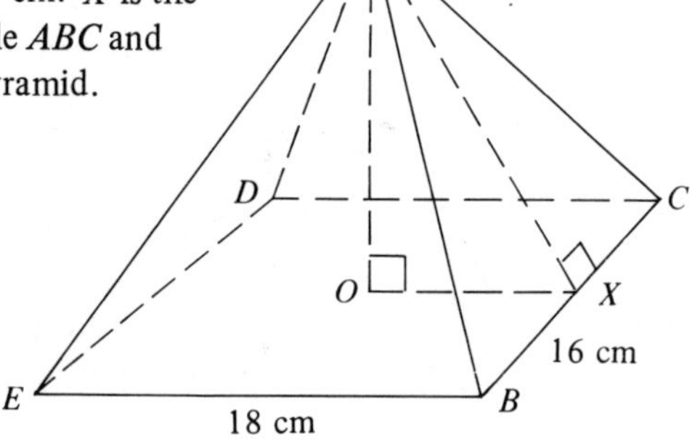

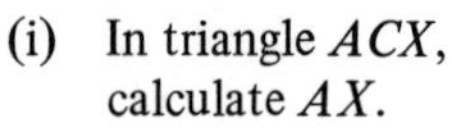

(i) In triangle ACX, calculate AX.

(ii) In triangle AOX, calculate AO, the perpendicular height of the pyramid.

(iii) Calculate the volume of *ABCDE*.

(iv) Write down tan angle OXA as a decimal.

(v) Use your tables to find angle OXA in degrees.

7 (i) Change the subject of the following equations to x.

(*a*) $x + a = b$
(*b*) $ax = b$
(*c*) $\frac{x}{a} = \frac{b}{c}$
(*d*) $\frac{3x}{4} = h$

(ii) (*a*) If $v^2 = u^2 + 2fs$, calculate v if $u = 4, f = 8$ and $s = 3$
(*b*) Change the subject of the formula to s.
(*c*) Calculate s if $v = 5, u = 3$ and $f = 2$

8 $\mathscr{E} = \{2,3,4,5,6,8,9,10,12,16\}$, A, B, and C are $\subset \mathscr{E}$

$A = \{\text{Square numbers}\}$, $B = \{\text{Factors of } 20\}$, $C = \{\text{Factors of } 12\}$

(i) Copy and complete the Venn diagram by entering the members of the sets in the appropriate region.

(ii) Find $n(A \cap B \cap C)$.

(iii) If a member is chosen at random from $\mathscr{E}$ what is the probability that it is a factor of 20?

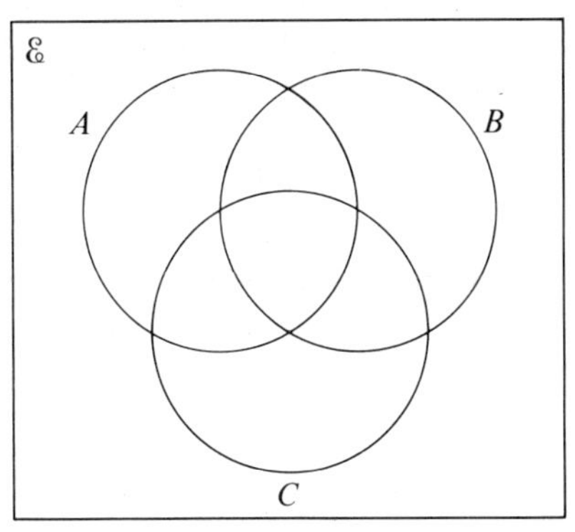

9 The diagram shows an ordinary school protractor, which is a semi-circle, radius 5 cm, and some of the points on it.

The arcs, AB, BC, CD, DE, EF, FG are equal.

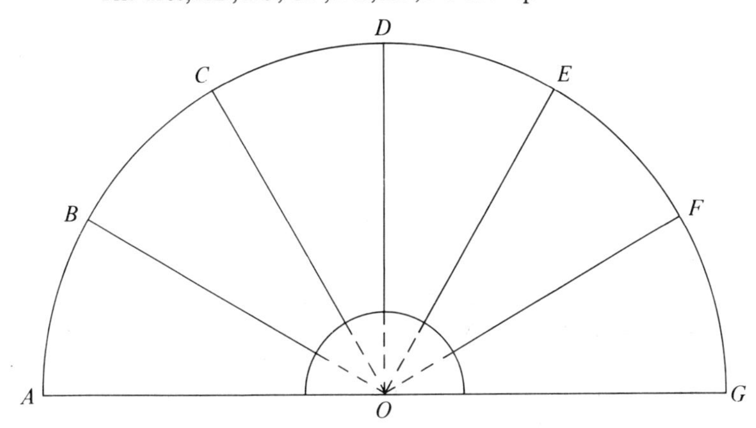

(i) Measuring anticlockwise from OG, write down the size of the angles:

(a) FOG (b) COG (c) BOG

(ii) Write down the value of:

(a) cos angle FOG
(b) sin angle EOG
(c) sin angle BOG
(d) cos angle COG

(iii) Calculate the length of the arc AD ($\pi = 3.14$).

(iv) Calculate the perimeter of the protractor.

(v) Calculate the area of the protractor.

(vi) Calculate the area of the sector COE.

10 (Squared paper) Draw x and y axes. Scale the x axis from 0 to 4 and the y axis from -2 to $+4$, using 1 cm : 1 unit on both axes.

(i) Plot the kite $ABCD$: $A(2,0)$, $B(1,-2)$, $C(1,-1)$, $D(0,-1)$.
(ii) Write the coordinates of $ABCD$ as a 2 by 4 matrix, K.
(iii) Multiply K, on the left, by the transformation matrix T $\begin{pmatrix} 0 & -2 \\ 2 & 0 \end{pmatrix}$ to give the coordinates of $A'B'C'D'$ the image of $ABCD$ under the transformation T.
(iv) Plot $A'B'C'D'$.
(v) Describe fully the transformation T.
(vi) Write down the ratio of the areas: $\dfrac{ABCD}{A'B'C'D'}$

11 ABC is the sector of a circle, centre C, radius 10 cm. AB is a chord and angle $ACB = 36°$.

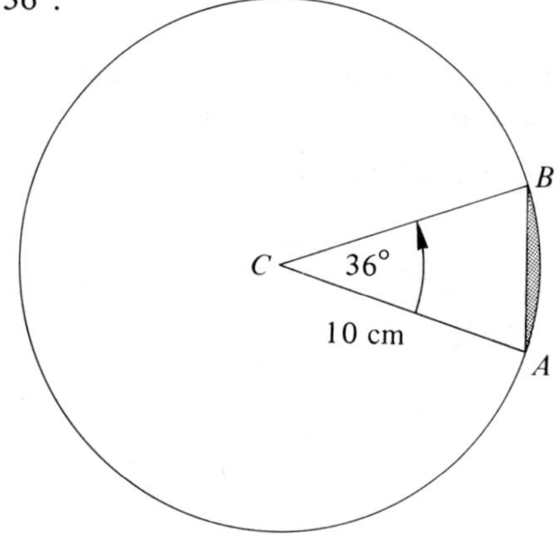

(i) Calculate the area of the sector ABC.
(ii) Use the formula Area $= \frac{1}{2}ab \sin C$, where $a = 10$, $b = 10$ and $C = 36°$ to find the area of triangle ABC.
(iii) From your answers to (i) and (ii) write down the area of the segment AB which is shaded on the diagram.
($\pi = 3.14$)

12 The diagram shows a cuboid whose length is $3x$ cm, breadth is $2x$ cm and height is x cm.

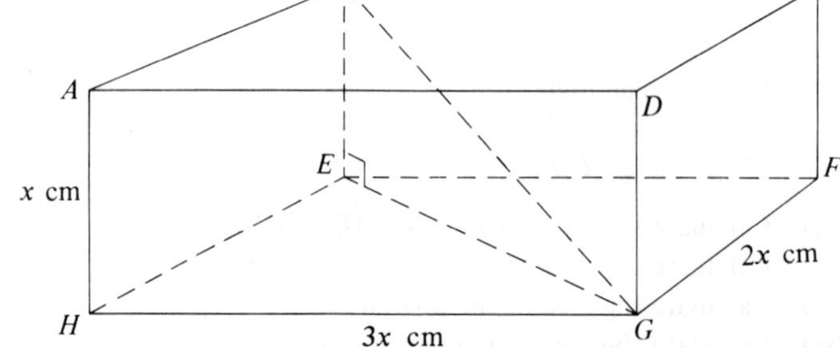

(i) Write down in terms of x:

(*a*) the total length of the edges
(*b*) the area of *ADGH*
(*c*) the area of *DCFG*
(*d*) the area of *EFGH*
(*e*) the total surface area of the cuboid
(*f*) the volume of the cuboid
(*g*) the length of *EG*, the base diagonal
(*h*) the length of *BG*, the space diagonal.

(ii) If the volume of the cuboid is 384 cm^3 find:

(*a*) the value of x
(*b*) the total surface area in square centimetres.

13 In the quadrilateral *PQRS*, *SR* is parallel to *PQ*.

The ratio $\frac{ST}{SQ} = \frac{1}{4}$

The ratio $\frac{PQ}{SR} = \frac{3}{4}$

S R T P Q

Given that $\vec{PQ} = 6\mathbf{a}$ and $PS = 2\mathbf{b}$

(i) Write in terms of **a** and **b**

(*a*) $\vec{SQ}$ (*b*) $\vec{ST}$ (*c*) $\vec{PT}$ (*d*) $\vec{QR}$

(ii) From your answers to (*c*) and (*d*)

(*a*) write down one fact about *PT* and *QR*

(*b*) complete the ratio: $\frac{PT}{QR}$

14 The following statistics were obtained in a survey of the number of children in a family.

Number of children in the family	1	2	3	4	5	6
Number of families	9	12	8	5	4	2

Draw a histogram (bar chart) to illustrate these statistics and then answer the questions.

(i) How many families took part in the survey?
(ii) Calculate the total number of children in all the families.
(iii) Write down the mode.
(iv) Calculate the mean, giving your answer correct to 1 decimal place.
(v) What percentage of the families in the survey

(*a*) had 4 children?
(*b*) had an only child?

(vi) Comment briefly on a similar survey, had it been taken 100 years ago.

15 (i) ABC is a right-angled triangle.
X is a point on AB.
$AC = 9.6$ and $BC = 7.9$ cm.

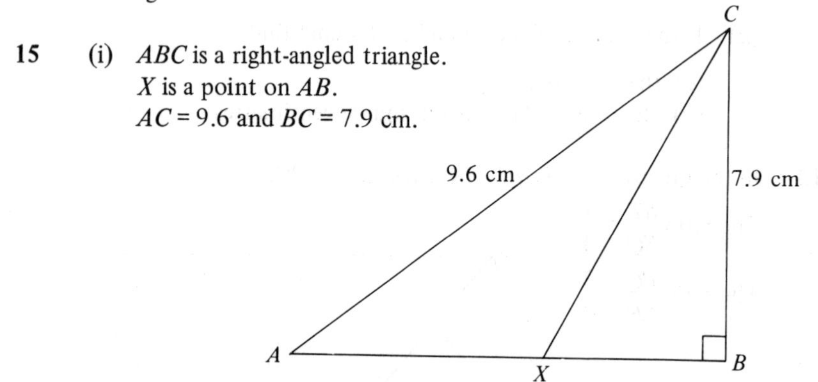

(*a*) Copy and complete the inequality: 7.9 ____ CX ____ 9.6
(*b*) If the value of CX is an integer, write down 2 possible values of CX.

(ii) Triangle ABC is right-angled at B and angle $ACB = 30°$.

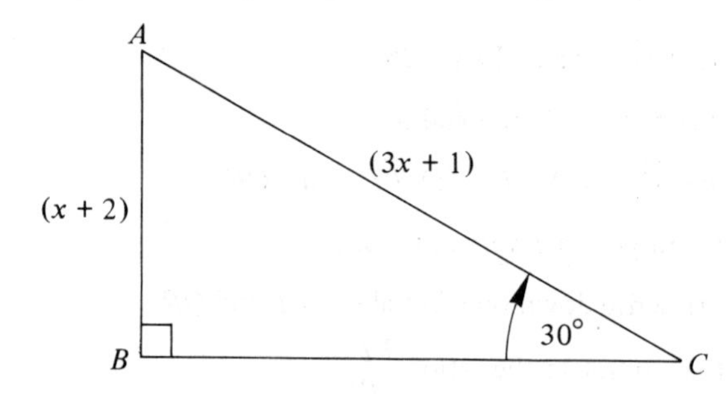

(*a*) By considering the relationship of AC to AB, form an equation in x.
(*b*) Solve the equation.
(*c*) Write down the length of AC and AB.
(*d*) Calculate the length of BC.

Set 10

1 Factorise:

(i) $x^3 + x^2 + x$
(ii) $9 - x^2$
(iii) $12x^2 - 27y^2$
(iv) $a^2 + 4a - 21$
(v) $2x^2 + x - 6$

2 The diagram shows the net of a closed cylinder, comprising a rectangle, 44 cm by 12 cm and two circles. When the cylinder is made, AB and DC will meet together exactly (use $\frac{22}{7}$ for the value of π).

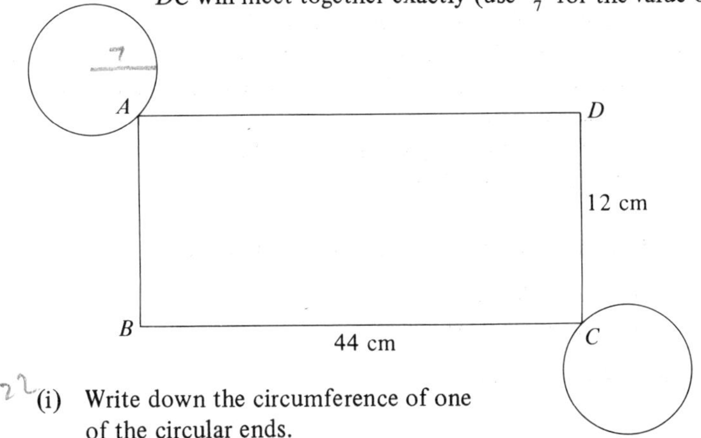

(i) Write down the circumference of one of the circular ends.
(ii) Calculate the radius of a circular end.
(iii) Calculate the area of a circular end.
(iv) Calculate the total surface area of the cylinder.
(v) Calculate the volume of the cylinder.
(vi) Calculate the volume of a cone on the same base and with the same height.

3 Mr Driver bought a car for £1800. Depreciation was 20% in year 1, 15% in year 2 and $12\frac{1}{2}$% in year 3. At the end of the 4th year the car was sold for £981.

Calculate:

(i) Depreciation in:

(*a*) year 1 (*b*) year 2 (*c*) year 3

(ii) The value of the car at the beginning of year 4.
(iii) The depreciation during year 4

(*a*) in money
(*b*) as a percentage of the original purchase price.

4 (i) Copy and complete the tables shown on values for the equations:

(*a*) $y = 2x - 3$

x	4	0	-1
y			

(*b*) $y = -\frac{1}{2}x + 2$

x	4	0	-2
y			

(ii) Draw the graphs of these equations.

(iii) From your graphs obtain the values of x and y which satisfy the equations simultaneously.

5 (i) Write 1001_8 in base 10.

(ii) Express $1\,000\,001_2$ as an octal number.

(iii) Evaluate $1000_2 + 234_5 + 567_8$ and give your answer in base 10.

(iv) The side of a square is 1000_2 cm. Calculate its area, giving your answer in base 2.

(v) If 15_8 multiplied by $x = 47_8$, find the value of x.

(vi) If $10_x = 1000_y$, find x and y.

6 The diagram shows 4 solids. For each of the solids:

(*a*) write down its name

(*b*) calculate its volume.

(Use 3.14 for the value of π.)

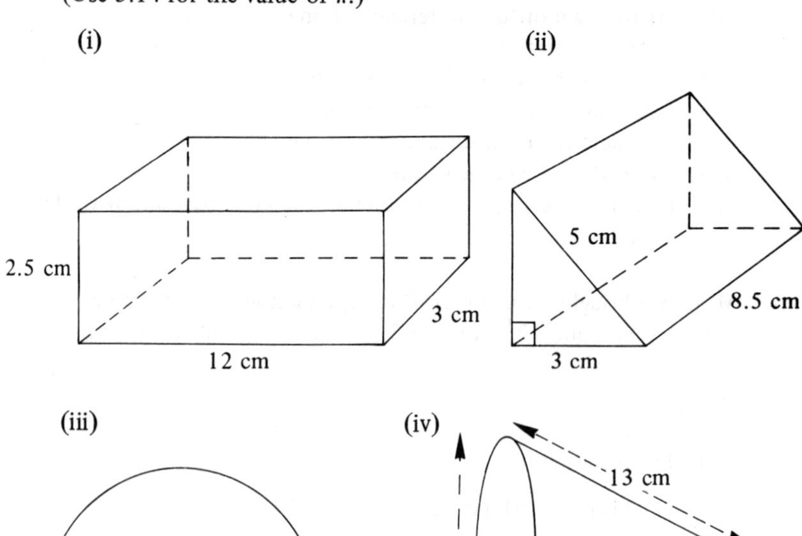

7 (i) Change the subject of the following formulae to x:

(*a*) $\frac{x}{4} = p$

(*b*) $2x - c = b$

(*c*) $a(x + b) = c$

(*d*) $ax + bx = c$

(ii) Given that $A = 2\pi rh$,

(*a*) calculate A when $\pi = \frac{22}{7}$, $r = 14$ and $h = 20$

(*b*) change the subject of the formula to r

(*c*) calculate r when $A = 528$, $\pi = \frac{22}{7}$ and $h = 12$

8 (i) Bill, Bob and Bertie share a Pools win of £45 000 in the ratio of their stakes, which is 3 : 5 : 7

Calculate their shares.

(ii) Calculate the sector angles of a pie chart showing their shares.

(iii) Two cubes have their volumes in the ratio 27 : 125
What is the ratio of their sides ?

9 (Squared paper) Draw x and y axes. Scale the x axis from -4 to $+4$ and the y axis from -2 to $+4$, using 1 cm : 1 unit.

(i) Plot the point A $(-1,3)$

(ii) B is the reflection of A in the y axis.
C is the reflection of B in $y = x$.
D is the reflection of C in the x axis.
E is the reflection of D in the y axis.
F is the reflection of E in the x axis.
Plot the points B, C, D, E, F, and join them up in order, finally joining F to A.

(iii) Describe fully the relation of F to A as a reflection.

(iv) Write down the equations of the lines:

(*a*) BC (*b*) ED (*c*) FA

(v) What is the equation of the line joining E and C ?

(vi) What is the gradient of the line joining F and D ?

(vii) Name the quadrilateral $ABCF$.

10 Copy the diagram on squared paper.

(i) Write down the coordinates of $ABCD$ as a 2 by 4 matrix, M.

(ii) Multiply M by $\begin{pmatrix} 0 & -1 \\ 1 & 0 \end{pmatrix}$ on the left.

(iii) Show on the diagram $A'B'C'D'$, the image of $ABCD$ under the transformation matrix $\begin{pmatrix} 0 & -1 \\ 1 & 0 \end{pmatrix}$

(iv) Describe the transformation fully.

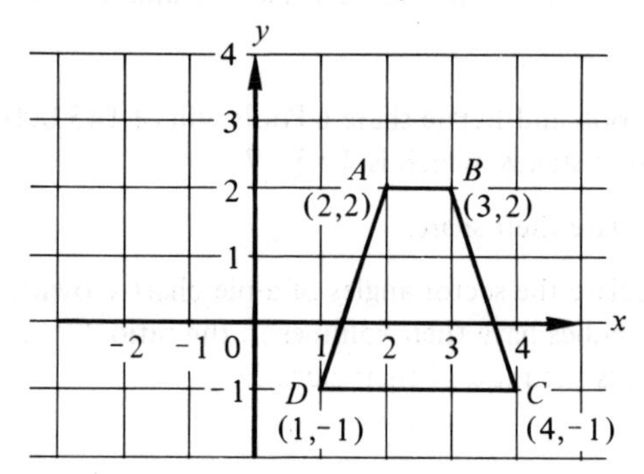

11 The diagram shows a circle, centre O, radius 5 cm. AB is a chord 6 cm long and C is a point on the circumference.

(i) Calculate angle AOB.

(ii) Write down the size of angle ACB.

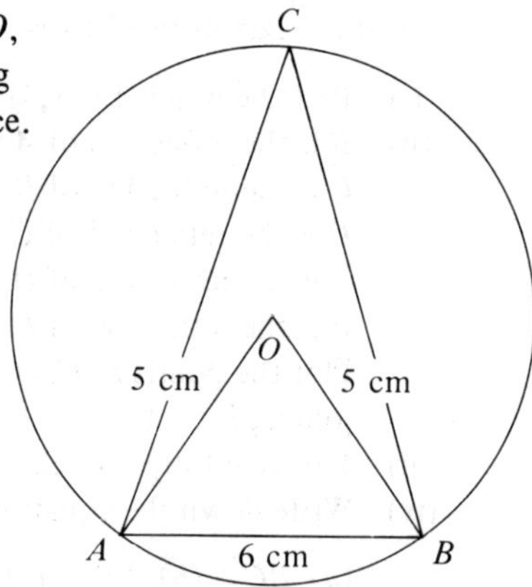

12 ABC is a right-angled triangle. $AB = (4x + 1)$ cm, $AC = (x + 2)$ cm and $BC = 4x$ cm.

(i) Use the theorem of Pythagoras to form an equation in x.

(ii) Solve the equation and find two sets of values for AB, AC and BC.

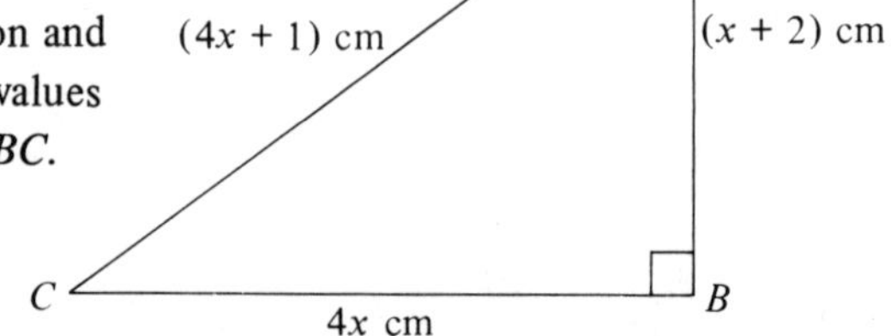

13 (i) In each diagram, calculate the bearing of A from B.

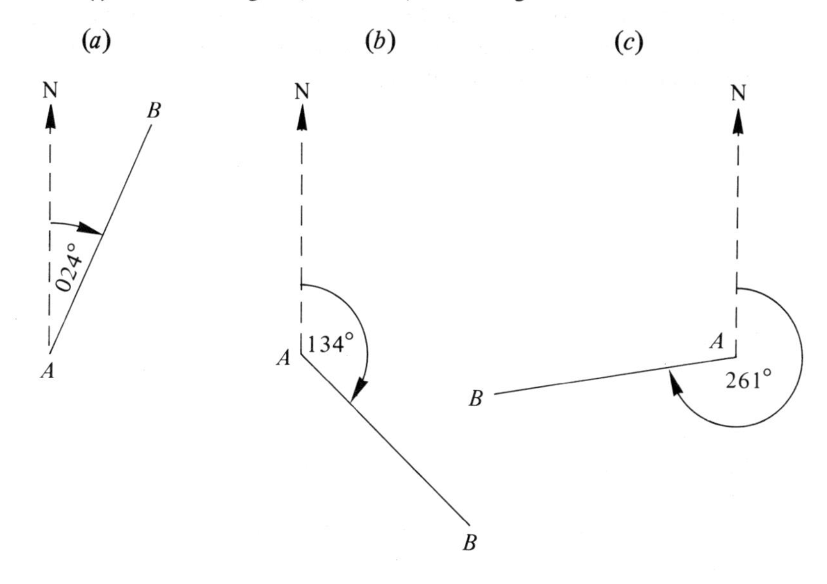

(ii) The diagram shows the positions of three points A, B and C. The bearing of B from A is 021° and angle $ABC = 92°$

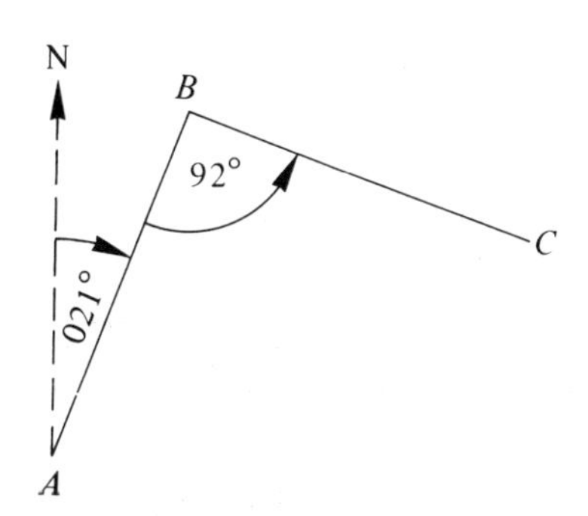

(*a*) Calculate the bearing of: (i) A from B (ii) C from B (iii) B from C.

(*b*) If $AB = BC$, find the bearing of C from A.

14 The three-dimensional diagram shows the rectangle $ABCD$ on a horizontal plane. $AB = 12$ m and $BC = 16$ m. AX is a vertical mast 5 m high, to the top of which wire stays have been connected from B, C and D, as shown by the dotted lines.

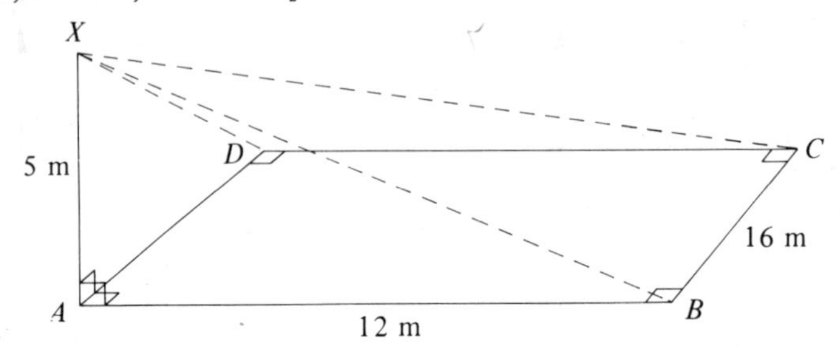

Calculate:

(i) BX
(ii) angle ABX, the angle of elevation from B to X
(iii) DX
(iv) angle ADX, the angle of elevation from D to X
(v) AC
(vi) angle ACX, the angle of elevation from C to X
(vii) the area of triangle ABX
(viii) the area of triangle ADX
(ix) the area of triangle ACX
(x) the area of triangle BXC
(xi) angle BCX.

15 (Squared paper) The partly completed table shows values of x and y for the graph of the equation: $y = -x^2 + 3$

x	-3	-2	-1	0	1	2	3
y	-6		2			-1	

(i) Copy and complete the table.
(ii) Draw x and y axes. Scale the x from -3 to $+3$ and the y axis from -6 to $+3$. Use 2 cm : 1 unit on the x axis and 1 cm : 1 unit on the y axis.
Plot the values from the table and join up the points to give the smooth curve graph of $y = -x^2 + 3$
(iii) Write down the values of x which satisfy the equation $-x^2 + 3 = 0$
(iv) On the same axes, draw the straight line whose equation is $y = 2x$
(v) Shade the region R, where $x \geqslant 0$, $y \geqslant 2x$ and $y \leqslant -x^2 + 3$
(vi) Make a brief statement about the symmetry of the graph of $y = -x^2 + 3$

Section 3

Set 1

1 A new car was priced at £4960 in the showroom but a customer who offered cash was allowed a $12\frac{1}{2}\%$ discount.
In the first year the car lost 18% of its original value as priced in the showroom and in the following year it lost 10% of its value at the beginning of that year.

(i) What price did the customer pay for the car?
(ii) What was the value of the car after 1 year?
(iii) What was the car valued at after 2 years?
(iv) Calculate the percentage loss to the customer if he sells the car at its value at the end of the second year. Give your answer correct to 3 significant figures.

2 (i) Solve the equations:

(*a*) $\dfrac{x - 2}{3} + 4 = \dfrac{52}{15} - \dfrac{2x - 3}{5}$

(*b*) $x - 5 = \dfrac{36}{x}$

(ii) (*a*) Factorise $3x^2 + 17x + 10$
(*b*) Use the factors of $3x^2 + 17x + 10$ to find two factors of the number 31 710 which lie between 100 and 400.

3 (i) A plastic drinking cup is in the shape of the frustrum of a cone as shown in the diagram. The diameter of the top of the cup is 6 cm and its base diameter is 4 cm. If the height of the cone is 15 cm, calculate the volume of the cup.

(ii) How many cups can be filled from a full urn which is a cylinder of diameter 30 cm and height 38 cm? ($\pi = 3$)

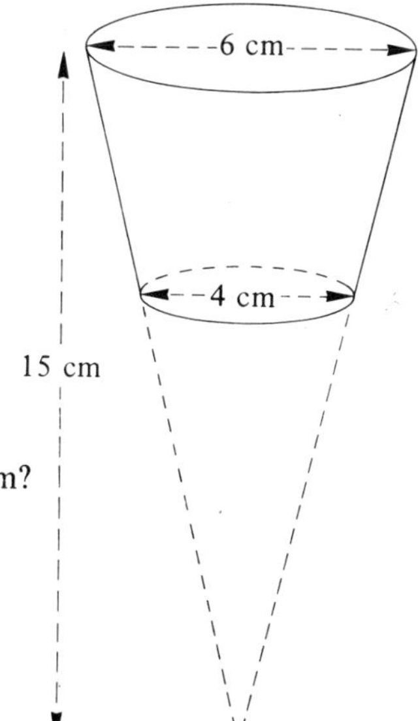

4 (i) Given a function $f : x \to x^2 + 5x - 4$:

Evaluate:

(*a*) $f(1)$
(*b*) $f(\frac{1}{2})$
(*c*) $f(-5)$
(*d*) $f(-6)$

(ii) Find the values of x such that $f(x) = 0$ for the function in (i).

(iii) Given that $p = \dfrac{r - s^2}{t}$, express s in terms of p, r and t.

5 A man at the top of a building of height 50 m observes a motor car below travelling away from the building along a straight level road. The first position of the car was P_1 and 5 seconds later the position of the car was P_2. The angles of depression of P_1 and P_2 were 52° and 22.5° respectively.

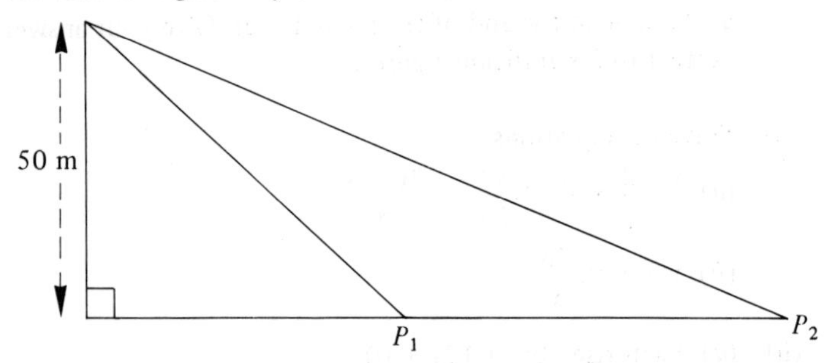

(i) Calculate in metres the distance P_1P_2 which the car had travelled in the 5 seconds. Give your answer correct to 3 significant figures.

(ii) Calculate the speed of the car, using your answer to (i):

(*a*) in metres per second
(*b*) in kilometres per hour.

6 (i) (*a*) Four men and three women are interviewed for a position as sales assistant. What is the probability that a man will be appointed?

(*b*) After being interviewed, one of the men and one of the women decide to withdraw their applications.
What is the probability that a woman will be appointed now?

(ii) Two dice are rolled. Find the probability of:

(*a*) a total score of 8
(*b*) a total score of less than 6.

(iii) Cards are drawn at random from a pack of playing cards and not replaced. Calculate the probability that the first two cards drawn are both hearts and the third is either a club or a spade.

7 Draw x and y axes. Scale the x axis from −5 to +8 and the y axis from −7 to +7, using 1 cm : 1 unit on both axes. Plot $A(1,2)$ $B(4,2)$ $C(3,3)$, $D(2,3)$ and join up the points to give the quadrilateral $ABCD$..

(i) Perform the transformation: $T\begin{pmatrix} x \\ y \end{pmatrix} \to \begin{pmatrix} -1 & 0 \\ 0 & -1 \end{pmatrix}\begin{pmatrix} x \\ y \end{pmatrix}$ on $ABCD$ and call the new points $A_1 B_1 C_1 D_1$
Plot these points on the diagram.

(ii) Plot $A_2(2,4)$, $B_2(8,4)$, $C_2(6,6)$ and $D_2(4,6)$ on the diagram and find the matrix which transforms $A_1B_1C_1D_1$ on to $A_2B_2C_2D_2$.

(iii) Perform the matrix transformation $X\begin{pmatrix} x \\ y \end{pmatrix} \to \begin{pmatrix} 1 & 0 \\ 0 & -1 \end{pmatrix}\begin{pmatrix} x \\ y \end{pmatrix}$ on $A_2B_2C_2D_2$ and show these new positions as $A_3B_3C_3D_3$.

(iv) What matrix will transform $A_3B_3C_3D_3$ on to $A_1B_1C_1D_1$?

8 Copy and complete the following table for values of x and y for the graph of: $y = \dfrac{10}{x} - x$

x	1	2	3	4	5	6	7	8	9
y	9	0.33				−4.33	−5.57		

(i) Draw the graph of $y = \dfrac{10}{x} - x$ for $1 \leqslant x \leqslant 9$ using 1 cm : 1 unit on both axes.

(ii) Draw the graph of $y = \dfrac{x}{2} + 2$ on the same axes and find from your graphs the solution to the equation: $\dfrac{10}{x} - x = \dfrac{x}{2} + 2$

(iii) By drawing the tangent to the curve at the point where $x = 5$, find the gradient of the curve at that point.

9 A livestock farmer goes to an auction prepared to spend a maximum of £900. He is interested in buying lambs and piglets which are priced at £5.00 and £7.50 respectively. He has sufficient space available for a total of 150 animals only and in order to have an efficient production output, neither the number of piglets, nor the number of lambs bought may exceed three times the other in number.

(i) From the above information, produce 4 inequalities if the number of lambs bought is x and the number of piglets is y.

(ii) Show these inequalities on a graph by shading the unwanted regions. Use 1 cm to represent 10 animals on both axes.

(iii) If the profit from a lamb and a piglet is £6.00 and £10.00 respectively after 20 weeks, find from your graph the number of each animal to give a maximum profit and calculate that profit.

10 In the diagram:

$\vec{AD} = 3\mathbf{a}$ and $AB = 4AD$
$\vec{AC} = 3\mathbf{b}$ and $CE = \frac{2}{3}CD$
$AE = \frac{1}{2}AF$

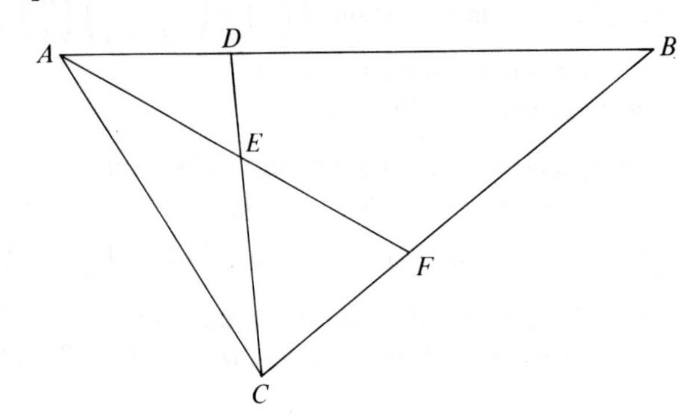

(i) Express AB in terms of $\mathbf{a}$.
(ii) Express in terms of $\mathbf{a}$ and $\mathbf{b}$:

(*a*) $\vec{CD}$ (*b*) $\vec{CE}$ (*c*) $\vec{AE}$ (*d*) $\vec{AF}$ (*e*) $\vec{CF}$ (*f*) $\vec{CB}$

(iii) Deduce the ratio $\frac{x}{y}$ if $\frac{CF}{CB} = \frac{x}{y}$

Set 2

1 (i) The sides of a rectangle measure 40 m by 20 m.
If the sides are increased in length by 15%, calculate the corresponding increase in area as a percentage.

(ii) The cash price of a camera was £96.64 but it could also be purchased under a hire purchase agreement with a deposit of £12.00 and 12 equal monthly payments.
If $12\frac{1}{2}\%$ is added to the cash price for the hire purchase agreement charge, calculate the monthly payment.

(iii) £1000 was invested for 2 years at $8\frac{1}{2}\%$ per annum compound interest. Calculate the amount at the end of 2 years, giving your answer correct to the nearest £.

2 (i) The values of x which satisfy the equation $x^2 + bx + c = 0$ are both equal to 5. Find the values of b and c.

(ii) Simplify by expressing as a single fraction in its lowest terms:

$$\frac{1}{x+1} + \frac{2}{x+2} - \frac{x+2}{x^2+3x+2}$$

(iii) Solve the equation: $\frac{1}{x} = \frac{5-x}{4}$

3 A is the point $(-4,4)$, B is the point $(0,2)$ and ABC is a straight line.

(i) Find the coordinates of C.
(ii) Write down the equation of the line ABC.
(iii) Find the area of $ABOD$.
(iv) Write down 3 inequalities to define the shaded triangle area BOC.

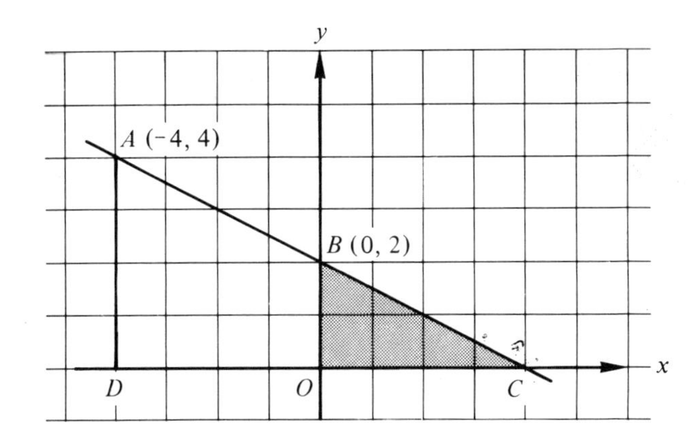

4 In the diagram $\overrightarrow{AC} = \mathbf{a}$ and $|\mathbf{a}| = 4$
$\overrightarrow{AB} = \mathbf{b}$ and $|\mathbf{b}| = 4$

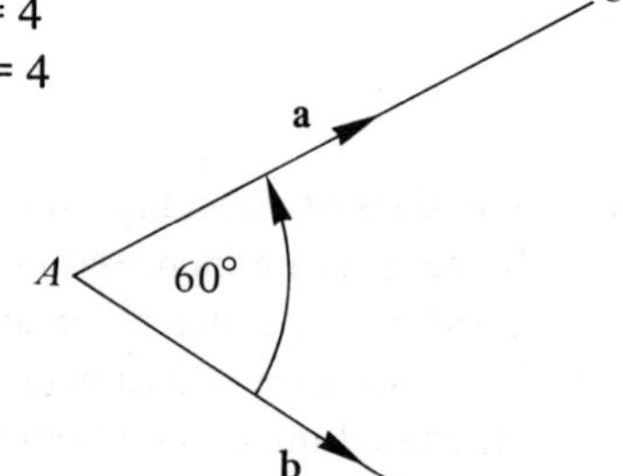

(i) Calculate $|\mathbf{a} - \mathbf{b}|$

(ii) Copy the diagram and then:

(*a*) Draw in a representative of $\mathbf{a} + \mathbf{b}$, letter it $\mathbf{c}$

(*b*) calculate $|\mathbf{c}|$.

5 (i) Two dice are rolled together. Calculate the probability that:

(*a*) they show the same number

(*b*) they show the same number and the square root of the total score is an integer.

(ii) Three dice are rolled together.

(*a*) What is the total number of possible outcomes?

(*b*) What is the probability they all show the same number?

(*c*) Calculate the probability of obtaining a total score of 10 when any 2 of the dice show the same number.

6 (i) S, the sum of the numbers $\{1, 3, 5, 7, 9, \ldots n\}$ is given by the formula:

$$S = \tfrac{1}{2}n(n + 1)$$

If $S = 190$, find the value of n.

(ii) The radius of a circular steel plate increases by 0.00157 cm when it is heated. If the radius is 5 cm before heating

(*a*) Calculate the increase in area in square centimetres, giving your answer correct to 3 decimal places. ($\pi = 3.14$)

(*b*) Express the increase in area in Standard Form notation, correct to 3 significant figures.

7 $y = 2x^2 + 4x - 7$.

x	-4	-3	-2	-1	0	1	2
y							

The table shows values of x for the graph of $y = 2x^2 + 4x - 7$

(i) Copy and complete the tables for values of y.
(ii) Draw x and y axes. Use 2 cm : 1 unit on the x axis and 1 cm : 2 units on the y axis.
(iii) Draw the graph, and from your graph write down the solutions of the following equations:

(*a*) $2x^2 + 4x - 7 = 0$
(*b*) $2x^2 + 4x = 0$
(*c*) $2x^2 + 4x = 11$

(iv) On the same axes draw the graph of $y = 4x - 2$ and write down the values of x at the point of intersection with the graph of $y = 2x^2 + 4x - 7$
(v) Find the single equation which these values satisfy.

8 The diagram shows the cross-section of a mathematical model of a cone, which is made in three sections.

$QR = 12$ cm, $PZ = 21$ cm, $PX = XY = YZ$

Calculate:

(i) the slant height of the cone, PR
(ii) XT and YV
(iii) the volume of cone PQR in terms of π

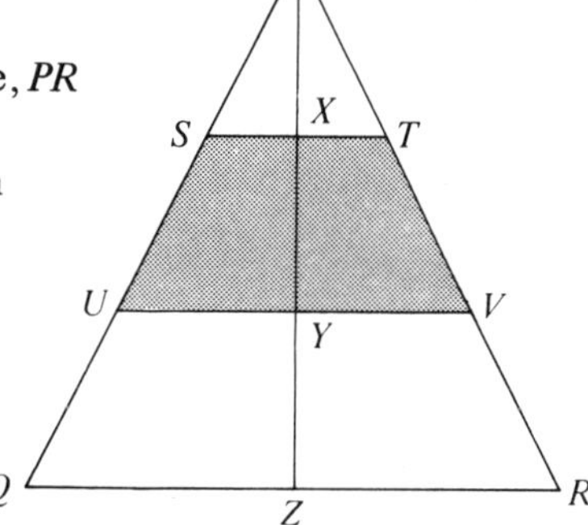

(iv) the ratio of the fraction; $\dfrac{\text{volume of cone } PST}{\text{volume of cone } PQR}$
and use this ratio to find the volume of cone PST in terms of π
(v) the ratio of the fraction; $\dfrac{\text{volume of cone } PUV}{\text{volume of cone } PST}$
and use this ratio to find the volume of cone PUV in terms of π
(vi) The volume of the shaded section of the model, in terms of π.

9 A farmer had 100 hectares in which to grow sugar beet and potatoes. He had a maximum of £3600 to cover costs and 360 labour units available.
The requirements for the two crops were as follows:

Sugar beet: Minimum area: 16 hectares
Labour units per hectare: 3
Cost per hectare: £45.00 .

Potatoes: Minimum area: 26 hectares
Labour units per hectare: 5
Cost per hectare: £40.00 .

Let x = the number of hectares under sugar beet
and y = the number of hectares under potatoes.

(i) Write down 5 inequalities which must be satisfied.
(ii) Use a scale of 1 cm : 10 units on both axes and draw the straight lines which will define the region within which (x,y) is to be found. Show the regions clearly by fringe-shading the regions *not* wanted.
(iii) From your graphs estimate the maximum total number of hectares which the farmer will be able to sow.

10 PQ is a vertical aerial on level ground, 20 m high. When the sun is Southwest, QS is the shadow of the aerial and is 35 m long. When the sun is due West, the angle of elevation of the sun is 21.8°.

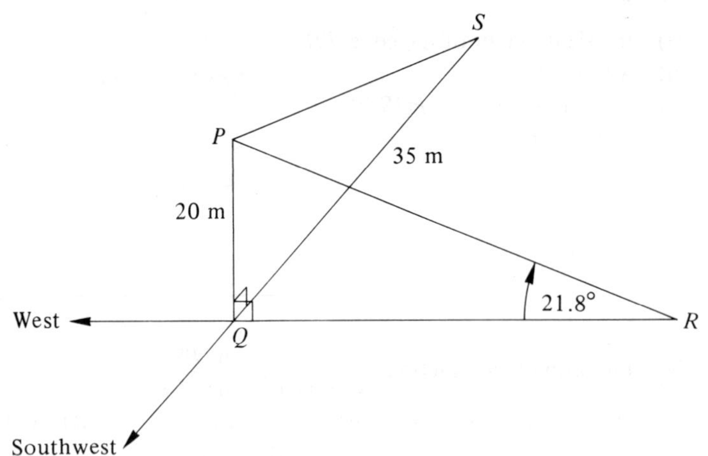

Calculate:

(i) the angle of elevation of the sun when it is Southwest
(ii) the length of the shadow QR when the sun is due West
(iii) the distance SR.

Set 3

1 (i) The sides of a 10 cm cube are lengthened by 10%.
Calculate the percentage increase:

(*a*) in total surface area
(*b*) in volume.

(ii) When a sum of money is divided in the ratio 1 : 2 : 4 among A, B and C respectively, B's share is £95.00.

Calculate:

(*a*) the sum of money
(*b*) A's share
(*c*) C's share.

(iii) During the first year of its life a car depreciated by 25% of its original value and during the second year by 20% of its value at the end of the first year. Express the total depreciation during the two years as a percentage of its original value.

2 (i) The solutions of the quadratic equation $x^2 + bx + c = 0$ are $x = -3$ or 4. Find the values of b and c.
(ii) If $f(x) = x^2 - x$, find the highest common factor of $f(5), f(6), f(10)$
(iii) Solve the equation: $\dfrac{x-3}{2} = \dfrac{5}{x}$

3 A sphere fits exactly inside a cube whose side measures $2x$ cm.

(i) Find a formula for D, the difference in volume between the cube and the sphere.
(ii) If $x = 10$ cm and $\pi = 3.14$, evaluate D, the difference in volume.

4 A and B are two points on a map drawn to a scale of 1 : 100 000
The distance in a straight line AB on the map is 5 cm, but A is on the 200 m contour and B is on the 600 m contour.

Calculate:

(i) the horizontal distance between A and B in metres
(ii) the vertical distance of B above A in metres
(iii) the gradient of the line AB as a decimal fraction
(iv) the angle of elevation of B from A.

5 The diagram shows a circle, centre O, radius 10 cm and a circle, centre P, radius 4 cm which touches the original circle at Q. APB is a straight line through P, perpendicular to the straight line OPQ. OAX and OBY are straight lines.

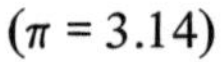

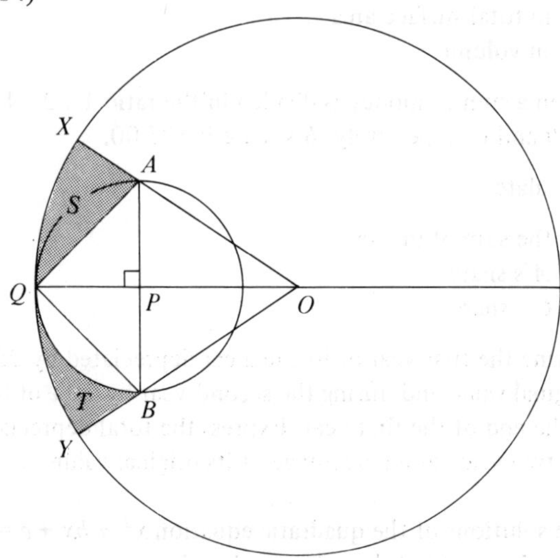

(i) (*a*) Name the quadrilateral $OAQB$
 (*b*) Find its area in square centimetres.

(ii) Calculate angle AOB.
(iii) Calculate the area of the minor sector XOY.
(iv) Calculate the shaded area S.
(v) Calculate the shaded area T.

6 (i) (*a*) Factorise the expression $a(x + 4) - 5(x + 4)$
 (*b*) Use your answer to (*a*) to solve the equation
 $ax + 4a - 5x - 20 = 0$

(ii) A circular wooden plate, whose radius is 'R' cm, has 100 small holes drilled in it, radius 'r'.

 (*a*) Write down a formula for the area of wood in the plate after the holes are drilled, in terms of π, R and r.
 (*b*) Factorise your formula fully.
 (*c*) Calculate the area left when $\pi = 3.14$, $R = 45$ and $r = 1.5$ cm

7

x	1	2	3	4	5	6
y						

$y = \frac{8}{x}$

The table shows values of x for the graph of $y = \frac{8}{x}$

(i) Copy and complete the table for values of y, giving your values correct to 2 decimal places if the value is not an integer.

(ii) Draw x and y axes. Scale the x axis from 0 to 6 and the y axis from 0 to 8, using 2 cm : 1 unit on both axes.

(iii) Draw on the same axes the graph of $y = \frac{1}{4}x$

(iv) Estimate from your graphs, the value of x at the point of their intersection.

(v) Write down the single equation which this value satisfies and hence show why the value is approximately $\sqrt{32}$

8 In the diagram, F and E are the mid points of AB and BC respectively. $GF = FE$ and GFE is a straight line.

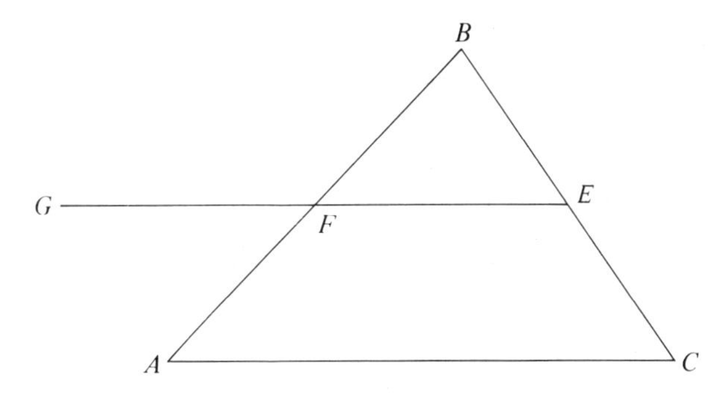

(i) If $\vec{AC} = \mathbf{a}$ and $\vec{AB} = \mathbf{b}$, write down: in terms of $\mathbf{a}$ and $\mathbf{b}$:

(*a*) $\vec{BC}$ (*b*) $\vec{BE}$ (*c*) $\vec{FB}$ (*d*) $\vec{FE}$ (*e*) $\vec{GE}$

(ii) What can you deduce about:

(*a*) GE and AC ?

(*b*) the quadrilateral $ACEG$?

9 The shaded part of the diagram shows the cross section of one end of the roof of a storage building. It is a segment of a circle, centre O and angle MON is 120°. The radius of the circle is 6 m and the roof is 15 m long.

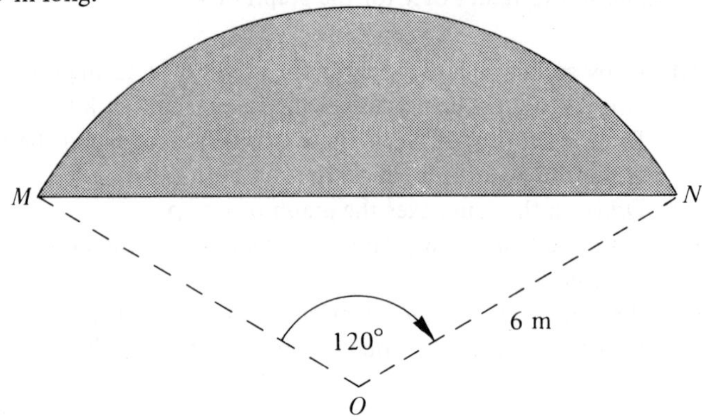

(i) Calculate correct to 3 significant figures:

(*a*) length MN in metres
(*b*) area of triangle MON
(*c*) area of sector MNO
(*d*) the shaded area of the roof end.

(ii) (*a*) Calculate exactly the area of the curved surface of the roof.
(*b*) Calculate the cost of covering the roof at £4.50 per square metre, giving your answer to the nearest £.

10 In the diagram,
$AD = 10, DC = 8, BC = 15$ cm.
Angle $ADB = 60°$, angle $BCD = 90°$

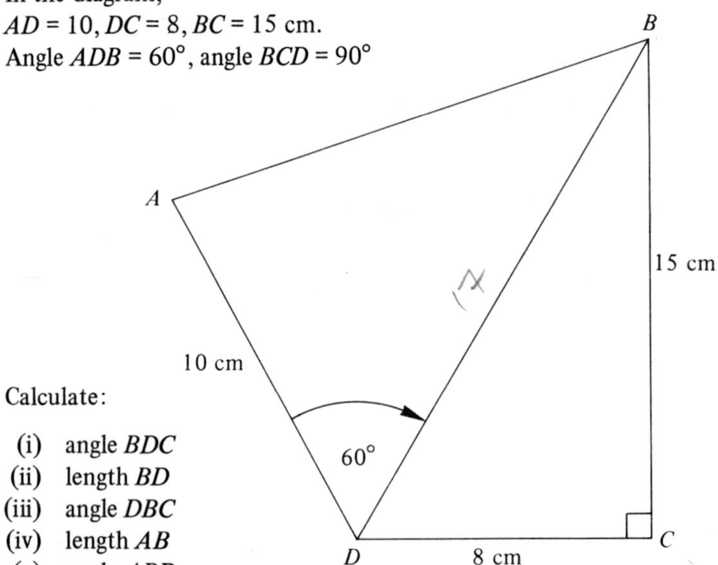

Calculate:

(i) angle BDC
(ii) length BD
(iii) angle DBC
(iv) length AB
(v) angle ABD
(vi) angle DAB.

Set 4

1 A car was purchased for £3600 and it depreciated as follows:
Year 1: 20%, Year 2: 15% of the value at the beginning of that year,
Year 3: $12\frac{1}{2}$% of the value at the beginning of that year. At the end of the 4th year it was sold for £1928.

Calculate:

(i) the value of the car

(*a*) after 2 years
(*b*) at the beginning of the 4th year

(ii) depreciation during the 4th year

(*a*) in money
(*b*) as a percentage of its value at the beginning of the 4th year, to the nearest whole per cent.

2 (i) If $f(x) = 2x^2 - 4x - 5$, evaluate $f(-3)$.

(ii) Solve the equation: $2x + \dfrac{1}{x} = 5 - \dfrac{1}{x}$

(iii) Given that $f(x) = (x + 4)(x + n)$ and $f(2) = 42$, find n.

3 A cylinder fits exactly inside a cube whose side is $4x$ cm.

(i) Find a formula for V, the difference in volume between the cube and the cylinder, in terms of x and π.

(ii) If $x = 7$ and $\pi = \frac{22}{7}$, evaluate the formula to find V, the difference in volume between the cube and the cylinder.

4 (i) If $A = \begin{pmatrix} 1 & 1 \\ 5 & 7 \end{pmatrix}$ write down A^{-1}

(ii) Express the simultaneous equations $x + y = 2$ and $5x + 7y = 4$ as a matrix equation.

(iii) Pre-multiply each side of the matrix equation by A^{-1}

(iv) Write down the values of x and y which satisfy the simultaneous equations $x + y = 2$ and $5x + 7y = 4$

5 The diagram shows a spherical metal cage which is made from 3 large hoops, radius 7.5 cm and 2 small hoops, radius 5 cm, welded together at A, B, C, D, E and F.

3 hoops are horizontal and 2 hoops are vertical and at right angles to each other.

O is the centre of the large horizontal hoop.

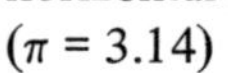

($\pi = 3.14$)

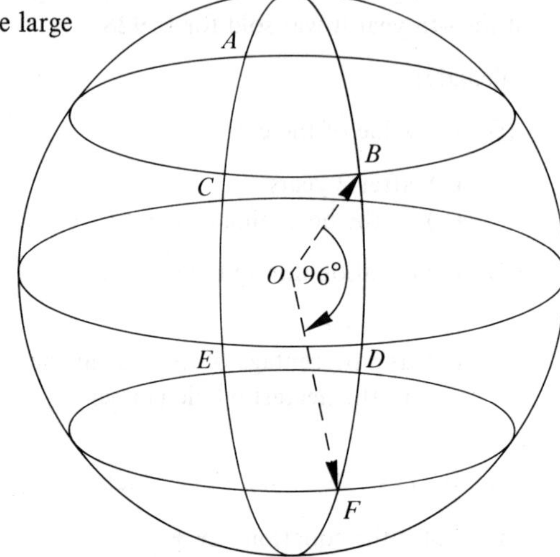

(i) Calculate the total length of wire needed to make the cage.

(ii) If angle $BOF = 96°$, calculate the distance from B to F along the arc BDF.

(iii) Calculate the volume of the spherical cage.

6 $y = x^2 + 5x$

x	-6	-5	-4	-3	-2	-1	0	1
y								

(i) The table shows values of x for the graph of $y = x^2 + 5x$. Copy and complete the table for values of y.

(ii) Draw the graph of $y = x^2 + 5x$, using a scale of 2 cm : 1 unit on the x axis and 1 cm : 1 unit on the y axis.

(iii) From your graph, obtain the solutions of the following equations:

(*a*) $x^2 + 5x = 0$
(*b*) $x^2 + 5x + 5 = 0$
(*c*) $x^2 + 5x - 2 = 0$

(iv) Estimate the gradient of the curve when $x = -4$

(v) The graph is symmetrical; write down the equation of the line of symmetry.

7 The dimensions of a cuboid are increased by 2 cm, 4 cm and 5 cm. This transforms the cuboid into a cube, whose volume is 440 cm^3 greater than that of the original cuboid.

Let the side of the cube be x cm, form an equation in x, solve it and from your solution find:

(i) the length of the side of the cube
(ii) the dimensions of the original cuboid
(iii) the volume of the cube.

8 (i) Find the transformation matrix and describe the transformation which:

(*a*) maps ABO on to $A'B'O$
(*b*) maps $A'B'O$ on to $A''B''O$
(*c*) maps $A''B''O$ on to $A'''B'''O$.

(ii) Plot and draw on your own diagram $A''''B''''O$, the image of ABO under the transformation matrix,

$$\begin{pmatrix} -2 & 0 \\ 0 & -2 \end{pmatrix}$$

Describe the transformation.

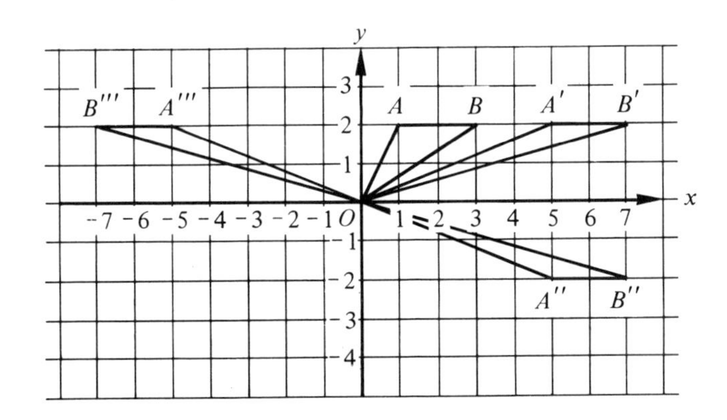

9 In the diagram, A, B and C are three points in a horizontal plane and a vertical radio mast RC is shown at C.
B is 80 m from A, bearing 120° and C is 60 m from B, bearing 030°. The angle of elevation of R, the top of the mast, from B is 16.7°.

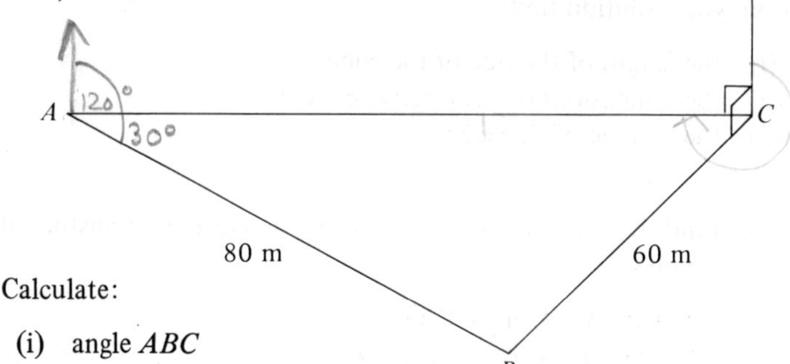

Calculate:

(i) angle ABC
(ii) the distance AC
(iii) the height of the mast RC
(iv) the angle of elevation of R from A, i.e. angle RAC
(v) the bearing of A from C.

10 The diagram shows a circle, centre O, radius 10 cm. OP and OQ are radii and angle $POQ = 72°$. ($\pi = 3.14$)

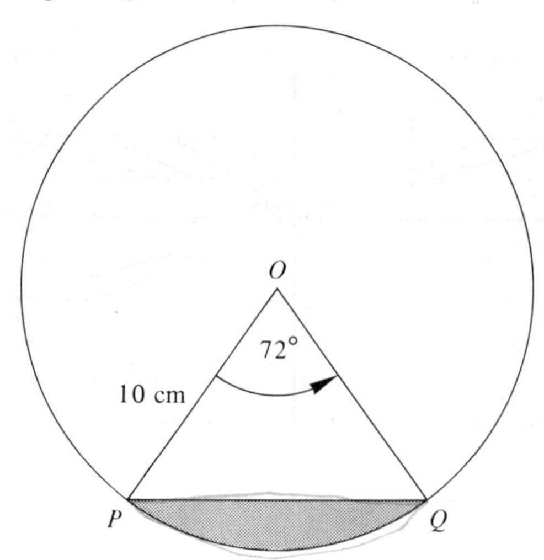

(i) Calculate

(*a*) the area of the minor sector POQ
(*b*) the area of the triangle POQ
(*c*) the area of the shaded segment
(*d*) the perimeter of the shaded segment.

(ii) If the radius of the circle were to be doubled and angle POQ remained unchanged, by what factor would your answers to (i) a, b and c be multiplied?

Set 5

1 (i) A building society agrees to make a loan of £11 764.00 and this represents 85% of the purchase price of the house. Calculate the purchase price.

(ii) Calculate the compound interest on £1000, invested for 3 years at $12\frac{1}{2}$% per annum. Give your answer correct to the nearest £.

(iii) In a mixed school, 60% of the pupils were girls. 40% of the girls cycle to school, 50% travel by bus and the rest walk. $37\frac{1}{2}$% of the boys cycle to school, 60% travel by bus and the rest walk. Find what percentage of the pupils walk to school.

2 (i) P is the solution set of $5x - 2 > 3(x + 4)$ and Q is the solution set of $7x - 9 > 4x + 3$ where x is an integer. List $P \cap Q$

(ii) Solve the simultaneous equations: $y = 5 - 3x$
$3x + 4y + 16 = 0$

(iii) Solve the equation: $(3x + 4)(2x + 15) = (2x + 3)(3x + 10)$

3 The diagram shows the top of a dining table which consists of a square and semi-circular ends. The side of the square is $4x$ m.

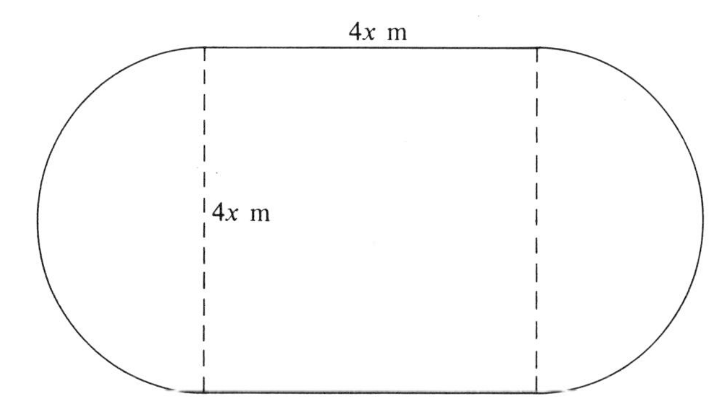

(i) Write down an expression for P, the perimeter of the table top and simplify it.

(ii) Write down an expression for A, the area of the table top and simplify it.

(iii) If $x = 0.35$ m and $\pi = \frac{22}{7}$

Evaluate:

(*a*) P, the perimeter
(*b*) A, the area of the table top.

4 $y = x^2 + 3x - 2$

x	-5	-4	-3	-2	-1	0	1	2
y								

The table shows values of x for the graph of $y = x^2 + 3x - 2$

(i) Copy and complete the table for values of y.

(ii) Draw the graph of $y = x^2 + 3x - 2$ using a scale of 2 cm : 1 unit on the x axis and 1 cm : 1 unit on the y axis.

(iii) From your graph write down the approximate values of x which satisfy the following equations:

(*a*) $x^2 + 3x - 2 = 0$
(*b*) $x^2 + 3x - 5 = 0$

(iv) (*a*) Draw on your graph the line whose equation is $y = x + 1$ and write down the values of x at the points of intersection with the graph of $y = x^2 + 3x - 2$
(*b*) Write down and simplify the single equation which these values satisfy.

5 (i) (*a*) A bag contains 16 counters, x of which are blue. Write down the probability that if a counter is chosen at random, it is blue.
(*b*) If 8 more blue counters are added to the bag, this probability is increased by $\frac{1}{4}$.
Write down an equation in x and solve it to find the number of blue counters originally in the bag.

(ii) A 2-digit number is formed by making a random choice of any 2 different digits from 1,3,5,7,9

(*a*) List the possible outcomes.
(*b*) Write down the probability that the number chosen is a prime number.

6 In the diagram TA is a tangent to the circle, centre O.
AB is a diameter and OC is parallel to AT.
C and D are points on the circumference and angle $AEO = 51°$.

Calculate the following angles:

(i) angle EOA
(ii) angle BDO
(iii) angle CBF
(iv) angle BFC
(v) angle ODA.

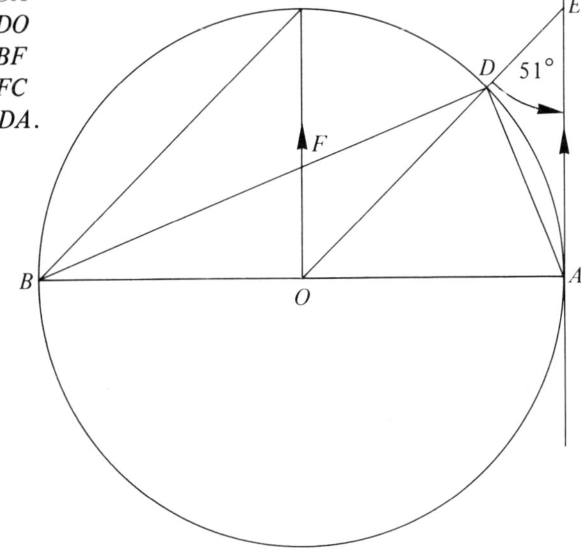

7 (i) A rectangle measures 17 cm by 11 cm. If these dimensions are both increased by x cm, the area is 432 cm^2. Form an equation and solve it to find the value of x.

(ii) D, the number of diagonals of a polygon may be found by the formula: $D = \dfrac{n(n-3)}{2}$ where n is the number of sides.
If a polygon has 54 diagonals

(*a*) Calculate the number of sides the polygon has.
(*b*) Write down its name.

8 (i) Factorise:

(*a*) $3a(2x + 3y) - 4b(2x + 3y)$
(*b*) $3t^2 + 5t - 2$
(*c*) $3b^2 - 48$
(*d*) $\dfrac{a^4}{81} - \dfrac{b^4}{16}$

(ii) Evaluate exactly by using factors, showing the method clearly:

$$\frac{6.96 \times 65 + 6.96 \times 35}{(8.48)^2 - (1.52)^2}$$

9 The diagram shows a triangular prism.
ABCD and *ADEF* are rectangles.

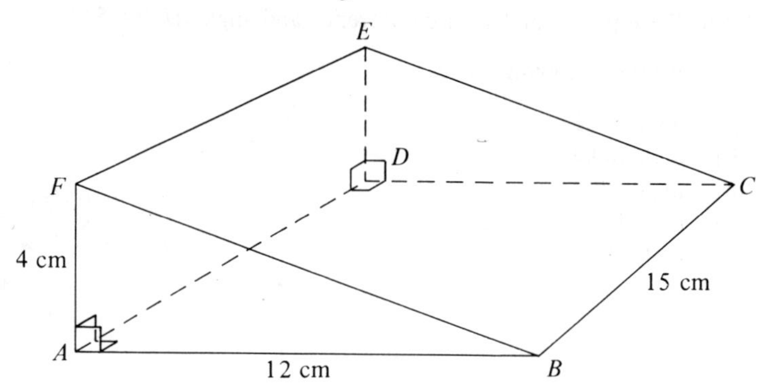

Calculate:

(i) the length *FB*
(ii) the total surface area
(iii) the volume of the prism
(iv) angle *ABF*
(v) the length *DB*
(vi) the length *EB*
(vii) angle *EBD*.

10 A ship sails from port on a bearing of 100° for 7.5 km, when it changes course to 205° to avoid a storm. But after sailing on this bearing for 5 km, the captain decided to return to port.

Calculate:

(i) the distance the ship is from port
(ii) the bearing needed for the return journey.

Set 6

1 A salt-cellar as shown in the diagram consists of a cylinder, radius r and height h, surmounted by a hemisphere of radius r.

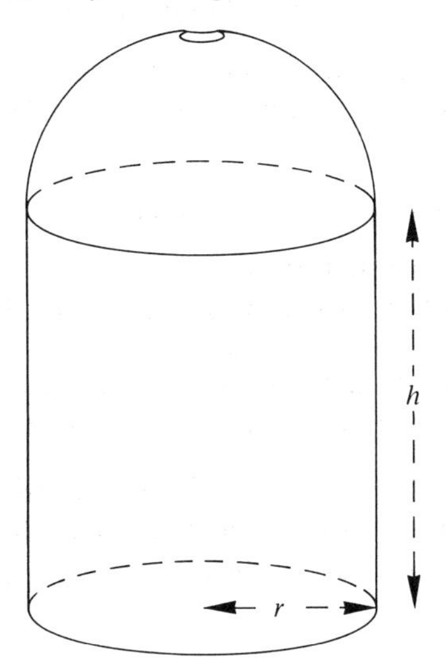

(i) Show that the total surface area of the salt-cellar is $\pi r(2h + 3r)$.

(ii) Show that the volume of the salt-cellar is $\frac{1}{3}\pi r^2 (2r + 3h)$.

(iii) If the external dimensions of the salt-cellar are: $r = 2$ cm and $h = 5$ cm and it is made from silver which is 1 mm thick, calculate the internal dimensions.

(iv) Using the formula from part (ii), calculate:

(*a*) the internal volume
(*b*) the external volume of the salt-cellar.

(v) From your answers to (iv), calculate:

(*a*) the volume of silver used
(*b*) the value of the silver used if the density is 10.5 g cm^3 and 1 g of silver costs 40p.

2 (i) $\mathscr{E} = \{x : x \text{ is an integer and } 8 < x < 20\}$
$A = \{x : x \text{ is an integer and } 10 < x < 16\}$
$B = \{x : x \text{ is an integer and } 12 < x < 19\}$

List the members of the following sets:

(*a*) $\mathscr{E}$ (*b*) A (*c*) B (*d*) $A \cap B$ (*e*) $(A \cup B)'$

(ii) 100 students were questioned regarding which of the three summer activities, cricket, tennis and athletics, they were taking part in.

The information was summarised as follows:

C is the set of students who play cricket.
A is the set of students who take part in athletics.
T is the set of students who play tennis.
$n(T \cap A) = 25, n(T \cap C) = 24, n(A \cap C) = 22$
$n(T) = 56, n(A) = 52, n(C) = 54$ and $n(A \cup C \cup T)' = 4$

Copy the Venn diagram and by letting $n(A \cap C \cap T) = x$, enter the information shown above and hence find the number of students who took part in all three activities.

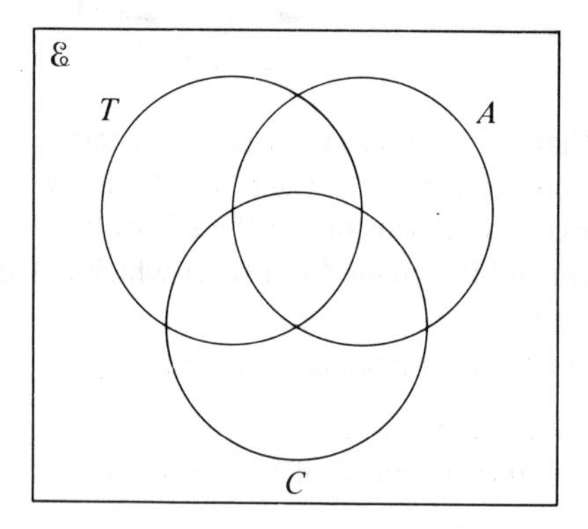

3 The table shows the results of a test taken by 600 students.

Mark	No. of students	Cumulative frequency
0 to 10	5	5
11 to 20	11	16
21 to 30	22	
31 to 40	47	
41 to 50	72	
51 to 60	103	
61 to 70	106	
71 to 80	156	
81 to 90	60	
91 to 100	18	

(i) Copy the table and complete the cumulative frequency column.

(ii) Draw a cumulative frequency graph. Use 2 cm to represent 10 marks and 2 cm to represent 100 students on the cumulative frequency axis.

(iii) From your graph, estimate:

(*a*) the median mark
(*b*) the upper quartile
(*c*) the lower quartile
(*d*) the interquartile range.

(iv) (*a*) If 60% of the students passed the test, what was the pass mark?
(*b*) What percentage of the students scored 50 marks or less?

4 An alloy A contains 3 parts tin to 7 parts lead by weight and an alloy B contains 3 parts tin to 5 parts copper by weight.

A new alloy C is made by mixing weights of A and B in the ratio 4 parts of A to 3 parts of B, by weight.

Tin costs £1250 per tonne, lead costs £100 per tonne and copper costs £500 per tonne.

(i) Calculate the cost of the metal required to make 1 tonne of the alloy C, giving your answer correct to the nearest £.

(ii) If the price of lead increases by 10% and the other two prices remain unchanged, what is the percentage increase in the cost of 1 tonne of the alloy C? Giving your answer correct to 3 significant figures.

5 (i) Factorise $(x + 4)^2 - (x - 3)^2$
How does your answer show that for all whole number values of x, the expression is exactly divisible by 7 ?

(ii) Give two positive values of x, one an integer and the other one not, for which the expression is divisible by 49.

(iii) Find a formula for all such values in the form $x = Pn + Q$ where P and Q are constants and n is any positive integer.

6 $ABCD$ is a square, side $2x$ units. O and Q are the mid points of BC and AD. $OPQR$ is a sector of a circle, centre O.

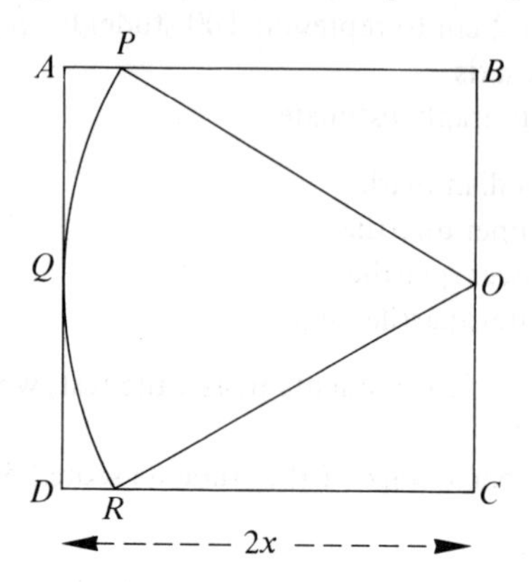

(i) Calculate in terms of π and x:

(*a*) the length of PQR, the arc of the sector $OPQR$
(*b*) the area of the sector $OPQR$
(*c*) the percentage of the area of the square occupied by the sector $OPQR$.

(ii) If the sector $OPQR$ is made into a cone, of height h, with O its apex and PQR the circular base, find the volume of the cone in terms of π, x and h.

7 Points X and Y are taken on the sides AB and AC respectively of triangle ABC so that $AX = \frac{1}{3}AB$ and $AY = \frac{1}{3}AC$. M is the mid point of BC. The directed segments $\overrightarrow{AX}$ and $\overrightarrow{AY}$ represent vectors $\mathbf{x}$ and $\mathbf{y}$ respectively.

(i) What vectors are represented by:

$\overrightarrow{XY}, \overrightarrow{AB}, \overrightarrow{AC}, \overrightarrow{BC}$ and $\overrightarrow{AM}$?

(ii) Find a vector represented by $\overrightarrow{XC}$.

(iii) If BY meets CX at L, use similar triangles to show that $\overrightarrow{XL}$ represents $\frac{1}{4}(3\mathbf{y} - \mathbf{x})$.

(iv) Hence show that $\overrightarrow{AL} = \frac{1}{2}\overrightarrow{AM}$. What conclusion can you draw about the relative positions of A, L and M ?

8 (i) Draw the graph of $y = x^2 - 2x - 3$ for values of x from -2 to $+4$, using 2 cm : 1 unit on the x axis and 1 cm : 1 unit on the y axis.

Use your graph to find:

(a) the least value of y
(b) the solutions to: (i) $x^2 - 2x - 6 = 0$; (ii) $x^2 - 2x - 1 = 0$

(ii) On the same axes, draw the graph of $3y = 4x - 9$ and:

(a) give the values of x for the intersection of the two graphs
(b) if set $A = \{(x,y): \frac{4}{3}x - 3 > y\}$
and set $B = \{(x,y): x^2 - 2x - 3 < y\}$
shade the region: $A \cap B$.

9 The total gives the heights in centimetres to the nearest centimetre of 100 children, aged 1 year.

Height in cm	65	66	67	68	69	70	71	72	73	74	75
Number of children	2	2	10	20	19	15	16	10	3	2	1

(i) It is suggested that the formula:

mean − mode = 3(mean − median)

is approximately valid for many sets of data.
By calculating the mean, median and mode of the data in the table, examine the validity of this statement for the above example.

(ii) If another sample of 500 children aged 1 year with a mean height of 72.03 was combined with this sample, what would be the mean height of the combined sample, correct to 3 significant figures?

10 (i) The transformation $\begin{pmatrix} x \\ y \end{pmatrix} \to T_1 \begin{pmatrix} x \\ y \end{pmatrix}$ is followed by

the transformation $\begin{pmatrix} x \\ y \end{pmatrix} \to T_2 \begin{pmatrix} x \\ y \end{pmatrix}$

If $T_1 \begin{pmatrix} x \\ y \end{pmatrix} = \begin{pmatrix} 0 & 1 \\ -1 & 0 \end{pmatrix}$ and $T_2 \begin{pmatrix} x \\ y \end{pmatrix} = \begin{pmatrix} 0 & 1 \\ 1 & 0 \end{pmatrix}$

find the matrix of the combined transformation and call it T_3.
Describe each of the transformations T_1, T_2, and T_3 in geometric terms.

(ii) A and B are sets of points given by:

$A = \{(x,y): y = 2x - 1\}$, $B = \{(x,y): y = x\}$

C is the image of A under transformation T_2.
Find C, expressing it in the same notation as A and B above and draw a clearly labelled Cartesian diagram, showing the sets A, B and C.

Part 2
Notes on solutions

Section 1

Set 1

2 *Number Bases:* Reference Section 9.

3 (i) Find the cost of 10 litres and then subtract the cost of a $\frac{1}{2}$ litre.

(iii) Check all subtractions by addition.

For example:

$$\begin{array}{r} 262 \\ -\underline{107} \\ \underline{155} \end{array} \Bigg\} \text{ Add these two lines to give you 262.}$$

When dealing with money, the 'Shopkeeper Method' of subtraction is often very useful (and reliable). For example, what change should you get from a £20 note after spending £14.78?

The shopkeeper gives you 22p to make the total of £15 and then counts out £5 (16,17,18,19,20) to make up the total to £20, i.e. change of £5.22

(v) Number of cubes $= \dfrac{\text{Large volume}}{\text{Small volume}} = \dfrac{V}{v}$

4 After rearranging the scores in order, check that you have 10 scores.

(i) The mode is the score which occurs most frequently.

(ii) The median is the middle score, but if there is an even number of items in the distribution, the middle pair are added and divided by 2.

For example: What is the median of 2,4, (5,8) ,10,13?

The median is: $\dfrac{5+8}{2} = \dfrac{13}{2} = 6.5$

(iv) x = the new total of the scores − the original total of scores i.e. $x = (11 \times 5) -$ (original total).

(v) read carefully: *more* than 7.

5 Remember to check your answers by substituting the value of your solution in the original equation.
e.g. Solve the equation: $3x = x - 12$

$$3x - x = -12 \quad \text{(subtracting } x \text{ from both sides)}$$
$$2x = -12$$
$$x = -6$$

Check:

	Left-hand side	Right-hand side
	$3 \times x$	$x - 12$
	$= 3 \times -6$	$= -6 - 12$
	$= -18$	$= -18$

If the left-hand side and right-hand side values do not correspond, check every step of your working.

(iv) The square root of a number may be positive or negative, i.e. $\sqrt{n^2} = +$ or $- n$ and $\sqrt{25} = +$ or $- 5$.

(v) If $a \times b = 0$, then either a or $b = 0$.
If $(b + 3)(b - 4) = 0$, then either $b + 3 = 0$, or $b - 4 = 0$.
If you solve these equations, they will give you the values of x which satisfy the quadratic equation.

6 (iv) (v) *Standard Form notation:* Reference Section 10.

7 (iii) Reference Section 2.3.

8 The questions are arranged in order so that you are guided from one to the other. Calculate answers in the order given.
(Reference Section 3.B.)

9 *Circle formulae:* Reference Section 2.8.

11 The graph of $y = x^2$ is symmetrical about the y axis. It should be possible to see this property as a pattern in your table of values for y.

(vi) You want the value of x such that $x = \sqrt{6}$, i.e. $x^2 = 6$. You have drawn the graph of $x^2 = y$, hence the line which will give you the values of $\sqrt{6}$ is $y = 6$.

12 (vi) C has three columns but B has only 2 rows, consequently there is no corresponding row for the 3rd column in C.

Note: Before two matrices can be multiplied together, the number of columns of the first one must be identical to the number of rows of the second one. This is illustrated by the 'domino rule', which also gives you the order of your answer matrix.

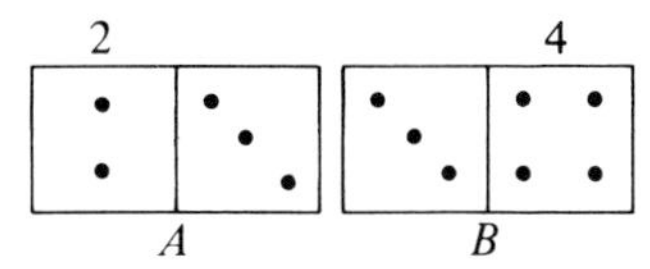

A is a (2 by 3) matrix, B is a (3 by 4) matrix, the multiplication AB will give you a (2 by 4) matrix.

13 *Position vectors:* Order of Counting.

1. Count right or left (right is positive, left is negative).
2. Count up or down (up is positive, down is negative).

In the diagram $\overrightarrow{AB} = \begin{pmatrix} 4 \\ -2 \end{pmatrix}$ i.e. 4 to the right, 2 down

but $\overrightarrow{BA}$ would be $\begin{pmatrix} -4 \\ 2 \end{pmatrix}$ i.e. 4 to the left, 2 up.

14 (i) The answer diagram shows full details of the enlargement.

(ii) Area of a kite: Reference Section 2.6.

Set 2

2 (iii) For an approximate answer use 8×5.

5 (v) Be careful with the square root of a fraction and always check your answer by squaring it, e.g.

$$\sqrt{0.09} = 0.30$$

Check: $0.3 \times 0.3 = 0.09$

Your answer may also be checked by changing the decimal fraction into an ordinary fraction as follows:

$$0.09 = \tfrac{9}{100} \text{ therefore } \sqrt{0.09} = \sqrt{\tfrac{9}{100}} = \tfrac{3}{10} = 0.3$$

(The square root of a fraction is obtained by taking the square roots of the denominator and numerator separately.)

6 Remember the first rule of factorising, which is to look for a possible common factor.

(iv) The common factor here is a ($a \div a = 1$).

8 (iii) Reference Section 2.8.

10 Matrix subtraction may be carried out by changing all signs in the matrix to be subtracted and then adding the matrices, e.g.

$$\begin{pmatrix} 3 & -4 \\ -2 & 1 \end{pmatrix} - \begin{pmatrix} -4 & 2 \\ 3 & -2 \end{pmatrix} \Rightarrow \begin{pmatrix} 3 & -4 \\ -2 & 1 \end{pmatrix} + \begin{pmatrix} 4 & -2 \\ -3 & 2 \end{pmatrix} = \begin{pmatrix} 7 & -6 \\ -5 & 3 \end{pmatrix}$$

11 (ii) An isosceles trapezium is a trapezium in which the non-parallel sides are equal.

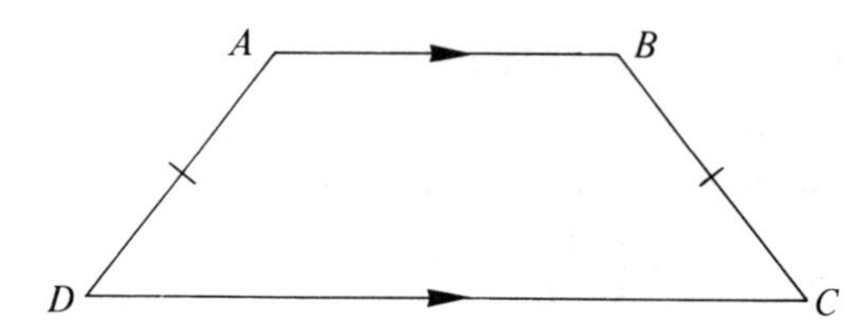

Isosceles trapezium: $AD = BC$

12 (v) $a^{\frac{1}{4}} = \sqrt[4]{a}$, e.g. $81^{\frac{1}{4}} = \sqrt[4]{81} = 3$

13 The answer diagram is explicit.

(iii) A rotation of +90° is an anticlockwise rotation: Tracing paper is very useful for rotations.

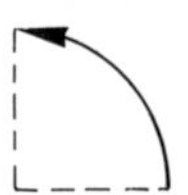

14 (i) To change a fraction to a percentage, multiply by 100%.
For example: express $\frac{3}{4}$ as a percentage

$\frac{3}{4} \times 100\% = 75\%$

(ii) To express a percentage as a fraction, write the percentage with a denominator of 100 and then simplify the fraction if possible.
For example: express 35% as a fraction

$$35\% = \frac{\overset{7}{\cancel{35}}}{\underset{20}{\cancel{100}}} = \frac{7}{20}$$

(iii) Make a fraction of the numbers and then use the method of (i).

(iv) To find a percentage of an amount.
For example: find 60% of 70

$$60\% = \frac{60}{100}$$

$$\text{thus } 60\% \text{ of } 70 = \frac{6\cancel{0}}{1\cancel{0}\cancel{0}} \times 7\cancel{0} = 42$$

(v) Profit % = $\dfrac{\text{Profit}}{\text{Cost price}} \times 100\%$.

For example: an article costing £2.40 was sold for £3.20, calculate the profit as a percentage of the cost price.

Cost price = £2.40
Selling price = £3.20
Profit = £0.80

$$\text{Profit } \% = \frac{0.80}{2.40} \times 100\%$$
$$= \tfrac{80}{240} \times 100\%$$
$$= \tfrac{100}{3}$$
$$= 33\tfrac{1}{3}\%$$

Note: When you make the fraction from: $\dfrac{\text{Profit}}{\text{Cost price}}$ numerator and denominator must be in the same units, in this case, both are in pence.

Set 3

1 (v) Use the method of 'trial and error'. You know that n must be greater than 5.

2 (iii) Check your answer by squaring it.

3 (iii) Double both sides of the equation.

4 (iv) Mean score = $\dfrac{\text{Total of all scores (i.e. total of column } (c))}{\text{Total number of trials (i.e. total of column } (b))}$

5 (iv) and (v) *Standard form notation:* Reference Section 10.

6 (iv) This is an expression which will factorise because it is the 'difference of 2 squares', i.e. $a^2 - b^2 = (a + b)(a - b)$, e.g. factorise: $x^2 - 25$

$$x^2 - 25 = x^2 - 5^2$$
$$= (x + 5)(x - 5)$$

(v) This is a trinomial. Study the following example:
Factorise $x^2 + 6x + 9$:

(*a*) $(x \quad)(x \quad)$ this gives x^2.
(*b*) $(x + 3)(x + 3)$ you need the factors of 9 which when added give 6, i.e. 3 x 3.

Always check factors by multiplying them out, in this case:

$$\begin{aligned}(x + 3)(x + 3) &= x(x + 3) + 3(x + 3)\\ &= x^2 + 3x + 3x + 9\\ &= x^2 + 6x + 9\end{aligned}$$

7 (i) and (ii) Use the theorem of Pythagoras, but remember that you are calculating shorter sides and hence the squares will be subtracted.

9 (iii) (*a*) The graph of $y = -x$ is the image of the graph of $y = x$ after reflection in the y axis. It has a gradient of -1 and passes through the origin.
The graph of $x + y = 2$ is quickly drawn by plotting the x and y intercepts (when $y = 0, x = 2$ and when $x = 0$, $y = 2$).

10 (vi) Given two matrices P and Q, if PQ or QP results in $\begin{pmatrix}1 & 0\\ 0 & 1\end{pmatrix}$ then Q is the multiplicative inverse of P.

11 A position vector of $\begin{pmatrix}-5\\ -2\end{pmatrix}$ means count 5 to the left, then 5 down.

(v) If $\vec{AB} = \vec{DC}$, then it follows that AB and DC are equal and parallel.

12 (iii) and (v) Brackets always have priority.

13 (ii) Under a translation vector, every point in the trapezium moves by the stated vector.

For example: $(-4,1) \rightarrow (1,3)$ under the translation $\begin{pmatrix}5\\ 2\end{pmatrix}$

14 (i) It is useful to learn that $12\frac{1}{2}\% = \frac{1}{8}$

(iv) and (v) What do you notice about your answers?

Set 4

1 *Language of sets:* Reference Section 4.

2 (iv) If you write down the operation described in (iii) as follows:

°F→ [−32] → [x5] → [÷9] →°C and then reverse the operation, you will have the method for converting °C into °F.

3 (iii) Multiply both sides of the equation by 4 first.

(v) Multiply both sides of the equation by 2 first.

4 Make sure that the total of column (c) is the same as the total number of scores in the distribution.

5 (ii) To change an ordinary fraction to a decimal fraction, divide the numerator by the denominator.

For example: change $\frac{3}{5}$ to a decimal fraction.

$$5\overline{)3.0} \quad \underline{0.6} \qquad \frac{3}{5} = 0.6$$

(v) A percentage may be regarded as a fraction with a denominator of 100. Thus $0.35 = \frac{35}{100} = 35\%$.

6 (iv) Study the following example of an expression which contains a common factor in brackets:

Factorise: $2a(x + y) - b(x + y)$.

The common factor is $(x + y)$ and taking this out we get:

$(x + y)(2a - b)$.

(v) Refer to the note on Q6(v), Set 2.

9 (i) The graph of $y = x + 2$ has a gradient of 1 and the y intercept is 2.

The graph of $y = x - 2$ has a gradient of 1 and the y intercept is -2.

These graphs are parallel (because they have the same gradient).

(ii) The graph of $x + y = -2$ is drawn by using the x and y intercepts.

If $y = 0, x = -2$ and if $x = 0, y = -2$.

11 (v) A comparison of the vectors shows (*a*) pairs of *equal* sides but (*b*) *no* parallel sides.

Note: Whenever the same pairs of numbers are used in a vector, the lengths of the vectors will be equal. The vectors in the following diagram are all equal in *length*.

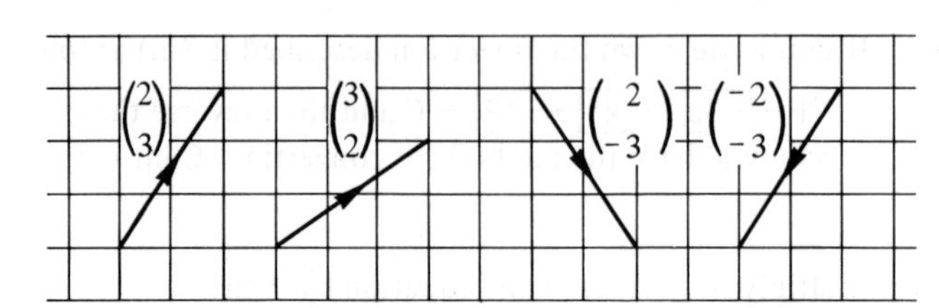

13 Since there is no change in the size of the shape, the transformation may be one of three: reflection, translation or rotation.

For reflections, state the equation of the line in which the shape has been reflected.

For translations, state the translation vector, e.g. the translation which maps G onto $D = \begin{pmatrix}4\\4\end{pmatrix}$

For rotations, state (i) the angle (ii) the direction and (iii) the centre of the rotation. (With a rotation of 180° it is unnecessary to state direction.)

Note: Direction may be positive or negative, e.g. +90° is anticlockwise and −90° is clockwise.

14 Calculations involving 15% are often easier if you first calculate 10% and then add on half of 10%.

For example: find 15% of £80.00

10% of £80.00 = £8.00
5% of £80.00 = £4.00
15% of £80.00 = £12.00

Set 5

1 It is a good idea to work out the intersections first, particularly (ii) (*b*).

2 (iii) If he runs 100 m in 12 seconds:

he will run 1000 m (1 km) in ? seconds
which is 1 km in ? minutes
and ? km in 1 hour.

(iv) $\text{Time} = \dfrac{\text{Distance}}{\text{Speed}}$

For example: if distance = 310 km and speed = 40 km/hr

then time $= \dfrac{31\not{0}}{4\not{0}} = 7\frac{3}{4}$ = 7 hours 45 minutes.

5 (iv) *Standard Form notation:* Reference Section 10.

6 (ii) The common factor is a^2 and $\dfrac{a^2}{a^2} = 1$

7 The following diagram shows the lengths to be calculated to give the x and y coordinates of the point Q.

QC is perpendicular to the x axis. The x coordinate of Q is the length of OC. The y coordinate of Q is the length of QC.

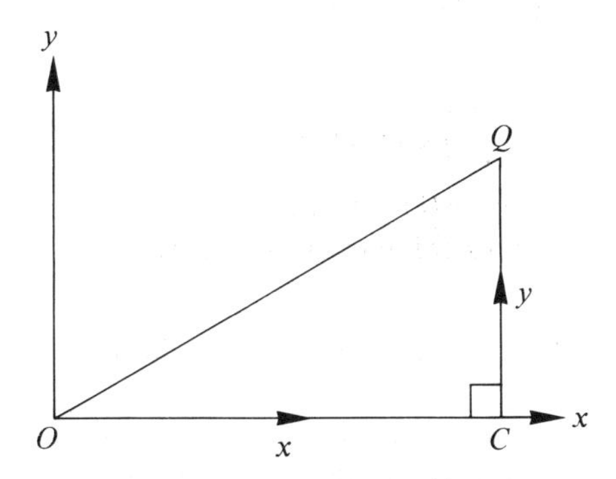

You will need basic trigonometry to calculate the y coordinate, which in this question should be calculated first; for reasons which you should quickly see.
The x coordinate may be found by using the theorem of Pythagoras but could be checked by basic trigonometry.

8 (i) Think in terms of letters. If you find this difficult, think what you would do if you were using numbers. Use the same process, but use the letters.

9 (i) The line whose equation is $y = 2x + 2$ has a gradient of 2 and the y intercept is 2.

(iii) Use the x and y intercepts to draw the graph of $x + y = 3$. The graph of $y = x - 3$ has a gradient of 1 and the y intercept is -3.

10 (viii) Ratio of corresponding sides = $\frac{8}{12} = \frac{2}{3}$

Therefore, the ratio of the areas = $\frac{2^2}{3^2} = \frac{4}{9}$

11 (vi)

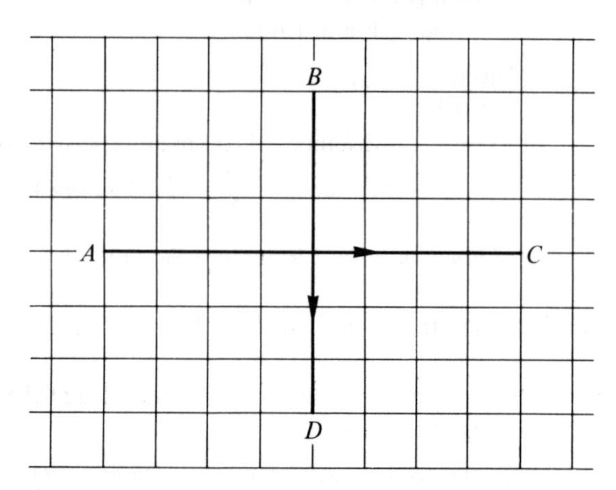

$\vec{AC} = \begin{pmatrix} 8 \\ 0 \end{pmatrix}$ and $\vec{BD} = \begin{pmatrix} 0 \\ -6 \end{pmatrix}$

The 0 in $\begin{pmatrix} 8 \\ 0 \end{pmatrix}$ shows that $\vec{AC}$ is horizontal and the 0 in $\begin{pmatrix} 0 \\ -6 \end{pmatrix}$ shows that $\vec{BD}$ is vertical.
If $\vec{AC}$ and $\vec{BD}$ intersect, then they must intersect at right angles.

13 See the full details of this question in the answer section.

14 (v) If we consider the pre-sale price as 100%, then:
£27.00 represents 100% − 10% = 90%

thus 90% = £27.00

$1\% = \frac{27.00}{90}$

and $100\% = \frac{27.00}{90} \times 100$ (or $27.00 \times \frac{10}{9}$)

Check your answer by finding 10% of it and then subtracting the 10% which should give you £27.00

Set 6

2 (v) Check your answer by cubing it.
For example, if the volume of the cube was 0.027 m^3 and your answer for the length of the edge of the cube was 0.3 m; check it by evaluating $(0.3)^3$:

$$(0.3)^3 = 0.3 \times 0.3 \times 0.3 = 0.027 \text{ m}^3.$$

4 The instructions are quite clear, follow them carefully.

(v) Theoretical probability is what ought to happen in theory. In other words, it is the result of dividing the number of 'favourable outcomes' by the number of 'possible outcomes'. As with many other theories, what happens in practice may give a slightly different result, depending upon the number of trials. The greater the number of trials, the more closely will theoretical probability relate to practical probability.

5 The laws of indices state that:

$(a)\ a^m \times a^n = a^{m+n}$
$(b)\ a^m \div a^n = a^{m-n}$ (Reference Section 12.)

(i) Evaluate the indices first.

(ii) Your answer is a power of x.

(iii) (iv) and (v) Reference Section 10.

6 (iii) The common factor is $2x$ and $\frac{2x}{2x} = 1$

(iv) The terms can be written as complete squares and then factorised as the 'difference of 2 squares', e.g.

$$\begin{aligned} 4a^2 - 25b^2 &= (2a)^2 - (5b)^2 \\ &= (2a + 5b)(2a - 5b) \end{aligned}$$

(v) You need the factors of 15 which add up to 8, the coefficient of b.

9 The diagram shows the graph of $y = \frac{1}{2}x$, which has a gradient of $\frac{1}{2}$ and passes through the origin.

(iii) The graph of $y = -\frac{1}{2}x$ is the image of the graph of $y = \frac{1}{2}x$ after reflection in the y axis.

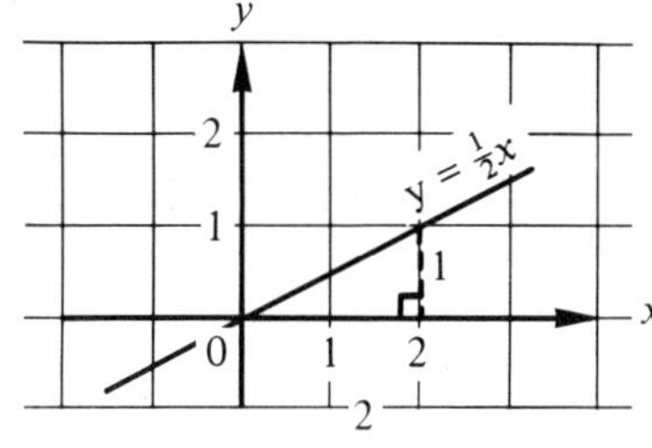

(iv) The diagram shows the graph of:

$y = 3x + 3$

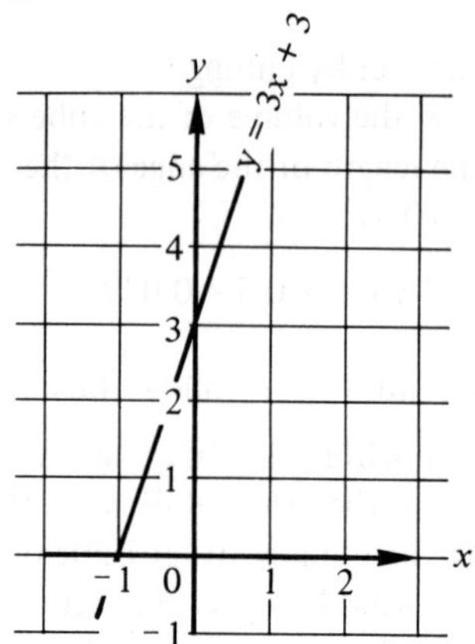

It has a gradient of 3 and the y intercept is 3.

The graph of $y = -3x + 3$ is the image of the graph of $y = 3x + 3$ after reflection in the y axis.

11 (v) *Addition of Vectors.* In the diagram $\vec{AB} = \begin{pmatrix} 6 \\ 4 \end{pmatrix}$, $\vec{BC} = \begin{pmatrix} 2 \\ -3 \end{pmatrix}$

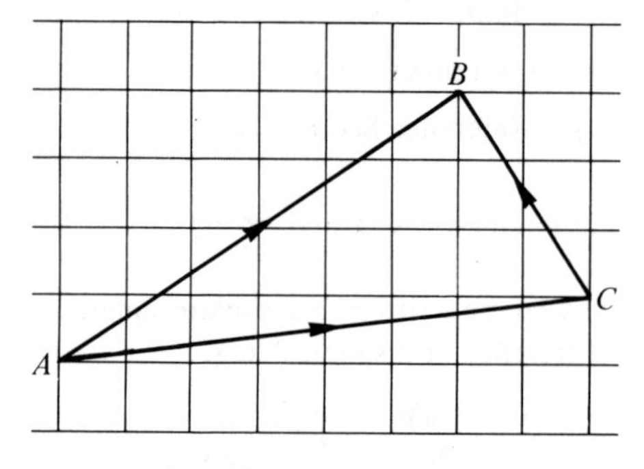

To evaluate $\vec{AB} + \vec{BC}$ we add the corresponding elements:

i.e. $\begin{pmatrix} 6 \\ 4 \end{pmatrix} + \begin{pmatrix} 2 \\ -3 \end{pmatrix} = \begin{pmatrix} 8 \\ 1 \end{pmatrix}$

Thus $\vec{AB} + \vec{BC} = \begin{pmatrix} 8 \\ 1 \end{pmatrix}$ and this may be checked on the diagram.

This also illustrates the fact that: $\vec{AB} + \vec{BC} = \vec{AC}$

13 $2x^2$ means $2 \times x^2$. To avoid mistakes, evaluate x^2 first and then multiply by 2.

14 Interest calculated like this is called compound interest. A similar example is shown below to give you a clear idea of method and setting down.
The question is identical apart from the deposit, which is £120.

			£	
	Year 1	*Amount:*	120.00	
(i)	*plus* 10% Interest:		12.00	(10% of £120.00 = £12.00)
(ii)	*Year 2*	*Amount:*	132.00	
(iii)	*plus* 10% Interest:		13.20	(10% of £132.00 = £13.20)
(iv)	*Year 3*	*Amount:*	145.20	
(v)	*plus* 10% Interest:		14.52	(10% of £145.20 = £14.52)
(vi)	*End of Year 3:*		£159.72	

Set 7

2 (iii) Calculate the area in square metres, then divide by 100.

(iv) Change the hectares to ares, then the ares to square metres.

3 (iv) Simplify the left-hand side, then divide both sides of the equation by 3.

(v) There are two methods of doing this.

Method 1

Rearrange the equation, giving $x^2 = 9$, and $x = \sqrt{9} = +$ or -3.

Method 2

Factorise $x^2 - 9$, giving the equation:

$(x + 3)(x - 3) = 0$
If $x + 3 = 0$, then $x = -3$
If $x - 3 = 0$, then $x = +3$, i.e. $x = +$ or -3

4 Even if a batsman fails to score, i.e. gets a 'duck', this is considered a completed innings.

6 (iv) $x^2 - 1$ is factorised as the 'difference of 2 squares'. At first glance the expression may appear to have no factors but $1^2 = 1$

(v) This trinomial involves negatives. Study the following example:
Factorise $x^2 - 2x - 15$

(*a*) $(x \quad)(x \quad)$ gives x^2.

(*b*) $(x + 3)(x - 5)$ you need the factors of 15, which when subtracted will give -2, i.e. $+3 \times -5$.

Check carefully by expanding your answer:

$$\begin{aligned} & x(x - 5) + 3(x - 5) \\ &= x^2 - 5x + 3x - 15 \\ &= x^2 - 2x - 15 \end{aligned}$$

Note: it is easy to make a mistake with signs but your check will reveal them.

9 (i) If O is the centre of the inscribed circle of triangle ABC, then CO is the bisector of angle ACB.

(ii) If x is the midpoint of BC, AX is perpendicular to BC and triangle OXC is right-angled.

(iii) Remember that $OA = OC$.

(vi) The shaded area is one of three such areas outside the circle and inside the triangle.

10 (i) When writing out matrix A, if there is no purchase of a particular item, write 0 in that column.

13 Full details are shown on the answer diagram.

14 (ii) Find the total marks he has scored (x) and the total marks which he could have scored (y). Now calculate the mean by evaluating

$$\frac{x}{y} \times 100\%.$$

(iii) The total possible mark in the 4 tests is $(40 + 25 + 75 + 60) = 200$
If his mean percentage mark was 74%, his total mark out of 200 would be 148. In English, Science and Maths he scored 28, 17 and 57, therefore his French mark was $148 - (28 + 17 + 57)$.

Set 8

1 (iii) There are two methods of doing this, one could be used to check the other.

Method (a): Change each number into base 10 first, then add.

Method (b): Add first in base 2, then change the resulting number into base 10. (Reference Section 9.)

2 (i) Check your answer by using the 'Shopkeeper Method' of subtraction (see the note on Q.3 (iii), Set 1).

5 (i) $10^4 + 10$ does *not* equal 10^5. Evaluate 10^4 and then add 10.

(iii) (iv) (v) *Standard Form notation:* Reference Section 10.

6 (iii) $(a + b)$ is the common factor.

(iv) $\frac{1}{2}$ is the common factor ($2 \div \frac{1}{2} = 4$ and $2a \div \frac{1}{2} = 4a$).

(v) Take out the common factor first.

7 The diagram shows the northerly and easterly components of the journey. PX is the northerly component. XS is the easterly component.

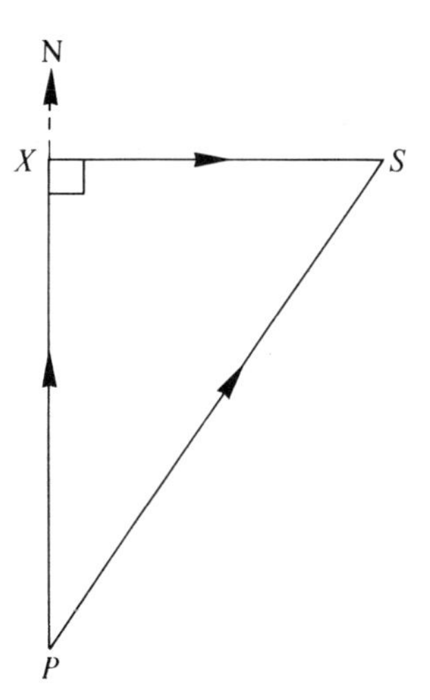

8 (i) (*b*) To change the subject to D, divide both sides of the equation by π.

(ii) (*a*) To change the subject to u, subtract ft from both sides of the equation.

11 (iv) See the note on Q.11 (v), Set 6.

12 (v) You could check this answer by first subtracting the fractions and then expressing the result as a percentage.

13 Be careful to use the scales exactly as given.

Set 9

2 (i) Write the length in centimetres before calculating the volume.

(ii) Number of small cubes in a large cube $= \dfrac{V\,(\text{Large volume})}{v\,(\text{Small volume})}$

3 (iii) Multiply both sides of the equation by 4.

(iv) Remember to multiply -4 by 2 when simplifying the left-hand side of the equation.

(v) Simplify the left-hand side of the equation and then divide both sides of the equation by 4.

4 (v) Read carefully: '*more* than 50%'.

6 (v) You need the factors of 6 which when subtracted give you 1. i.e. 2×3. The coefficient of x is -1, thus the signs will be $+2$ and -3.

7 (i) A good diagram is most important, showing all the information given as in the following example:

(*a*)

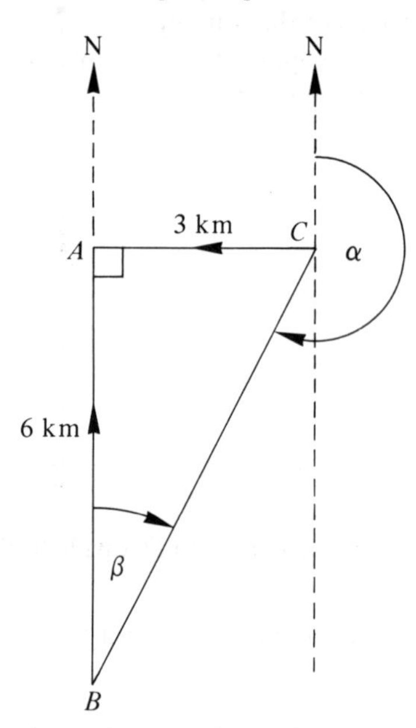

(*b*) The angle marked β is the angle you need to calculate to find the bearing of C from B.

(*c*) The return bearing is marked α in the diagram.

(ii) The diagram shows the northerly and easterly components of the journey.
BX is the northerly component and XA is the easterly component.
Angle $ABX = 45°$, therefore the triangle ABX is a right-angled isosceles triangle.

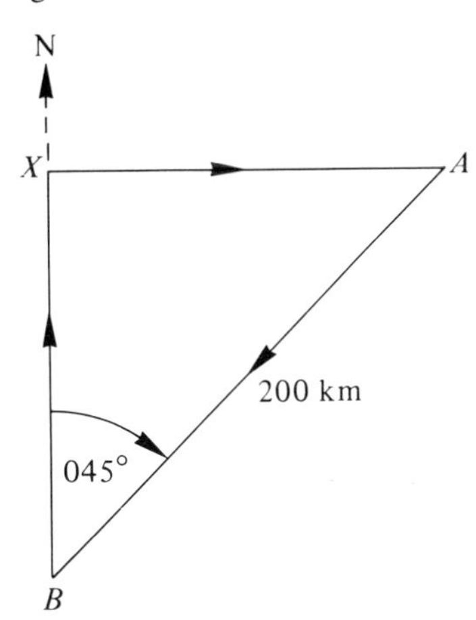

8 *Fact:* The diagonals of a rhombus bisect each other at right-angles.

9 (i)

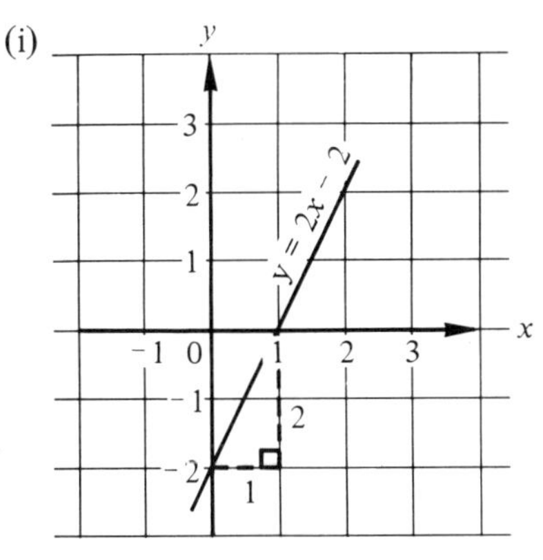

The diagram shows the graph of

$y = 2x - 2$.

It has a gradient of 2 and the y intercept is -2.

The graph of $y = -2x - 2$ is the image of the graph of $y = 2x - 2$ after reflection in the y axis.

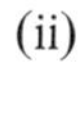

(ii)

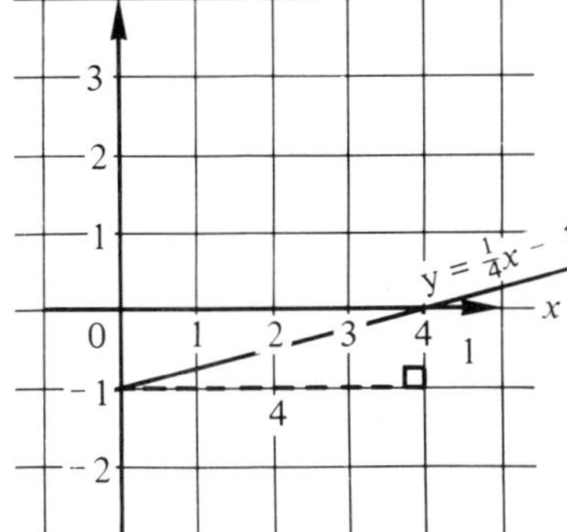

The diagram shows the graph of

$y = \frac{1}{4}x - 1$

gradient $\frac{1}{4}$, y intercept -1.

The graph of $y = -\frac{1}{4}x - 1$ is the image of $y = \frac{1}{4}x - 1$ after reflection in the y axis.

(iii) The graph of $y = 2x$ has a gradient of 2 and passes through the origin.

The graph of $y = -2x$ has a gradient of -2, passes through the origin and is the image of $y = 2x$ after reflection in the y axis.

13 (i) *Reflection:* State the equation of the line in which T is reflected.

(ii) *Enlargement:* State the Scale Factor and the centre of enlargement.

(iii) See (i).

(iv) *Translation:* State the translation vector.

(v) *Rotation:* Check with tracing paper for the centre. State the angle of rotation and the coordinates of the centre of rotation.

14 (i) Depreciation in the value of a car means the loss in value. Remember that 15% = 10% + 5%.

Set 10

2 (iii) Greenwich Mean Time (G.M.T.) is the time in London. (In fact, Cairo time is 2 hours ahead of G.M.T.)

3 (iii) Multiply both sides of the equation by $\frac{4}{3}$ (check your answer).

(iv) If $x(x - 5) = 0$, then either $x = 0$ or $x - 5 = 0$. Hence one value of x is 0 and the other value is obtained from the equation $x - 5 = 0$

(v) If $(3a - 9) \times (2a + 7) = 0$, then either $3a - 9 = 0$ or $2a + 7 = 0$

Solve these equations for the values of x which satisfy the quadratic equation. Check that your values of x make the values of the brackets 0.

4 (iii) The mean is found by dividing the total of column (c) by the total of column (b).

5 *Standard Form notation:* Reference Section 10.

6 (v) Take out the common factor first.

7 Consider the following facts:

(i) Angle BAD = angle BAC + angle DAC.

Triangle ACB, containing angle BAC is isosceles and triangle ACD containing angle DAC is a right-angled isosceles triangle.

(ii) Use the theorem of Pythagoras (X is the midpoint of AC).

(iii) Consider triangle ADX, what kind of triangle is it?

(iv) Use either the theorem of Pythagoras or trigonometry, (sin or cos ratio of angle ADX).

(vi) The quadrilateral is a kite (Reference Section 2).

8 (iii) Changing the subject to r (Reference Section 2).

Section 2

Set 1

1 (ii) (*a*) There are three pairs of faces which are equal in area, thus the total surface area may be found by:
2(area of side + area of top + area of end).

3 *Standard Form notation:* Reference Section 10.

4 (iii) Solve by factorising the left-hand side of the equation,

e.g. $x^2 - 5x = 0$ (take out the common factor)

giving: $x(x - 5) = 0$ and if x multiplied by $(x - 5) = 0$ then

either $x = 0$ or $(x - 5) = 0$
therefore $x = 0$ or 5. (Check your answer by substitution.)

(v) Solve by factorising. Study the following example:
Solve the equation:

$x^2 + 3x - 18 = 0$ (factorise the left-hand side)
giving: $(x - 3)(x + 6) = 0$

If $(x - 3) \times (x + 6) = 0$, then either $(x - 3) = 0$ or $(x + 6) = 0$

If $x - 3 = 0$ and if $x + 6 = 0$
$x = 3$ $x = -6$

Answer: $x = 3$ or -6.

Check: (i) if $x = 3$, $x^2 + 3x - 18 = 9 + 9 - 18 = 18 - 18 = 0$.
(ii) if $x = -6$, $x^2 + 3x - 18 = 36 - 18 - 18 = 36 - 36 = 0$.

5 Orderly arrangement of the calculation is of the utmost importance. Study the following worked example which is identical to Q.5 except for the purchase price of the car, which is £4000.

		£	
	Purchase price:	4000.00	
	less 25%:	1000.00	(25% of £4000 = £1000)
(i) (*a*)	Value at the end of Year 1:	3000.00	
	less 15%:	450.00	(15% of £3000 = £ 450)
(*b*)	Value at the end of Year 2:	2550.00	
	less $12\frac{1}{2}$%:	318.75	($12\frac{1}{2}$% of £2550 = £ 318.75)

(*c*) Value at the end of Year 3: £2231.25 Total: £1768.75.

(ii) Total depreciation for 3 years = £4000.00
− 2231.25
£1768.75

Check this with the total shown above.

Depreciation as a percentage of purchase price:

$$\frac{1768.75}{4000} \times 100\% = \frac{1768.75}{40} = \frac{176.875}{4} = 44.21\%$$

to the nearest whole per cent: 44%.

6 (v) The mean is calculated by dividing answer (ii) by answer (i).

(vii) '*Less* than 4 passengers'.

7 (ii) (*a*) A singular matrix is one which has no inverse and may be identified by the fact that the determinant is 0.

In the matrix $\begin{pmatrix} a & b \\ c & d \end{pmatrix}$, if $(a \times d) - (b \times c) = 0$ the matrix is singular, e.g.

if $A = \begin{pmatrix} 3 & 6 \\ 1 & 2 \end{pmatrix}$

$$\begin{aligned} \text{the determinant} &= (3 \times 2) - (6 \times 1) \\ &= 6 - 6 \\ &= 0 \end{aligned}$$

and A is a singular matrix. (Reference Section 6.E.)

8 (i) First calculate the height of triangle ABC. A diagram is always useful:

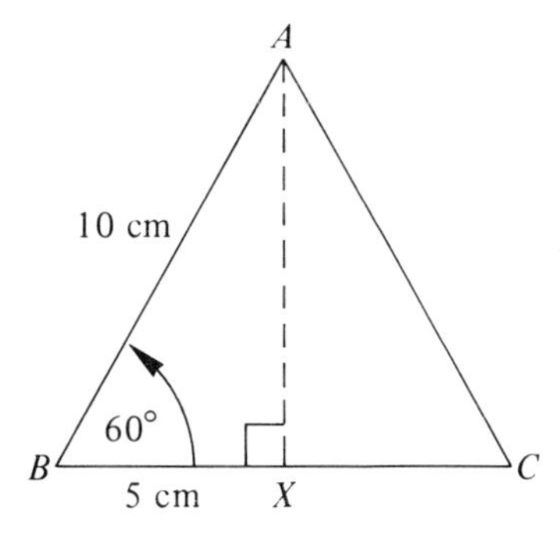

AX, the height of the triangle may be calculated by either of 2 methods:

(*a*) by using the theorem of Pythagoras

or (*b*) by using the sine ratio of angle ABX.

(Why not use both as a check?)

(ii) The height of triangle XYZ may be similarly calculated, but there is a quicker method to calculate the area of triangle XYZ, using the fact that the triangles ABC and XYZ are similar.

$XY = \frac{1}{2}AB$, i.e. the ratio $XY : AB$ is $1 : 2$ or $\frac{1}{2}$
If the ratio of the sides of the triangles is $1 : 2$, then the ratio of the areas will be $1^2 : 2^2$ i.e. $1 : 4$ or $\frac{1}{4}$ and

$$\text{the area of triangle } XYZ = \frac{\text{area of triangle } ABC}{4}$$

Note: If you are familiar with the trigonometrical formula for the area of a triangle: $A = \frac{1}{2}ab \sin C$ then it is unnecessary to calculate the height of the triangle. (Reference Section 3.C.)

(vi) The distance BD may clearly be seen in the diagram as the diagonal of rectangle $BCDF$.

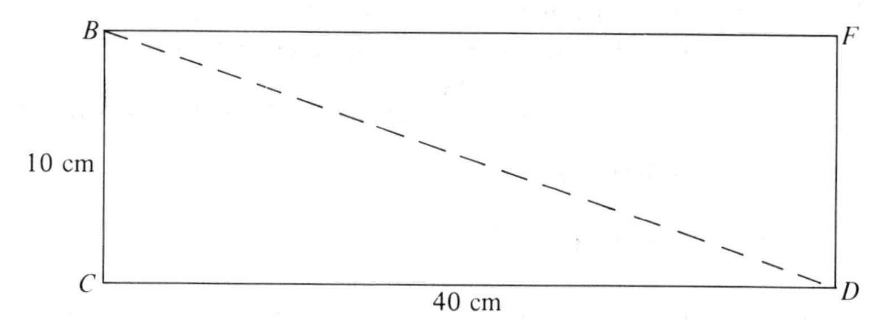

9 (i) Refer to the note on Q.4 (v).

(ii) Quadratic equations may also be solved by using the formula:

$$x = \frac{-b \pm \sqrt{b^2 - 4ac}}{2a}$$

and this method is particularly useful in quadratics which are either difficult or impossible to factorise.

Study the following worked example:

Solve $3x^2 - 5x - 3 = 0$

$$x = \frac{-b \pm \sqrt{b^2 - 4ac}}{2a}$$

a = the coefficient of $x^2 = 3$
b = the coefficient of $x = -5$
c = the constant term $= -3$

Substitute the values of a, b, and c in the formula:

$$x = \frac{-(-5) \pm \sqrt{(-5)^2 - 4(3 \times -3)}}{2 \times 3}$$

$$x = \frac{5 \pm \sqrt{25 + 36}}{6}$$

$$x = \frac{5 \pm \sqrt{61}}{6} = \frac{5 + 7.81}{6} \text{ or } \frac{5 - 7.81}{6}$$

$$x = \frac{12.81}{6} \text{ or } \frac{-2.81}{6}$$

$x = 2.135$ or -0.468

$x = 2.14$ or -0.47 (correct to 2 decimal places).

(It is usual to give answers correct to 2 decimal places where necessary.)

10 (iii) (*b*) To change m/s to km/h. The operation involves dividing the metres by 1000 to change them to kilometres and multiplying the result by 3600 (seconds in 1 hour). The operation may be simplified by combining the two processes:

i.e. $\frac{\times\ 3600}{\div\ 1000}$ is equal to multiplying by 3.6.

Method: Multiply the speed in m/s by 3.6.

11 (i) Use the 'gradient, intercept' method to identify the equation of the lines. If $y = mx + c$, m is the gradient of the line and c is the y intercept (if a line passes through the origin, then $c = 0$).

Study the following examples:

(*a*)

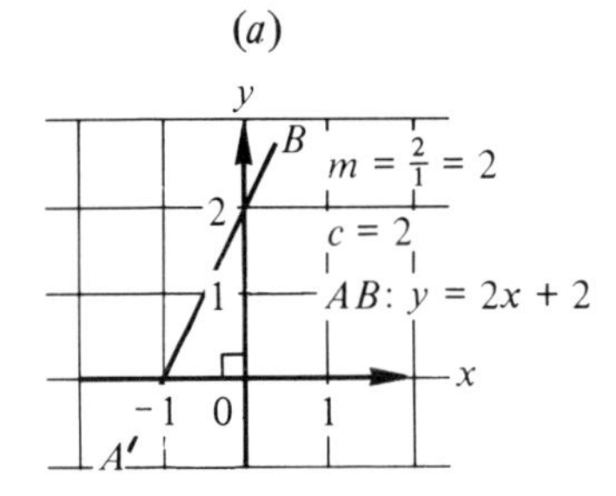

(*b*)

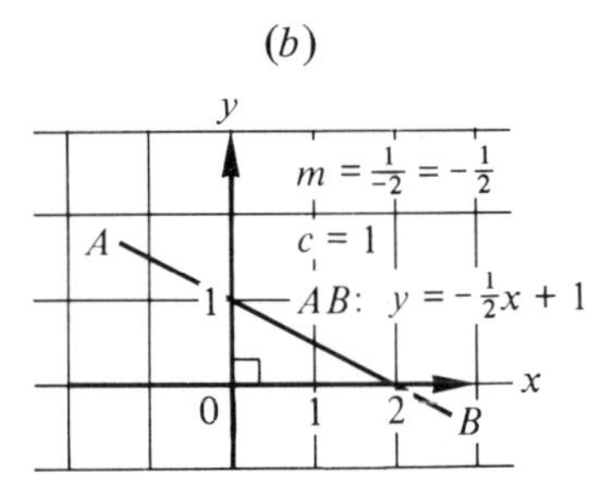

12 (i) $y = \frac{1}{4}x^2$ means $y = \frac{1}{4} \times x^2$ or $y = \frac{x^2}{4}$, therefore first evaluate x^2 and then divide by 4.

(iii) The graph of $y = \frac{1}{4}x + 1$ has a gradient of $\frac{1}{4}$ and the y intercept is 1.

13 (ii) and (iii) You are given 2 angles and the non-included side. See the Reference Section 3.C. for the necessary formula.

(iv) The shortest distance from C to AB is the perpendicular distance from C to AB, shown in the diagram as CX.

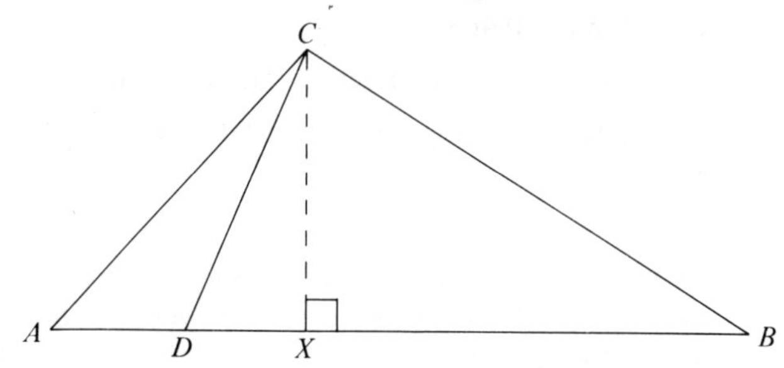

(v) In a 3-dimensional situation, always draw a diagram to give you a clear idea of the problem. The right-angled triangle PCD shows the angle of elevation of P from D, $DC = 5$ m and the required height is PC.

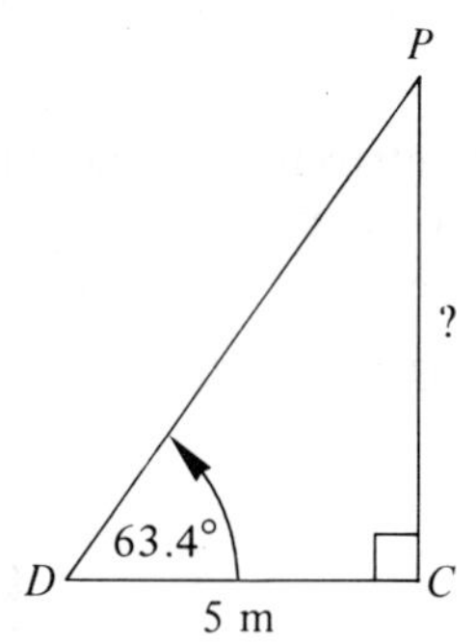

14 See the Reference Section 6.F for matrix transformations.

Set 2

1 (iii) Be sure to factorise fully. When you have taken out the common factor you can factorise the expression inside the bracket.

For example: Factorise $4x^2 - 100$

$$4x^2 - 100 = 4(x^2 - 25)$$
$$= 4(x + 5)(x - 5)$$

2 (i) (*b*) The quadrilateral $BCFD$ has 1 pair of parallel sides and 1 pair of equal sides. It is, therefore, an isosceles trapezium.

(ii) (*d*) You could check your answer here by using the formula $A = \frac{1}{2}ab \sin C$ to find the area of the triangle BDE.

3 (i) (*c*) Evaluate $(Q + R)$ first.

(ii) A^{-1} is the inverse of matrix A (see the Reference Section 6.E for the standard method of calculating A^{-1}).

6 (i) Be careful to use the scales given.

(ii) $y = 3x$ has a gradient of 3 and passes through the origin. Use the x and y intercepts to draw the graph of $x + y = 15$ and $x + y = 20$. For example, in $x + y = 15$, if $y = 0, x = 15$ and if $x = 0, y = 15$.

(iv) Remember that x and y are integers.

7 *Standard Form notation:* Reference Section 10.

(iii) Study the following example:

$$9\,000\,000 \div 4000$$
$$= (9.0 \times 10^6) \div (4.0 \times 10^3)$$
$$= \frac{9.0}{4.0} \times (10^{6-3})$$
$$= 2.25 \times 10^3$$

9 (i) (*a*) Start by using the angle sum of a triangle in triangle OAT (angle $OAT = 90°$).

(*b*) Triangle OAB is isosceles (equal radii).

See the Reference Section 7 for (1) the tangent facts and (2) the relation between the angle 'at the centre' and the angle 'at the circumference'.

10 See the Reference Section 6.F for matrix transformations and the answer section for the diagram of the transformations.

11

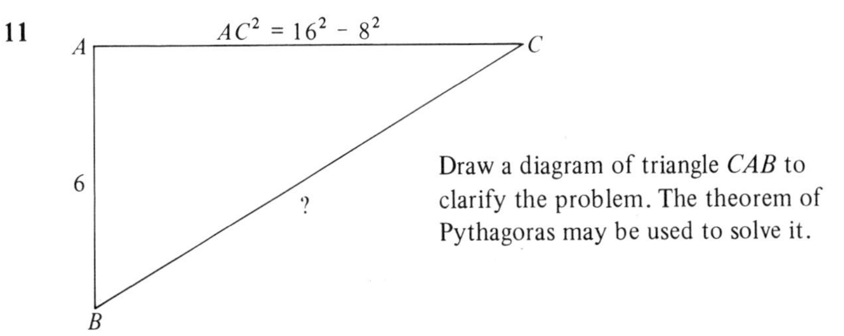

Draw a diagram of triangle CAB to clarify the problem. The theorem of Pythagoras may be used to solve it.

14 The answer section shows the scale drawing.

15 (vi) (*a*) If $\overrightarrow{DE} = \mathbf{b}$ and $\overrightarrow{AC} = 2\mathbf{b}$, then DE and AC are parallel and $AC = 2DE$. The diagram illustrates a similar situation numerically.

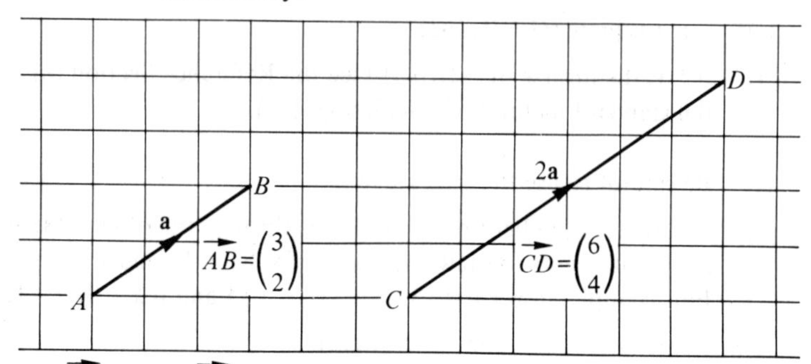

If $\overrightarrow{AB} = \mathbf{a}$ and $\overrightarrow{CD} = 2\mathbf{a}$ then AB and CD are parallel and $CD = 2AB$.

(*b*) The same conclusion may be reached with AB and FE.

Set 3

1 (i) Do you remember that $12\frac{1}{2}\% = \frac{1}{8}$?

(ii) 60% of the number is 24. The number is represented by 100%. 100% may be found by the *unitary method* of proportion, i.e.

If 60% $= 24$ or *more directly by:* $24 \times \frac{100}{60}$
then 1% $= \frac{24}{60}$
and 100% $= \frac{24}{60} \times 100$

(iii) In this question 108% = £27 and 100% = £27 $\times \frac{100}{108}$

2 See the Reference Section 2.8 for circle formula.

4 (i) Solution of simultaneous equations. There are several methods in use, but the easiest one in the case of this question is to eliminate one of the unknowns by subtraction. This results in a value which is then substituted to evaluate the other unknown.

For example: Solve the simultaneous equations:

(i) $2x + 3y = 5$
(ii) $x + 3y = 7$

Method (a) (i) $2x + 3y = 5$
(ii) $x + 3y = 7$ subtract (ii) from (i)
$x = -2$

(b) Substitute this value for x in equation (i)

$$2x + 3y = 5$$
$$-4 + 3y = 5$$
$$3y = 9$$
$$y = 3$$

(c) *Answer:* $x = -2$ and $y = 3$

(d) Check these values by substituting them in equation (ii):

$$x + 3y$$
$$= -2 + 9$$
$$= 7$$

4 (ii) See the note on Q.9 (ii) Set 1.

7 (i) Factorise both numerator and denominator.

9 (i) (*a*) A tangent to a circle is perpendicular to the radius at the point of contact. Use this fact in quadrilateral $ATOY$.

(ii) OA bisects angle TAY. (Reference Section 3.C.)

10 Reference Section 6.F: Matrix transformations.
See the answer section for the diagram.

11 The cosine ratio of an *obtuse* angle is *negative*.

13 Draw a sketch first showing the information given and the information you have to find. Then draw the accurate scale diagram and be careful to use the scale given. The sketch should be neat and an example is drawn for you.

Make sure you have a sharp pencil point. 2H pencils are recommended for scale drawings. Show all distances and bearings clearly.

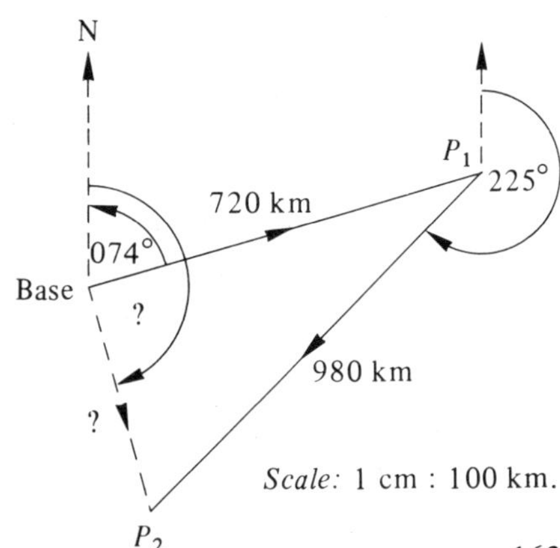

14 Be sure to use the scales given.

15 If $\frac{AX}{XC} = 3$, then AX will be equal to $\frac{3}{4}AC$.

(vii) (*a*) If 2 adjoining line segments share the same vector, then they must be part of the same straight line.
For example:

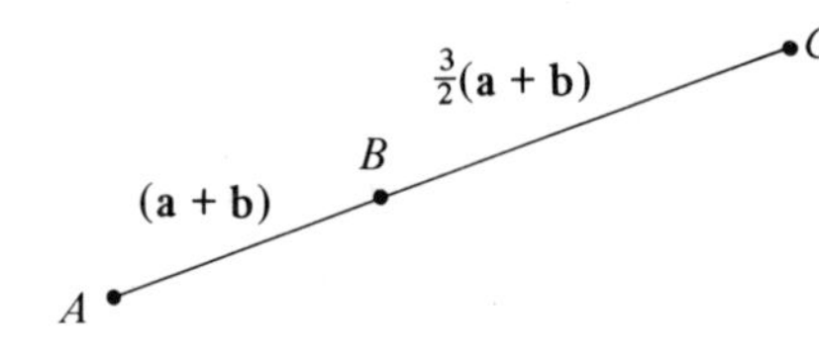

If $\vec{AB} = (\mathbf{a} + \mathbf{b})$ and $\vec{BC} = \frac{3}{2}(\mathbf{a} + \mathbf{b})$ they share the same basic vector $(\mathbf{a} + \mathbf{b})$, thus ABC must be a straight line, i.e. A, B and C are colinear.

(*b*) If 2 line segments have the same basic vector, one may be expressed as a ratio of the other, e.g.

$\vec{AB} = (\mathbf{a} + \mathbf{b}) \quad \vec{BC} = \frac{3}{2}(\mathbf{a} + \mathbf{b})$

Since the vectors are the same then: $\frac{AB}{BC} = \frac{1}{\frac{3}{2}}$

(multiply numerator and denominator by 2)

$= \frac{2}{3}$

i.e. $AB = \frac{2}{3}BC$

Set 4

1 (v) The coefficient of k^2 is 5 and this complicates the factorising process, but as 5 is a prime number there is no great difficulty.

(i) $(5k \quad)(k \quad)$ $5k \times k = 5k^2$

(ii) The factors of 2 are 2 and 1 and this produces as possible factors,

either (*a*) $(5k + 1)(k - 2)$
or (*b*) $(5k + 2)(k - 1)$

but (*a*) produces $5k^2 \; -\!\textcircled{9k} - 2$ and (*b*) produces $5k^2 \; -\!\textcircled{3k} - 2$. Thus to produce the required middle term $-3k$ we must arrange the factors as in (*b*) and
$5k^2 - 3k - 2 = (5k + 2)(k - 1)$

2 (i) A regular hexagon is composed of 6 equilateral triangles. Use this fact.

(iv) The trigonometry formula $A = \frac{1}{2}ab \sin C$ is very useful here.

4 You might find it useful to expand the diagram. For example, if from X to Y there are 3 routes and from Y to Z, 2 routes.

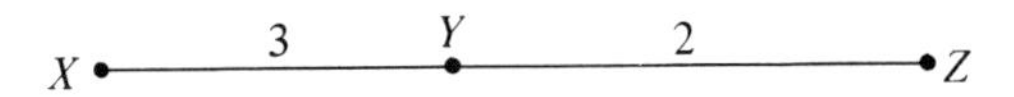

This could be drawn as:

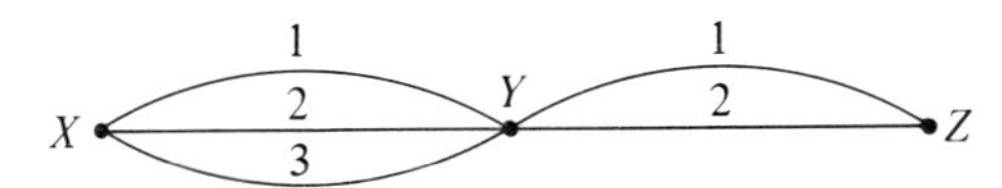

and the number of different routes from X to Z passing through Y is $3 \times 2 = 6$

6 (ii) (*b*) Before trying to factorise, re-arrange the equation to give: $x^2 - x - 6 = 0$

7 What is the size of angle XPY?

8 (ii) (*b*) Look for gradients and y intercepts. In graphs of $y = mx + c$, m is the gradient, c is the y intercept.

9 See the answer diagram and Reference Section 6.F.

10 (iii) Use PS and QS in triangle PQS to calculate angle y.

11 Solution of simultaneous equations by Matrix Method. Study the following example:

Solve the simultaneous equations:

(i) $2x + y = 12$
(ii) $x - y = 3$

1. Write the equations in matrix form: $\begin{pmatrix} 2 & 1 \\ 1 & -1 \end{pmatrix} \begin{pmatrix} x \\ y \end{pmatrix} = \begin{pmatrix} 12 \\ 3 \end{pmatrix}$

2. Multiply both sides of the matrix equation by the inverse of the left-hand side matrix. Note that the inverse is not fully evaluated at this stage.

$$-\tfrac{1}{3} \begin{pmatrix} -1 & -1 \\ -1 & 2 \end{pmatrix} \begin{pmatrix} 2 & 1 \\ 1 & -1 \end{pmatrix} \begin{pmatrix} x \\ y \end{pmatrix} = -\tfrac{1}{3} \begin{pmatrix} -1 & -1 \\ -1 & 2 \end{pmatrix} \begin{pmatrix} 12 \\ 3 \end{pmatrix}$$

3. Simplify the left-hand side and evaluate the right-hand side:

$$\begin{pmatrix} 1 & 0 \\ 0 & 1 \end{pmatrix} \begin{pmatrix} x \\ y \end{pmatrix} = -\tfrac{1}{3} \begin{pmatrix} -15 \\ -6 \end{pmatrix}$$

$$\begin{pmatrix} x \\ y \end{pmatrix} = \begin{pmatrix} 5 \\ 2 \end{pmatrix}$$

which gives the solution set: $\{x = 5, y = 2\}$

It may be seen from the example that great care must be taken with negative values in the evaluation of the right-hand side of the equation.

Remember to check your answers by substituting the values you have found in *both* equations, as follows:

Check: $x = 5, y = 2$

(i) $2x + y$	(ii) $x - y$
$= 10 + 2$	$= 5 - 2$
$= 12$	$= 3$

13 Rewrite the scale using the units the question involves and simplify as follows:

1 cm : 200 cm, i.e. 1 cm represents 2 m.

You should now find the conversion a great deal easier.

(v) The ratio of areas will be as the squares of the ratio of the lengths, e.g. if 1 cm : 5 m, then 1 cm^2 : 25 m^2.

14 (*g*) See the note on Q.15, Set 3.

Set 5

1 (iii) $12\frac{1}{2}\% = \frac{1}{8}$, therefore 30p = $\frac{1}{8}$ of the total pocket money.

(v) *Compound interest.* Study the following example:
Calculate the compound interest on £900 for 2 years at 10% per annum.

	£	
Principal:	900.00	(The *Principal* is the sum of money on which the interest is calculated)
plus 10%:	90.00	
End of Year 1: *Amount*:	990.00	(The *Amount* is the *Principal + Interest*)
plus 10%:	99.00	(Interest is calculated on the *Amount* of £990)
End of Year 2: *Amount*:	£1089.00	

The compound interest is found by £1089 − £900 = £189.
(Check by adding £90 and £99.)

2 Draw your own diagram and draw in the other diagonal AC. (What do you know about the diagonals of a rhombus?)

(iv) A useful formula for the area of a rhombus is:

$$A = \frac{\text{Long diagonal} \times \text{Short diagonal}}{2}$$

3 (v) You may use the laws of indices to simplify $A^2 \times A^{-1}$

4 (i) See the note on Q.4, Set 3. The same method may be used.

(ii) See the note on Q.9, Set 1.

5 (ii) See the Reference Section for length of arc and area of sector formulae (Section 2.A.9, 10).

8 (ii) Write the fractions with the same denominator, i.e.

$\frac{1}{9} = \frac{10}{90}$ and $\frac{1}{10} = \frac{9}{90}$

There is an infinite number of fractions between $\dfrac{9}{90}$ and $\dfrac{10}{90}$

but an obvious one is $\dfrac{9\frac{1}{2}}{90} = \dfrac{19}{180}$

10 *Matrix Transformations:* Reference Section 6.F. See the answer section for the detailed diagram.

11 (i) Triangle BCD is isosceles and angle CBD may be found as shown in the following diagram.

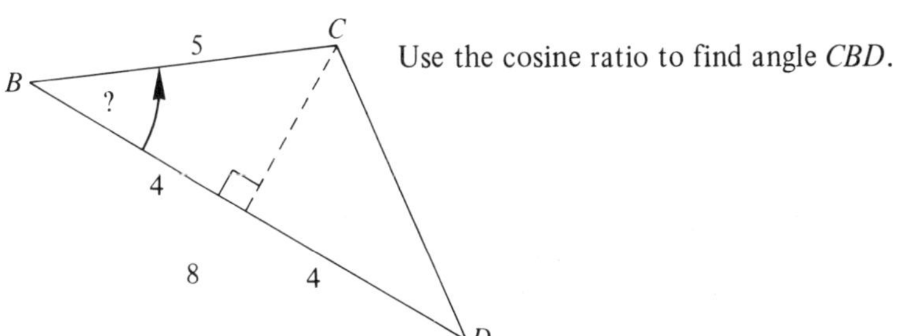

Use the cosine ratio to find angle CBD.

12 (i) Add up the ratios. It can then be seen that B's share was $\frac{5}{18}$ of the sum of money, i.e. $\frac{5}{18}$ = £90.

If you calculate $\frac{1}{18}$, you can then find A's share and C's share.

The sum of money is represented by $\frac{18}{18}$. Check that the sum of the three shares is equal to your answer to (c).

(ii) 1 km = 100 000 cm.

(iii) The volumes will be in the ratio of $2^3 : 5^3$

13 Check the scale along the time axis carefully to find how many minutes 1 small square represents.

14 See the note on Q.15, Set 3.

15 (iii) Change the subject of the equation to either x or y.

(iv) Remember that x and y are integers.

Set 6

1 (iii) Take out the common factor first.

(v) In this trinomial, the coefficient of a^2 is 9 and factors of 9 are either (9×1) or (3×3), both of which must be considered. After inspection, (9×1) cannot produce the desired coefficient of 3 for the middle term, as the only factors of 2 are (2×1) or (1×2). Starting then with $(3a \quad 2b)(3a \quad b)$, you must decide the signs which will give you -3 after addition and -2 by multiplication. After inspection it will be seen that $(3a - 2b)(3a + b)$ will give the required result.

5 (i) OX bisects angle YXZ.

6 (iv) (vi) See the Reference Section 2 for the necessary formulae.

8 (ii) Refer to the note on Q.12 (i), Set 5.

10 The answer diagram is explicit.

12 (iv) This may be done by substituting values of x to see which values satisfy the inequality (start with $x = 1$). An alternative approach is to factorise the left-hand side of the inequation, i.e. $(x + 4)^2 \leqslant 36$.

(v) Multiply both sides by $(x - 4)$.

13 (i) Take care with the negative values. The graph of a quadratic is symmetrical and this symmetry is often to be seen in the values for y in the table. A wrong value or a wrongly plotted value will also be evident if the plotted points do not produce the smooth curve of a parabola.

(iv) When you draw the graph of $y = x - 2$, remember that the axes are scaled differently.

(vi) At the points of intersection of the curve and the straight line:

$$x^2 - 2x - 2 = x - 2$$

(Rearrange and simplify the equation.)

14 *ABCD* is a trapezium in which *AB* is parallel to *CD*, therefore: $\vec{DC} = \mathbf{a}$.

(ii) Consider $\vec{XY}$ and $\vec{CD}$.

15 Study the following example of calculating 'in terms of π'.
Area of a circle $= \pi r^2$. If $r = 5$, then $A = \pi 5^2 = 25\pi$ (in 'terms of π').
(Reference Section 2 for circle and cylinder formulae.)

Set 7

1 (iv) £1280 represents 80%, to find the price he paid for it, calculate 100%.

(v) Refer to the note on Q.1 (v), Set 5.

2 (ii) If you regard triangle *PSQ* as the base of the pyramid, then *PR* is the height of the pyramid and the volume is found by $\frac{1}{3}$ area of base x height.

3 (i) (*b*) A singular matrix is one which has no inverse and its determinant is 0.

(ii) The operation of multiplication must be carried out before the operation of addition.

5 See the Reference Section 7 for facts about angles in circles.

(iii) Calculate *AC* first and then $XC = AC - AX$

6 See the Reference Section 2 for the necessary formulae.

9 These are graphs of $y = mx + c$. To find their equations, look for m, the gradient of the line and c, the y intercept. Remember that in mathematics, a gradient may be negative, as shown in the following diagram.

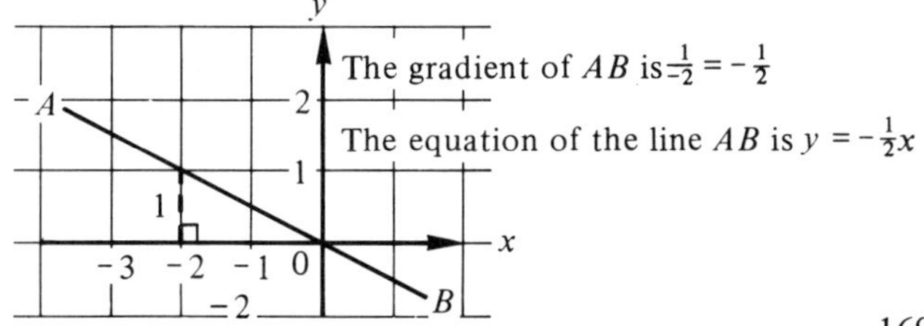

10 Consider the basic vectors $\begin{pmatrix}1\\0\end{pmatrix}$ and $\begin{pmatrix}0\\1\end{pmatrix}$

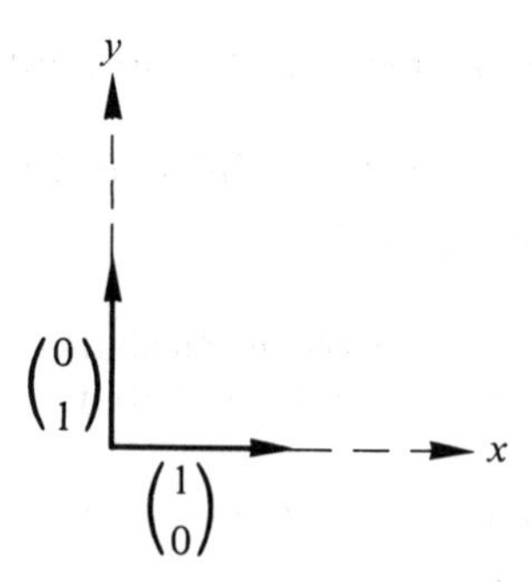

$\begin{pmatrix}1\\0\end{pmatrix} \to \begin{pmatrix}0\\1\end{pmatrix}$ and $\begin{pmatrix}0\\1\end{pmatrix} \to \begin{pmatrix}-1\\0\end{pmatrix}$

(Each has been rotated by what angle and in which direction about the origin?)

11 (ii) AC may also be calculated by using the cosine formula which would provide a very good check on your answer.

12 When copying the diagram, in order to avoid mistakes, it is a good idea to use position vectors to plot the positions of B, C and D. For example, the position of B may be considered as a journey from A of $\begin{pmatrix}8\\6\end{pmatrix}$ i.e.

$\vec{AB} = \begin{pmatrix}8\\6\end{pmatrix}$; $\vec{AC} = \begin{pmatrix}11\\2\end{pmatrix}$ which you can check by $\vec{BC} = \begin{pmatrix}3\\-4\end{pmatrix}$

(iv) Average speed = $\dfrac{\text{Total distance}}{\text{Total time}}$

13 (i) If you are using the method of eliminating one of the unknowns, first of all multiply the first equation by 3 and then y can be eliminated by addition. Do not forget to multiply the whole equation by 3. A similar worked example is shown:

Solve the simultaneous equations:

(i) $x - 2y = 10$
(ii) $2x + y = 10$

Multiply (ii) by 2, giving $4x + 2y = 20$.
Now add the equations:

(i) $x - 2y = 10$
(ii) $4x + 2y = 20$

$5x = 30$

and $x = 6$

Substitute $x = 6$ in (i):

$$6 - 2y = 10$$
$$-2y = 4$$
$$y = -2$$

Answer: $x = 6, y = -2$

Check in (ii):

$$(2 \times 6) + (-2)$$
$$= 12 - 2$$
$$= 10$$

14 (iii) To arrive at your answers, consider:

(*a*) $\overrightarrow{BX}$ and $\overrightarrow{YD}$

(*b*) $\overrightarrow{BY}$ and $\overrightarrow{XD}$.

(iv) Study your answers to (iii).

Set 8

1 (v) The common factor of this trinomial is 6 and the factorising difficulties are greatly reduced when you divide the whole expression by 6, giving: $6(x^2 + 4x + 4)$ and leaving inside the bracket a straightforward trinomial. Do not forget to show the 6 in your answer.

2 (i) Consider the sum of the angles at B.

(iii) Use either $A = \frac{1}{2}bh$ or $A = \frac{1}{2}ab \sin C$. (Or use one to check the other!)

(vii) A diagram of the completed pyramid will help to solve this problem. Study the following diagram:

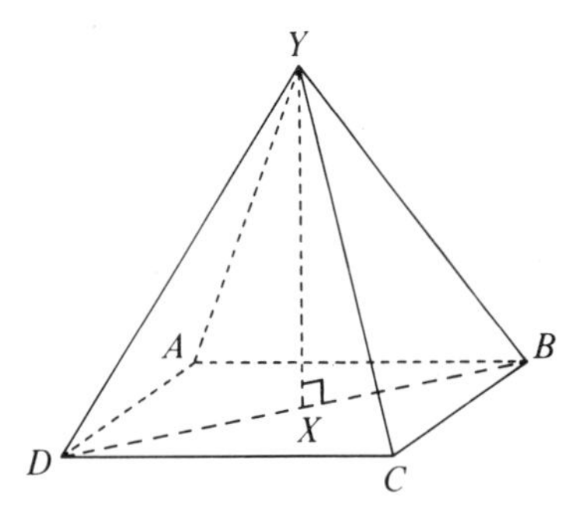

In the diagram, the point Y is the meeting point of the points E, F, G and H. XY is the perpendicular height of the pyramid and may be calculated in the right-angled triangle BXY. $BY = 3$ cm and $BX = \frac{1}{2}DB$ (calculate DB first).

3 (ii) Be careful in this question. The depreciation in the second year is given as a % of the car's value at the end of the first year, which is 75% of its original value. Study the following similar example:

Depreciation in the 1st year is 20% of its original value and in the 2nd year it depreciates by 15% of its value at the end of the first year.

1st year: Depreciation is 20%, therefore the car's value as a percentage of the original value is 80%.

2nd year: Depreciation is 15% of 80% = 12%

Total Depreciation as a percentage of original value

= 20% + 12% = 32%

4 (i) Addition of the equations will eliminate y.

(ii) Refer to the note on Q.9 (ii) Set 1.

6 (iii) $\pi r^2 \times 2r = 2\pi r^3$

(iv) Comparing $\frac{4}{3}\pi r^3$ with $2\pi r^3$, $\frac{4}{3}$ has to be calculated as a fraction of 2, i.e.

$$\frac{\frac{4}{3}}{2} = \frac{2}{3}$$

12 (ii) Study $\vec{CA}$ and $\vec{GH}$ carefully.

15 Refer to the note on Q.13, Set 6.

Set 9

1 (i) 15% is quickly found by finding 10% first and then adding on half of 10%.

(ii) It is possible to have percentages greater than 100%. If, for instance, an article was bought for £1 and sold for £3, a profit of 200% would have been made.
Thus $1\frac{1}{2}$ represents 150% and $3\frac{1}{4}$ represents 325%.

3 Be careful here. As a result of the fact that you are subtracting the squares of the two longer sides, the *minimum* value of the third side is found by using the *minimum* length of the hypotenuse and the *maximum* length of the other longer side.

The *maximum* value of the third side is found by using the *maximum* length of the hypotenuse and the *minimum* length of the other longer side.

Remember that if AB, a line segment, is measured as 5 cm, correct to the nearest centimetre, the maximum length is 5.5 cm and the minimum length is 4.5 cm.

5 (i) The outcome of each toss of a coin is independent of the previous outcome.

6 It is a good idea to extract and sketch any triangles used in different parts of the problem unless, of course, they are clearly shown in the perspective diagram. For example, in (i) the triangle ACX is used and it may be shown as follows with the information you need for the calculation:

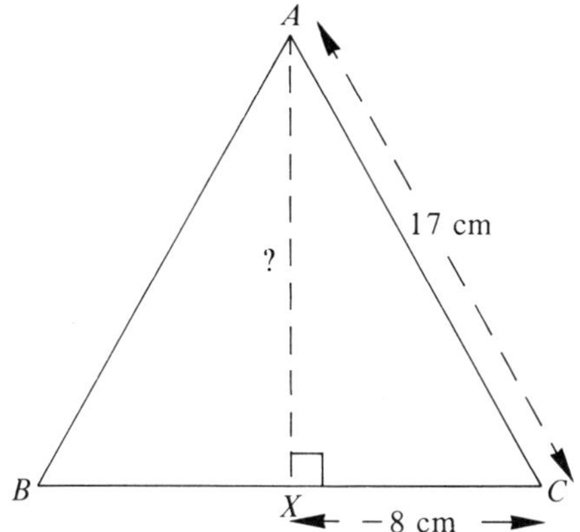

10 (v) The matrix $\begin{pmatrix} 0 & -2 \\ 2 & 0 \end{pmatrix}$ can be written as: $2\begin{pmatrix} 0 & -1 \\ 1 & 0 \end{pmatrix}$

Considering the base vectors $\begin{pmatrix} 1 \\ 0 \end{pmatrix} \rightarrow \begin{pmatrix} 0 \\ 1 \end{pmatrix}$ and $\begin{pmatrix} 0 \\ 1 \end{pmatrix} \rightarrow \begin{pmatrix} -1 \\ 0 \end{pmatrix}$, both have been rotated anticlockwise 90° about (0, 0). Therefore, the complete transformation is a rotation as described, followed by an enlargement, Scale Factor 2, centre 0.

13 (ii) Study $\vec{PT}$ and $\vec{QR}$.

14 Use the horizontal axis for the 'Number of children in a family' and the vertical axis for the 'Number of families'.

(iv) The mean is found by: $\dfrac{\text{Total number of children in all families}}{\text{Number of families}}$.

15 (ii) (*a*) Consider the sine ratio of angle ACB.

Set 10

2 See the Reference Section 2 for the necessary formulae.

3 See the note on Q.5, Set 1.

4 *Graphical solution of simultaneous equations.*

(iii) The solution set of the equations is found by the x and y coordinates at the point of intersection of the graphs.

5 (ii) There is a quick method for conversion of base 2 to base 8 and vice versa. Study the following examples:

(*a*) Convert $101\ 111_2$ to base 8.
Divide the number into groups of three, marking off the groups of three from right to left, and treat each group in binary notation:

4	2	*U*	4	2	*U*
1	0	1	1	1	1
	5			7	

Answer: $101\ 111_2 = 57_8$

(*b*) Convert 47_8 into base 2.
Convert each digit of 47 into a base 2 number:

		4	7		
4	2	*U*	4	2	*U*
1	0	0	1	1	1

Answer: $47_8 = 100\ 111_2$

The reason you can perform this operation is because 8 is a column in the binary system.

6 See the Reference Section 2 for the necessary formulae.

8 (iii) If 2 cubes have their volumes in the ratio $x : y$, then their sides are in the ratio $\sqrt[3]{x} : \sqrt[3]{y}$

11 (i) The diagram shows how the angle AOB may be calculated.

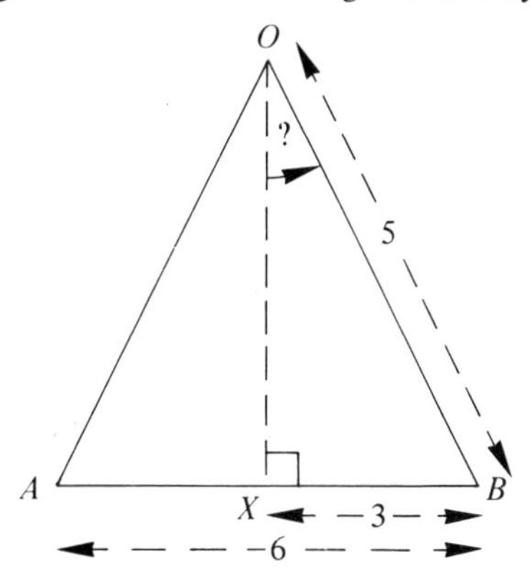

As triangle AOB is isosceles, OX is perpendicular to AB and also bisects AB.
Use triangle BXO and calculate angle BOX.
Angle $AOB = 2BOX$.

(ii) Reference Section 7.B: Angles in a circle.

12 (ii) You should have a quadratic equation which will factorise, giving 2 positive values of x from which you can produce 2 sets of values for AB, AC and BC.

13 In this question you are *calculating*, not *measuring*.

(i) Draw neat diagrams, putting in North lines at B, parallel to the one at A. It is also helpful to extend this North line in a southerly direction as in the following example. Mark the angle you need to calculate for the return bearing.

The return bearing is marked β in the diagram and is made up of two angles, one of which is 180° and the other is an alternate angle (parallel North lines) which is equal to the outward bearing angle of 44°. Hence, the bearing of A from $B = 180° + 44° = 224°$

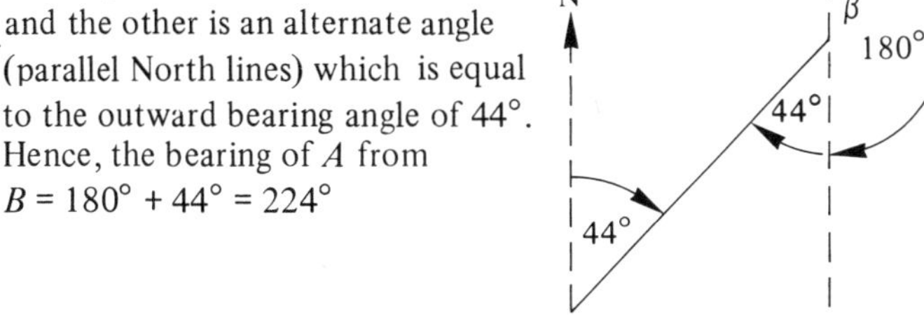

(ii) A similar use of North and South lines will help you in this question.

(ii) (*a*)

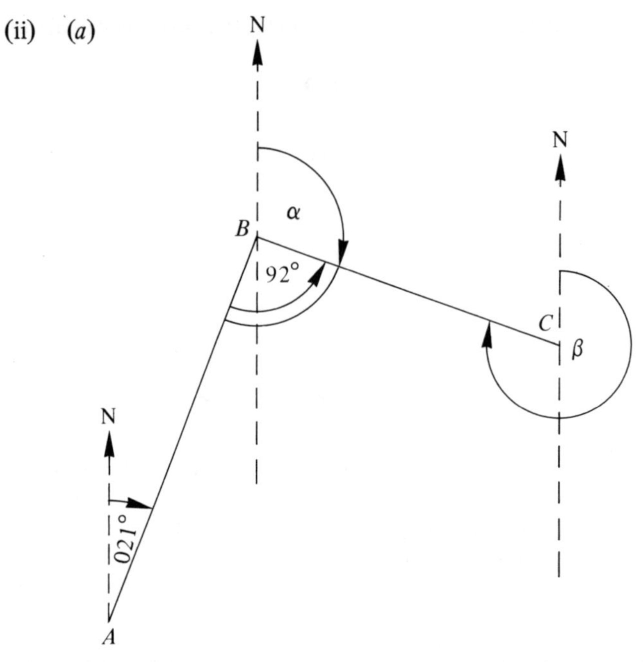

(i) Calculate A from B as shown in the first part of the question.

(ii) The bearing of C from B, marked α in the diagram will be answer (i) $-92°$.

(iii) The bearing of B from C will be angle $\alpha + 180°$.

(ii) (*b*) If $AB = BC$, triangle ABC is isosceles and angle BAC can be found. The bearing of C from A will be $021°$ + angle BAC.

14 Where necessary, draw diagrams of the separate triangles involved in the calculations as shown below:

(iii)

(iv)

(viii)

(v)

(vi)

(ix)

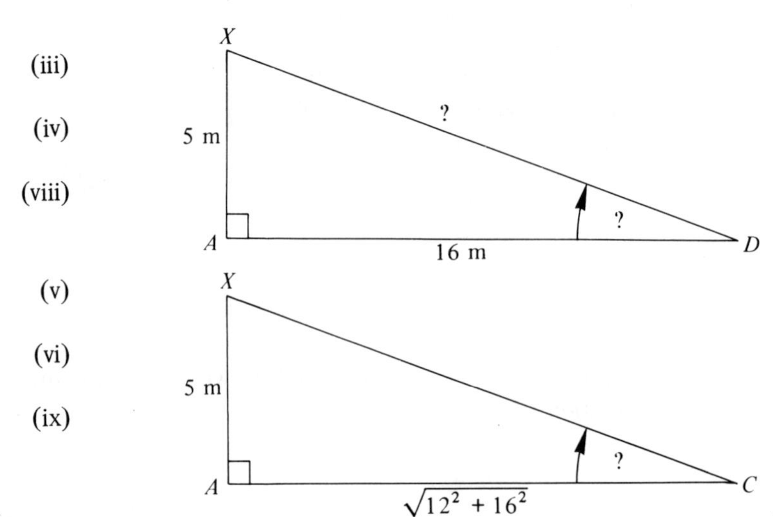

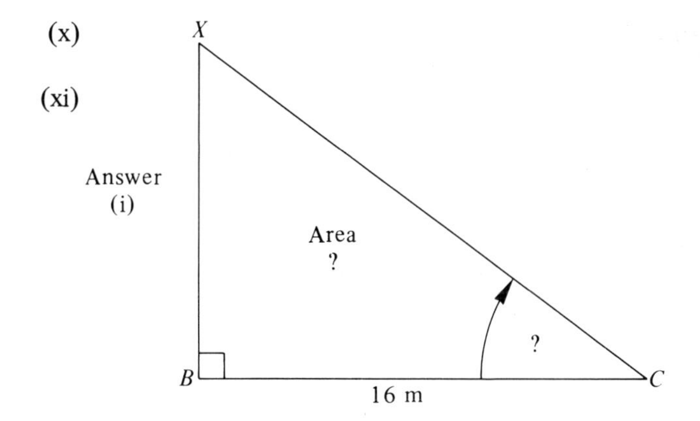
(x)
(xi)
X
Answer
(i)
Area
?
?
B
C
16 m

Section 3

Set 1

1 (i) Find $12\frac{1}{2}\%$ of £4960 and subtract it from £4960.

(ii) Value after one year = £4960 − (18% of £4960).

(iii) Value after 2 years = answer (ii) − 10% of answer (ii).

(iv) Find the loss to the customer by subtracting answer (iii) from answer (i).

The percentage loss will be: $\frac{\text{loss to the customer}}{\text{answer (i)}} \times 100\%$

2 (i) (*a*) Firstly bracket your fractions as shown:

$$\frac{(x-2)}{3} + 4 = \frac{52}{15} - \frac{(2x-3)}{5}$$

Then, to remove your fractions, multiply every element, including 4, by the L.C.M. of your fractions, i.e. 15.

Don't forget to check your answer by replacing x by the value you have obtained for it.

(*b*) Multiply every element by the L.C.M. of your fraction, i.e. x. You will produce a quadratic equation, which will produce two solutions when solved by factorising.

For example:

$$x + 3 = \frac{10}{x}$$

becomes $x^2 + 3x = 10$

$\Rightarrow$ $x^2 + 3x - 10 = 0$

Factorising: $(x+5)(x-2) = 0$

and $x = 2$ or -5

(ii) (*b*) $3x^2 + 17x + 10 = 31710$

$= (3 \times 10000) + (17 \times 100) + 10$

i.e. $17x = 17 \times 100$

$x = 100$

Substitute the value of 100 for x in each of your factors of $3x^2 + 17x + 10$ and you will have found the required factors.

3 (i) Find the height of the small cone, then subtract the volume of the small cone from the large one to find the volume of the frustrum.

If the volume of the large cone = $\frac{1}{3}\pi R^2 H$ and the volume of the small cone = $\frac{1}{3}\pi r^2 h$, then

$$\text{the volume of the frustrum} = \tfrac{1}{3}\pi R^2 H - \tfrac{1}{3}\pi r^2 h$$
$$= \tfrac{1}{3}\pi(R^2 H - r^2 h)$$

(ii) Find the volume of the cylinder and divide it by the volume of each cup.

4 (i) If $f(x) = x^2 + 5x - 4$, the value of $f(2)$ is obtained by substituting $x = 2$ in the expression, as follows:

$$f(2) = 2^2 + (5 \times 2) - 4$$
$$= 4 + 10 - 4$$
$$= 10$$

(ii) If $f(x) = 0$, then $x^2 + 5x - 4 = 0$ and as this quadratic does not factorise, you will need to use the formula:

$$x = \frac{-b \pm \sqrt{b^2 - 4ac}}{2a}$$

(iii) Remove the fraction by multiplying both sides of the equation by t.

5 (i) $TP_2 = 50 \tan \text{angle } TXP_2$

$TP_1 = 50 \tan \text{angle } TXP_1$

$$P_1P_2 = TP_2 - TP_1$$
$$= 50 \tan TXP_2 - 50 \tan TXP_1$$
$$= 50(\tan TXP_2 - \tan TXP_1)$$

Complete the calculation.

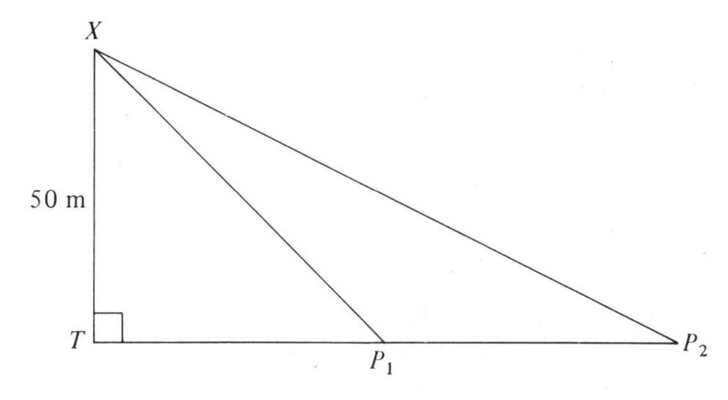

(ii) (*a*) To find the speed in metres per second, divide the answer to (i) by 5.

(*b*) To change metres per second to kilometres per hour, use the following method:

$$1.8 \text{ m/s} = \frac{1.8}{1000} \text{ km/s}$$

$$= \frac{1.8}{1000} \times 60 \times 60 \text{ km/h.}$$

This gives the general rule for changing metres per second into kilometres per hour, i.e. multiply by $\frac{3600}{1000}$ or more simply, *multiply by* 3.6

6 (i) Write your probability as a fraction:

$$\frac{\text{Number of men}}{\text{Number of possible outcomes}}.$$

(ii) (*a*) There are two methods of approaching this question:

Method 1

The lattice shows all the possible outcomes of rolling two dice. Those which result in a total score of 8 are ringed, resulting in a probability of $\frac{5}{36}$

6	*	(*)	*	*	*	*
5	*	*	(*)	*	*	*
4	*	*	*	(*)	*	*
3	*	*	*	*	(*)	*
2	*	*	*	*	*	(*)
1	*	*	*	*	*	*
	1	2	3	4	5	6

Method 2

Make up a table to show all possible outcomes.

Total score	Scores on each die	Number of ways
2	(1,1)	1
3	(1,2), (2,1)	2
4	(1,3), (2,2), (3,1)	3

(ii) (*b*) Either of the above methods may be used.

(iii) The tree diagram is a useful method.

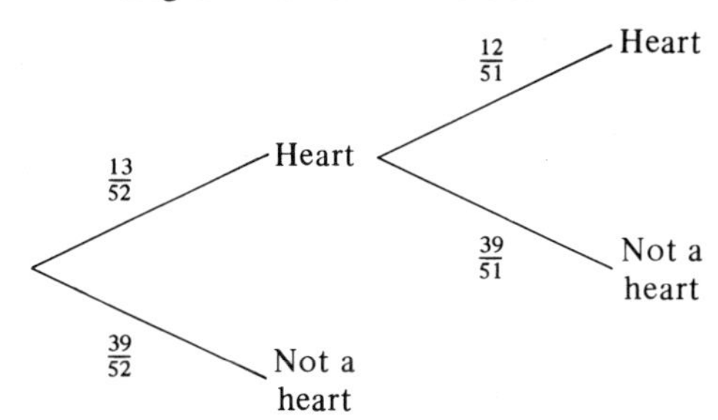

The probability that the first two cards drawn are hearts is:

$\frac{13}{52} \times \frac{12}{51}$

For the third to be a club or a spade, the probability is:

$$\frac{\text{Number of clubs and spades}}{\text{Number of cards left}}$$

7 A useful method of checking is to observe what happens to the two basic vectors $\begin{pmatrix}1\\0\end{pmatrix}$ and $\begin{pmatrix}0\\1\end{pmatrix}$ as shown in the diagram: e.g. for a $\frac{1}{2}$ turn about (0,0):

$$\begin{pmatrix}1\\0\end{pmatrix} \rightarrow \begin{pmatrix}-1\\0\end{pmatrix} \text{ and } \begin{pmatrix}0\\1\end{pmatrix} \rightarrow \begin{pmatrix}0\\-1\end{pmatrix}$$

therefore, the matrix which performs a $\frac{1}{2}$ turn about (0,0) is $\begin{pmatrix}-1 & 0\\0 & -1\end{pmatrix}$

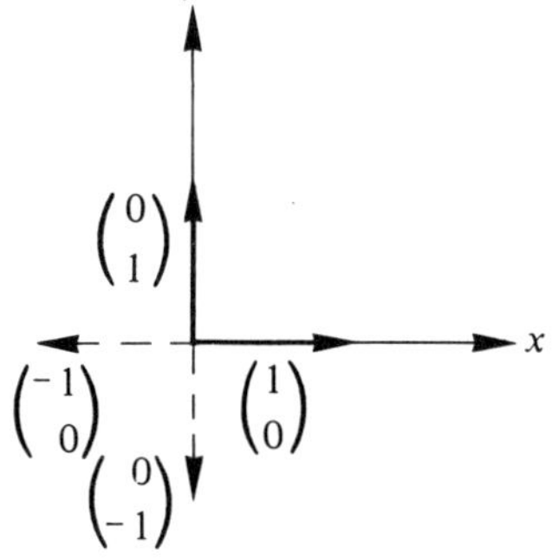

8 (i) Join your points with a smooth curve and you should be able to tell whether the points you have calculated are correct.

(ii) The solution to the equation will occur where the two graphs intersect.

(iii) The gradient is $\frac{y}{x}$ but be careful to check both the scales.
PT is a tangent to the curve.

T

y

P

x

9 (i) An example of how to form your first inequality is:

x lambs cost £$5x$

y piglets cost £$7\frac{1}{2}y$

Total cost of x lambs and y piglets = £$(5x + 7\frac{1}{2}y)$.
Since this must not exceed £900, the inequality becomes:

$$5x + 7\tfrac{1}{2}y \leqslant 900$$

(ii) Take care when shading the unwanted regions. A useful check is to take any point (x,y) in the unshaded region and after substituting the values for x and y into your inequality ensure that it gives a satisfactory result.

10 (i) Make sure that you use the correct sign because direction is all important in vectors, e.g. if $\overrightarrow{AB} = \mathbf{a}$ then $\overrightarrow{BA} = -\mathbf{a}$.

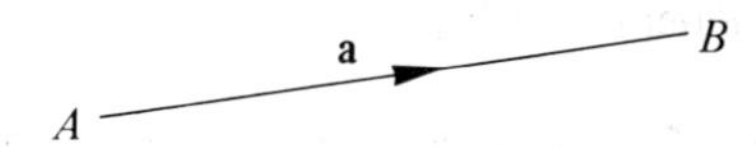

(iii) Since CFB is a straight line, $\overrightarrow{CB}$ should be the same basic vector as $\overrightarrow{CF}$ except for a different scalar multiplier.
It is advisable to factorise: e.g. if $\overrightarrow{AC} = 2\mathbf{a} + \mathbf{b}$ and ACB is a straight line and $\overrightarrow{AB} = 6\mathbf{a} + 3\mathbf{b}$.

By factorising: $\overrightarrow{AB} = 3(2\mathbf{a} + \mathbf{b})$ which shows that $\overrightarrow{AB}$ and $\overrightarrow{AC}$ have the same basic vector and $\frac{AC}{AB} = \frac{1}{3}$

Set 2

1 (i) Find the original area and then increase each side by 15% to find the new area. The percentage increase in area is found by:

$$\frac{\text{New area} - \text{Original area}}{\text{Original area}} \times 100\%$$

An alternative method is to use the enlargement approach. Since the scale factor is $\frac{115}{100}$ for each side, the new area will be the Original area $\times \left(\frac{115}{100}\right)^2$

(ii) Be careful with the order of procedure.

(*a*) Increase the cash price by $12\frac{1}{2}\%$.

(*b*) Subtract the down payment from this new price.

(*c*) Divide the remaining sum of money by 12.

(iii) With compound interest, you receive interest upon your interest, e.g. to find what £100 amounts to after 2 years at 10% per annum compound interest.

Interest for 1st year = £10.00
Total after 1 year = £110.00
Interest for 2nd year = £11.00 (10% of £110.00)
Amount after 2 years = £121.00

2 (i) To find the values of x which satisfy $x^2 - 8x - 16 = 0$, we factorise the left-hand side of the equation, giving:

$$(x - 4)(x - 4) = 0$$

i.e. both values of x are 4.

Conversely, if the two solutions to a quadratic equation are identical, say $x = a$, then the equation will be

$$(x - a)(x - a) = 0,$$

i.e. $x^2 - 2ax + a^2 = 0$

(ii) By factorising $x^2 + 3x + 2$, you will see that it is the L.C.M. of all your factors.

(iii) See the note for Q.2 (i) (*b*) Set 1 (the L.C.M. of your fractions is $4x$).

3 (ii) The general equation of any straight line is $y = mx + c$, where m is the gradient of the line and c is the y intercept. The diagram shows you the y intercept and reminds you how to consider a suitable right-angled triangle from which to obtain the gradient.

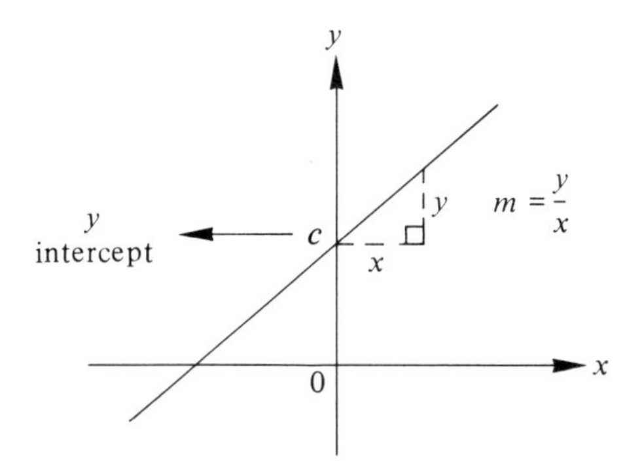

4 (i) Calculate $|\mathbf{a} - \mathbf{b}|$ is another way of asking you to calculate the length of *CB*.

(ii) (*a*) To obtain a representative of $(\mathbf{a} + \mathbf{b})$ you need to draw **b** from *C* parallel to your original vector as shown in the answer diagram.

(*b*) $|\mathbf{c}|$ i.e. *AD*, is the diagonal of the rhombus *ACDB*.

5 (i) See the note on Q.6 (ii), Set 1.

(ii) (*a*) Each die has 6 different ways of landing, so that 3 dice rolled together have $(6 \times 6 \times 6)$ different ways of landing.

(ii) (*c*) Here are some of the possible ways of scoring 10 as a total when any 2 of the dice show the same number: (2,2,6), (2,6,2), (6,2,2,) . . .

6 (i) Form a quadratic equation by expanding the formula. Solve the equation by the method of factorisation.

(ii) (*a*) Write down an expression for the increase in area. Use 'the difference of two squares' to factorise the expression before substituting the values of the radii.

7 (iii) (*a*) The solutions to $2x^2 + 4x - 7 = 0$ occur where $y = 0$, i.e. where the graph cuts the x axis.

(*b*) (*c*) Rewrite the equation in the form of the original equation: e.g. to solve $2x^2 + 4x - 5 = 0$, rewrite it as:

$$2x^2 + 4x - 5 - 2 = -2$$ (adding -2 to both sides of the equation)

i.e. $2x^2 + 4x - 7 = -2$

and the solutions occur where the graph of $2x^2 + 4x - 7 = -2$, in other words at the points of intersection of the graph and the straight line whose equation is $y = -2$

8 (i) Use the theorem of Pythagoras, $PR^2 = PZ^2 + RZ^2$.

(ii) Triangles PXT, PYV and PZR are similar so that XT, YV, and ZR are in proportion.

(iii) Volume of a cone $= \frac{1}{3}\pi r^2 h$. (Reference Section 2.B.6.)

(vi) The volume of the shaded section = volume of cone PUV – volume of cone PST.

9 (i) Here are some examples of how to form your inequalities: If there are to be x hectares of sugar beet with a minimum of 16 hectares, then $x \geqslant 16$ and similarly $y \geqslant 26$.
To form the labour units inequality: for sugar beet it is 3 units per hectare and for potatoes, 5 units per hectare. Therefore, the total units for x hectares of sugar beet and y hectares of potatoes will be $(3x + 5y)$ but since this must not exceed 360 units, the inequality will be $3x + 5y \leqslant 360$.

Note: Check that your inequality is expressed in the simplest possible form, for example: $18x + 12y \leqslant 240$ can be simplified to give: $3x + 2y \leqslant 40$.

(ii) Check your region carefully, if necessary refer to the note on Q.9 (ii), Set 1.

10 Extract triangle PSQ from the diagram as shown below and you will see that the angle you require is angle PSQ. Use the tangent ratio of angle PSQ.

(i)

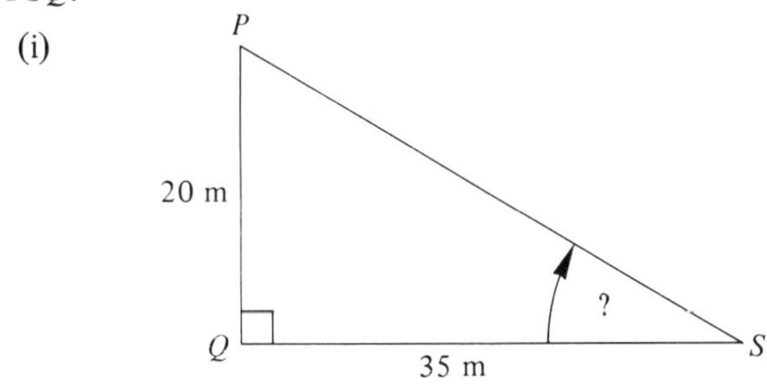

(ii) Triangle PQR is the one you require. The tangent ratio may be used to calculate QR and a division may be avoided by using angle QPR.

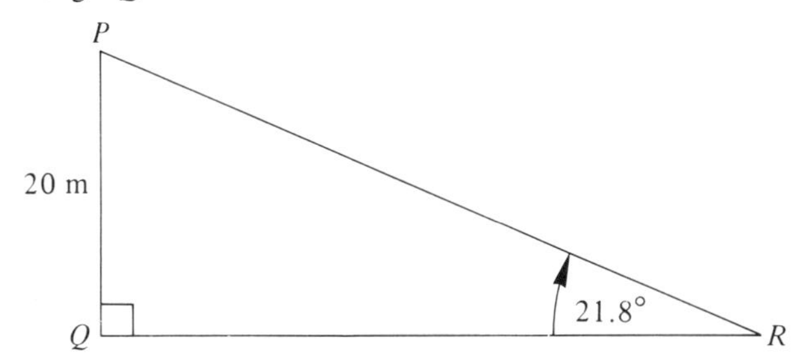

(iii) In triangle SQR, angle SQR can be found because it is the angle between the directions NE and E.

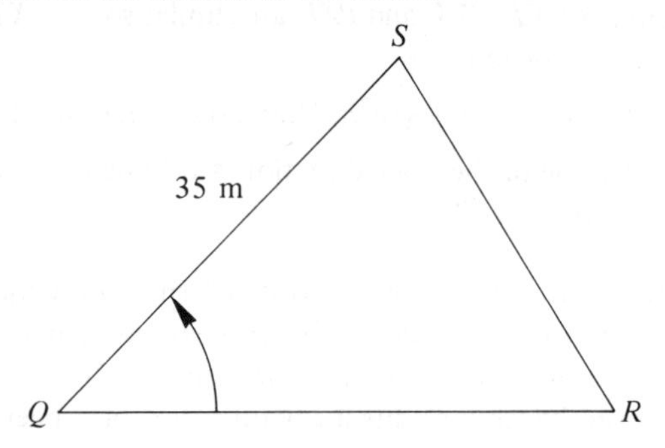

You now have two sides and an included angle, so you will need to use the cosine formula to calculate SR.

Set 3

1 (i) Find the surface area and the volume of the 10 cm cube. Then increase the length of the side by 10% to find the length of the side of the new cube.

$$\text{Percentage increase} = \frac{\text{Increase}}{\text{Original}} \times 100\%$$

(ii) B's share as a fraction of the total is $\frac{2}{7}$, i.e.

$$\frac{2}{7} = £95$$

$$\frac{1}{7} = \frac{£95}{2}$$

$$\text{The sum of money: } \frac{7}{7} = \frac{£95}{2} \times 7$$

(iii) See the note for Q.1, Set 1.

2 (i) If the solutions to a quadratic equation are 2 or -1, it means that

either $x = 2$, i.e. $(x - 2) = 0$

or $x = -1$, i.e. $(x + 1) = 0$

Therefore, the factors of the quadratic are: $(x - 2)(x + 1) = 0$ and the quadratic is $x^2 - x - 2 = 0$

(ii) Find the values of $f(5)$, $f(6)$ and $f(10)$ as explained in the note on Q.4 (i), Set 1 and then find the highest number which will divide into all three values.

(iii) Multiply both sides of the equation by $2x$, the L.C.M. of 2 and x.

3 A diagram showing the front elevation of the cube is helpful in this question.

(i) The volume of the cube is $(2x)^3$ cm^3.
The volume of the sphere is $\frac{4}{3}\pi x^3$ cm^3.

Subtract these volumes to find D, the difference in volume between the cube and the sphere.

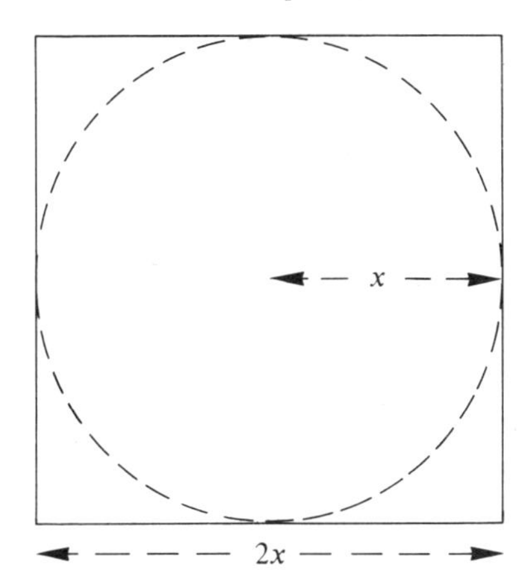

4 (i) If the scale is 1 : 100 000, 1 cm represents $\frac{100\,000}{100}$ metres.

(ii) The vertical distance = the difference in contours.

(iii) Gradient = $\frac{\text{Vertical height}}{\text{Horizontal distance}}$ (express in the same units).

(iv) The angle of elevation is the angle marked α in the diagram.

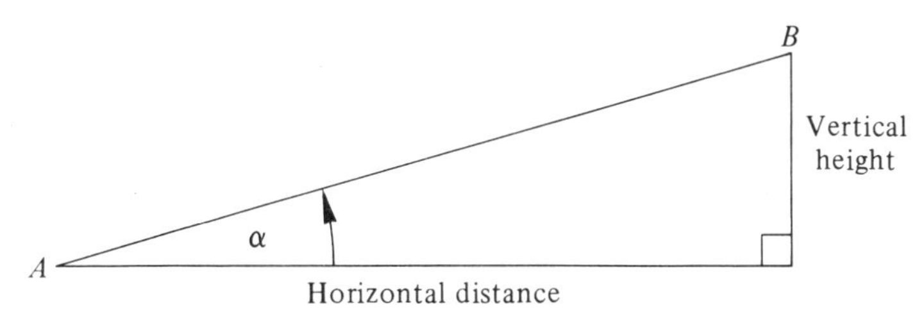

Angle a may be calculated because tan angle α = the gradient of AB.

5 (*a*) (ii) Reference Section 2.A.6.

(*b*) Calculate angle AOP in the right-angled triangle APO. Angle AOB is twice angle AOP.

(*c*) Area of a sector: Reference Section 2.A.10.

(*d*) Shaded area $S = \frac{1}{2}$(Area of sector XOY − area $OAQB$).

(*e*) To calculate area T, first calculate the area of the segment in the quadrant QPB by subtracting the area of triangle QPB from the area of the quadrant QPB. Shaded area T = (Shaded area S − area of segment).

6 (i) (*a*) $(x + 4)$ is the common factor.

(*b*) Here is an example of how to solve this type of equation:

$$ax - 3a + 4x - 12 = 0$$

$$a(x - 3) + 4(x - 3) = 0$$

$$(x - 3)(a + 4) = 0$$

either $(x - 3) = 0$ or $(a + 4) = 0$

and $x = 3$ and $a = -4$

Check these answers by substituting the values in the equation.

(ii) (*b*) Take out the common factor first, then use the difference of 2 squares.

7 (i) Another way of writing the equation $y = \frac{8}{x}$ is $xy = 8$.

This provides a quick method of checking your table of values by making sure that the value of xy is always 8.

(iii) The graph of $y = \frac{1}{4}x$ passes through the origin and has a gradient of $\frac{1}{4}$.

(v) The single equation occurs at the point of intersection of the graphs: i.e. $\frac{8}{x} = \frac{1}{4}x$ or $\frac{8}{x} = \frac{x}{4}$. By cross-multiplication: $x^2 = 32$, and $x = \sqrt{32}$

8 (ii) Check vectors $\vec{AC}$ and $\vec{GE}$ and also $\vec{AG}$ and $\vec{CE}$.

9 (i) (*a*) $MN = 2XN$ (see the Reference Section 7.D.2). The diagram shows how XN may be calculated.

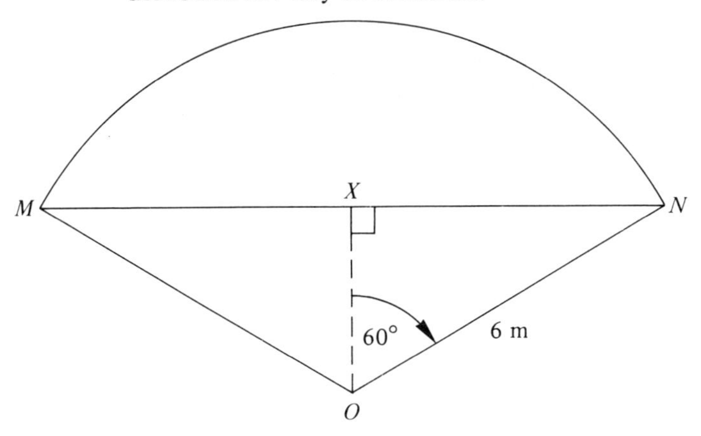

(*b*) Area of triangle $MON = \frac{1}{2}ab \sin C$

$= \frac{1}{2} \times 6 \times 6 \times \sin 120°$.

(ii) (*a*) The area of the curved surface of the roof is the area of the rectangle shown in the diagram:

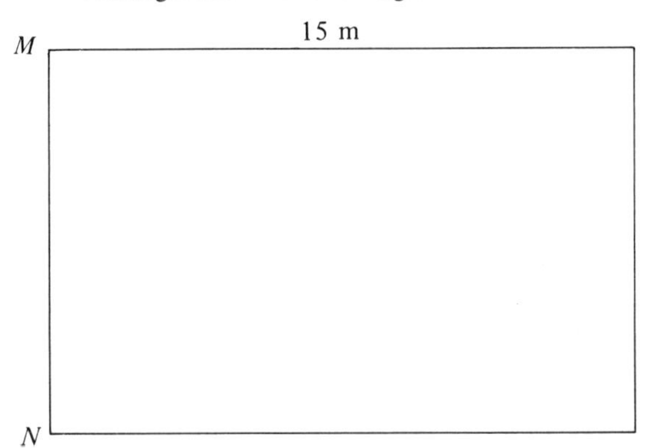

The distance MN is the length of the arc MN and must be calculated first. (Reference Section 2.A.9.)

10 (iv) You know 2 sides and the included angle. Use the cosine formula (Reference Section 3.C).

(v) Use the sine formula (Reference Section 3.C).

Set 4

1 (ii) (*b*) Depreciation during the 4th year as a percentage of its value at the beginning of the year is found by:

$$\frac{\text{Depreciation in that year}}{\text{Value at the beginning of 4th year}} \times 100\%$$

2 (ii) Multiply every term by x.

(iii) Here is a similar example:
If $f(3) = 28$, find n

$$f(3) = (3 + 4)(3 + n)$$

therefore $(3 + 4)(3 + n) = 28$

and $7(3 + n) = 28$

$$n = 1$$

3 See the note for Q.3, Set 3.

4 See the Reference Section 6.E.

5 (ii) See the Reference Section 2.A.9.

(iii) Reference Section 2.B.7.

6 (iii) To find solutions from your graph of $x^2 + 5x = y$, retain the L.H.S. of the equation and adjust the R.H.S. in order to find a value for y, e.g:

(*a*) To solve $x^2 + 5x = 2$

Draw the straight line whose equation is $y = 2$ on your graph and read off the values of x at the points of intersection.

(*b*) To solve $x^2 + 5x - 1 = 0$

Rewrite the equation as $x^2 + 5x = 1$, draw the straight line whose equation is $y = 1$ and read off the values of x at the points of intersection.

(iv) To find the gradient, draw a tangent to the curve at the point where $x = -4$ and then complete a suitable right-angled triangle, as shown in the diagram:

The gradient of the curve is the gradient of the tangent: $\frac{PA}{XA}$

Take care when finding PA and XA because the axes are scaled differently; if XA is a negative distance then the gradient will be negative.

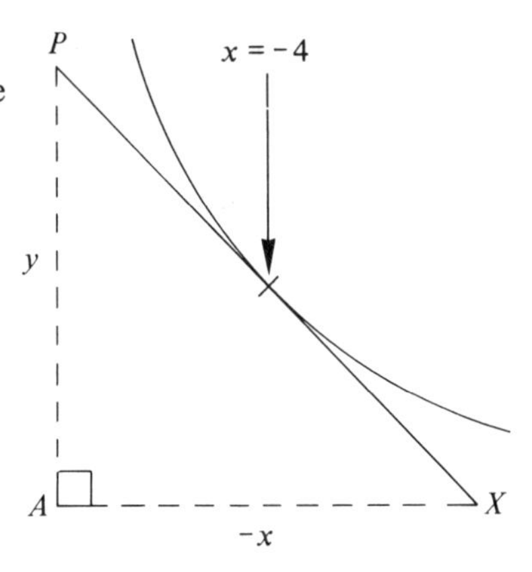

7 If the side of the cube is x cm, then the sides of the cuboid will be $(x - 2)$, $(x - 4)$ and $(x - 5)$ cm.
The volume of the cuboid = $(x - 2)(x - 4)(x - 5)$ cm^3, and the volume of the cube = x^3 cm^3.
The difference in volume = $x^3 - [(x - 2)(x - 4)(x - 5)]$ cm^3, i.e.

$$x^3 - [(x - 2)(x - 4)(x - 5)] = 440$$

When this equation is simplified, you will have a quadratic equation to solve by factorisation.

8 (i) (*a*) Reference Section 6.F.
It is a shear transformation with $y = 0$ invariant and the matrix is of the form $\begin{pmatrix} 1 & k \\ 0 & 1 \end{pmatrix}$

(*b*) This is a reflection in the x axis ($y = 0$). Study the basic vectors in the diagram and notice their position after reflection in the x axis:

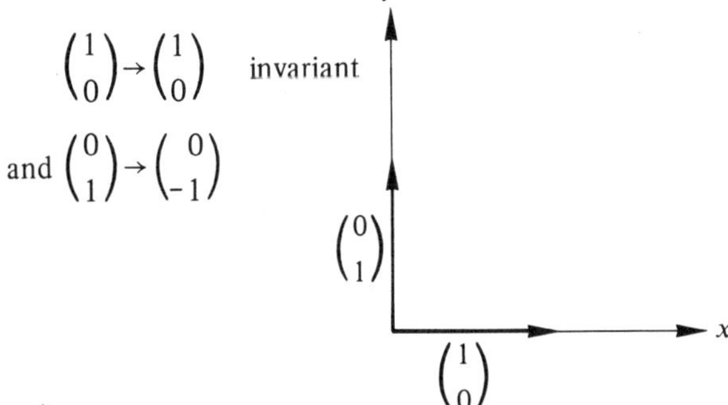

thus, the transformation matrix is $\begin{pmatrix} 1 & 0 \\ 0 & -1 \end{pmatrix}$

(*c*) This is a rotation of 180° about (0,0). Once again study the positions of the two basic vectors after this rotation.

$$\begin{pmatrix}1\\0\end{pmatrix}\rightarrow\begin{pmatrix}-1\\0\end{pmatrix} \text{ and } \begin{pmatrix}0\\1\end{pmatrix}\rightarrow\begin{pmatrix}0\\-1\end{pmatrix}$$

thus, the transformation matrix is $\begin{pmatrix}-1 & 0\\0 & -1\end{pmatrix}$.

(ii) See the Reference Section 6.F.

9 The diagram shows the bearings and North lines. The angle marked a is (180°–120°) as a result of parallel North lines.

(i)

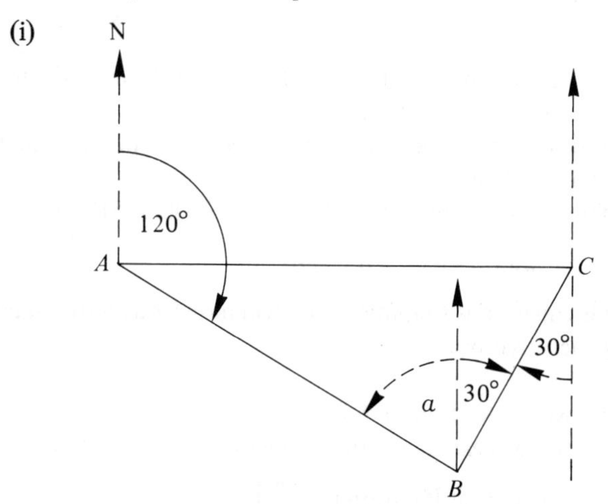

(iii) Draw a diagram of the triangle you require as shown below. Now choose the necessary trigonometrical ratio.

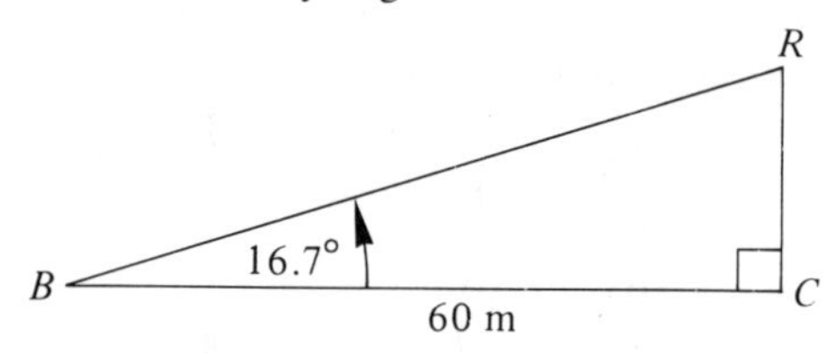

(iv) The angle of elevation of R from A is shown in the following diagram, angle RAC.

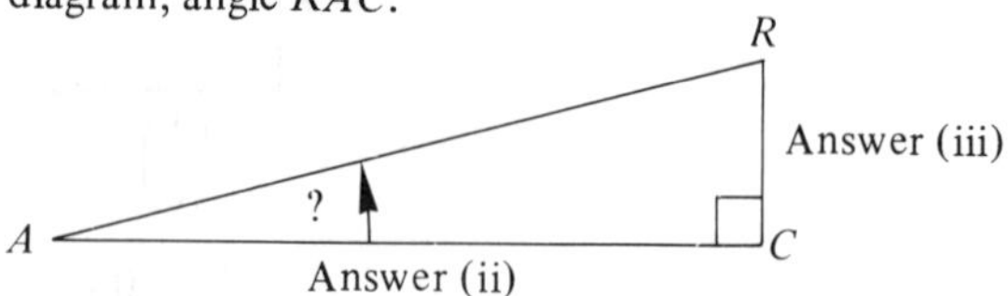

(v) Study the diagram for (i). You will need to calculate angle ACB in triangle ACB by trigonometry, then the bearing of A from C will be (180° + 30° + angle ACB).

10 (i) (*a*) Reference Section 2.A.10.

(*b*) Area of triangle $POQ = \frac{1}{2}ab \sin C$

$= \frac{1}{2} \times 10 \times 10 \times \sin 72°$.

(*c*) Area of shaded segment = area of sector − area of triangle.

(*d*) Perimeter of shaded segment = line segment PQ + arc PQ.

Line segment $PQ = 2XQ$ and XQ is found by using the appropriate trigonometrical ratio.

The diagram clarifies the problem. To find the length of arc PQ see the Reference Section 2.A.9.

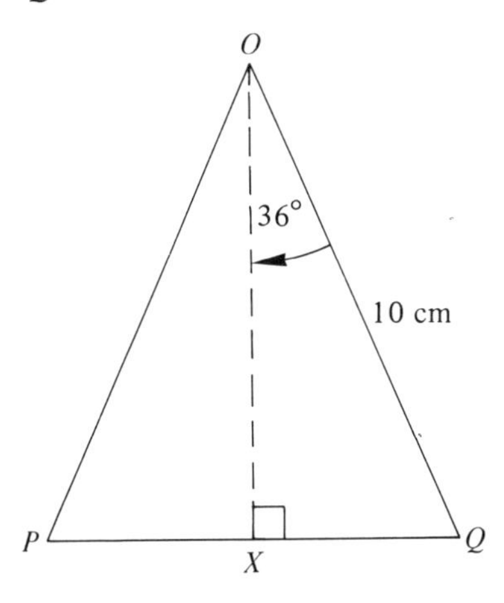

(ii) If the lengths are doubled, i.e. multiplied by a scale factor of 2, then the areas will be multiplied by the square of the scale factor.

Work them out and test the results for yourself.

Set 5

1 (i)

85% of the purchase price = £11 764

$$1\% \text{ of the purchase price} = \frac{£11\,764}{85}$$

$$\text{Therefore the purchase price } (100\%) = \frac{£11\,764}{85} \times 100.$$

(ii) Study the following example of compound interest. Find the Compound Interest on £1000 for 3 years at 10% per annum.

	£	
Principal:	1000	(the sum of money on which interest is calculated)
plus 10%:	100	(10% of £1000)
End of Year 1: *Amount*:	1100	(Amount=Principal+Interest)
plus 10%:	110	(10% of £1100)
End of Year 2: *Amount*:	1210	
plus 10%:	121	(10% of £1210)
End of Year 3: *Amount*:	£1331	

Compound interest for 3 years = £100 + £110 + £121 = £331.

This may be checked by: Amount after 3 years − Original principal, i.e.

£1331 − £1000 = £331

2 (i) When solving inequalities treat them as equations: e.g.

$$7x - 3 > 3(x + 5)$$
$$7x - 3 > 3x + 15$$
$$7x - 3x > 15 + 3$$
$$4x > 18$$
$$x > \tfrac{18}{4}$$

If x is an integer, then $\{x = 5, 6, \ldots\}$

$P \cap Q$ means the members of set P which are also members of set Q.

(ii) Re-arrange both equations, giving

(i) $3x + y = 5$

(ii) $3x + 4y = -16$

Now subtract the equations, thus eliminating x and substitute your value for y in equation (i) to find a value for x. Check your answers by substituting the values you have found in equation (ii).

(iii) Simplify both sides of the equation by expansion and you will then be able to subtract $6x^2$ from both sides, leaving a simple equation to solve.

3 (i) The perimeter consists of the circumference of a circle, radius $2x$ m together with two straight lengths of $4x$ m each.

(ii) The area consists of the area of a square of side $4x$ m together with the area of a circle, radius $2x$ m.

4 (iii) See the note for Q.6, Set 4.

(*b*) Rewrite $x^2 + 3x - 5 = 0$ as $x^2 + 3x - 2 - 3 = 0$, and then: $x^2 + 3x - 2 = 3$

(iv) The single equation which these values satisfy is:

$x^2 + 3x - 2 = x + 1$ (now simplify it).

5 (i) (*a*) $$\text{Probability} = \frac{\text{Number of ways of selecting a blue counter}}{\text{Total number of possibilities}}$$

$$= \frac{x}{16}$$

(*b*) If 8 more blue counters are added, the probability of choosing a blue one $= \dfrac{x+8}{24}$

This increases the previous probability by $\frac{1}{4}$, which results in the equation: $\dfrac{x+8}{24} = \dfrac{x}{16} + \dfrac{1}{4}$

Multiply every element by the L.C.M. of 24, 16 and 4, which is 48 and then solve for x.

(ii) (*a*) Be methodical when making the list; begin by writing down the 2 digit numbers whose first digit is 1, i.e. 13, 15, 17, 19 and then use 3 as the first digit and so on.

6 (i) Use triangle AOE and remember that angle OAE is 90° (the angle between the radius and the tangent), see the Reference Section 7.C.1.

(ii) Angle EOA = angle DOA and angle $DBA = \frac{1}{2}$ angle DOA (Reference Section 7.B.1).

(iii) Angle COD = angle OEA (alternate angles, CO is parallel to EA) therefore angle $COD = 51°$.

(iv) Triangle BOC is a right-angled isosceles triangle, therefore angle BCO (angle BCF) = 45°.
In triangle BFC, angle $BFC = 180° -$ (angle $CBF + 45°$).

(v) Triangle ODA is isosceles, therefore angle $ODA = \frac{1}{2}(180° -$ angle $EOA)$.

7 (i) It is helpful to draw a diagram showing the new length as $(17 + x)$ cm and the new breadth as $(11 + x)$ cm. The new area will be seen to be $(17 + x)(11 + x)$ cm^2.

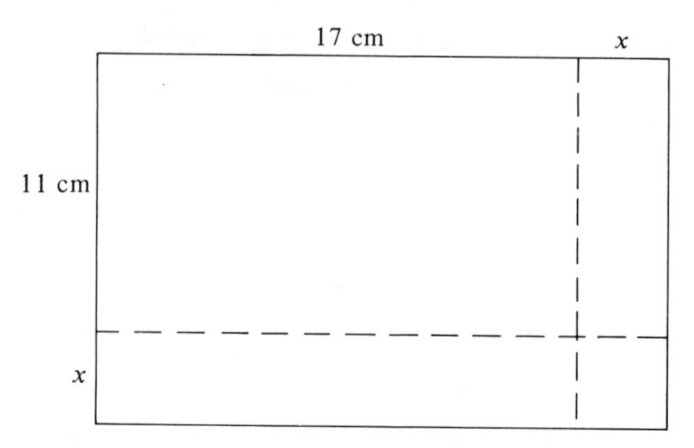

You can now form the equation $(17 + x)(11 + x) = 432$
Simplify and solve the resulting quadratic equation by factorising.

(ii) Substitute $D = 54$ into the equation $D = \frac{n(n-3)}{2}$ and solve for n.

8 (i) (*a*) The common factor is $(2x + 3y)$

(*c*) Take out the common factor first and then factorise further using the difference of two squares.

(*d*) $\frac{a^4}{81} - \frac{b^4}{16}$ can be written as $\left(\frac{a^2}{9^2}\right)^2 - \left(\frac{b^2}{2^2}\right)^2$

Now use the factors of the difference of two squares:

giving: $\left(\frac{a^2}{3^2} - \frac{b^2}{2^2}\right)\left(\frac{a^2}{3^2} + \frac{b^2}{2^2}\right)$ (the first factor may now be factorised again)

giving: $\left(\frac{a}{3} + \frac{b}{2}\right)\left(\frac{a}{3} - \frac{b}{2}\right)\left(\frac{a^2}{3^2} + \frac{b^2}{2^2}\right)$

(ii) Factorise the numerator by taking out the common factor and factorise the denominator by using the difference of two squares.

9 (iii) Volume = Area of cross section × Length

= Area of triangle $ABF \times 15$ cm^3.

(v) BD is the diagonal of rectangle $ABCD$ and may be found by using the theorem of Pythagoras.

(vi) EB is the diagonal of rectangle $BCEF$, you have already calculated FB and $BC = 15$ cm.

10 It is essential to draw a clear diagram. Study the one drawn for you and then draw your own.

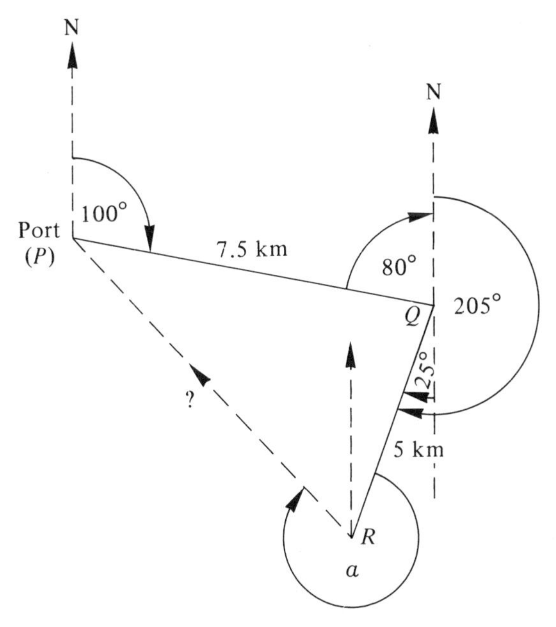

(i) To calculate *PR*, the distance the ship is from port, you need to find angle *PQR* and then you will know 2 sides and the included angle in triangle *PQR* and the cosine formula may be used. (Reference Section 3.C.3.)

(ii) The return bearing is marked a in the diagram. Before you can find angle a you will need to calculate angle *PRQ*, using the sine formula (Reference Section 3.C.2).

Set 6

1 (i) Total surface area = Curved surface area of the cylinder + Area of circular base + Area of hemisphere.
(Reference Section 2.B.4 and 7 for the appropriate formulae.)

(ii) Volume of salt-cellar = Volume of cylinder + Volume of hemisphere.
Hint: $x + \frac{2}{3}y$ can be written as $\frac{3x + 2y}{3}$
(Reference Section 2.B.4 and 7 for the appropriate formulae.)

(iii) Internal radius = 2 cm − Thickness of silver.
Internal height = 5 cm − Thickness of base.

(iv) Substitute the respective values in the formula. Remember to evaluate the contents of the bracket first.

(v) (*a*) Volume of silver used = External volume − Internal volume.

(*b*) Value of silver = Mass of silver in grams multiplied by 40p per gram.

2 (i) Be careful with the inequality signs when listing the members of the sets.

(*d*) $A \cap B$ means those members which are in *both* A and B.

(*e*) $(A \cup B)'$ means those members which are in *neither* A or B.

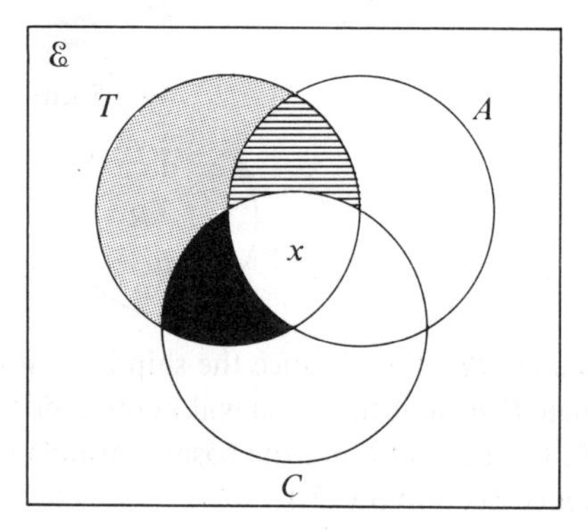

(ii) To find the value to be entered in the part of the diagram shaded with horizontal lines:

$n(T \cap A) = 25$ but x members are already in the intersection, therefore the value required is $(25 - x)$.

$n(T \cap C) = 24$, therefore the value to be entered in the solid black part of the diagram will be $(24 - x)$.

The value to be entered in the dotted part of the diagram, i.e. the number of students who play tennis only, is found by subtracting x together with the values of the lined and solid parts of the diagram from $n(T)$, i.e.

$$56 - (x + 25 - x + 24 - x)$$
$$= 56 - (49 - x)$$
$$= 56 - 49 + x$$
$$= 7 + x$$

If the remainder of the Venn diagram is completed in this way, a value for x can be found by adding all the various parts and forming an equation. Do not overlook the 4 students who do not take part in any activities.

3 (iii) The diagram below shows how to obtain the median, the quartiles and the interquartile range.

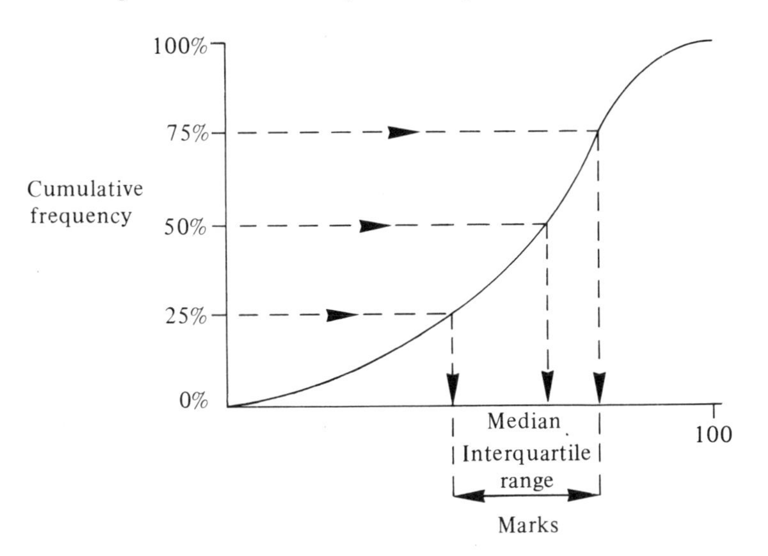

(iv) (*a*) The students who pass form the top 60% on the cumulative frequency axis, i.e. the bottom 40% will fail.

(*b*) Read off from the graph the cumulative frequency number which corresponds to 50 marks and then express this as a percentage of 600 (your reading may be checked by referring to your cumulative frequency table).

4 (i) Alloy A consists of tin and lead in the ratio 3 : 7
Alloy B consists of tin and copper in the ratio 3 : 5
Alloy C consists of alloy A and alloy B in the ratio 4 : 3
To make 1 tonne of C you need $\frac{4}{7}$ of a tonne of A + $\frac{3}{7}$ tonne of B.
Weight of tin in $\frac{4}{7}$ tonne of $A = (\frac{4}{7} \times \frac{3}{10})$ tonne.
Weight of lead in $\frac{4}{7}$ tonne of $A = (\frac{4}{7} \times \frac{7}{10})$ tonne.
Weight of tin in $\frac{3}{7}$ tonne of $B = (\frac{3}{7} \times \frac{3}{8})$ tonne.
Weight of copper in $\frac{3}{7}$ tonne of $B = (\frac{3}{7} \times \frac{5}{8})$ tonne.
To find the cost of 1 tonne of C, multiply each of the above by the cost per tonne and find the total, correct to the nearest £.

(ii) The percentage increase in cost $= \frac{\text{Increase}}{\text{Original}} \times 100\%$

5 (i) Use the difference of 2 squares to factorise the expression:

$$[(x+4)+(x-3)]\ [(x+4)-(x-3)]$$

This becomes $(2x + 1) \times 7$ after simplifying the brackets. The 7 is significant for the second part of the question, because whatever whole number value x takes, if $(2x + 1)$ is multiplied by 7, the new number must be divisible by 7.

For example:

if $x = 4$, $(2x + 1) = 9$	if $x = -6$, $(2x + 1) = -11$
$(2x + 1) \times 7 = 63$	$(2x + 1) \times 7 = -77$
which is divisible by 7.	which is divisible by 7.

(ii) For $(2x + 1) \times 7$ to be divisible by 49, $(2x + 1)$ must be a multiple of 7, e.g.

if $2x + 1 = 7$	and if $2x + 1 = 14$
$x = 3$	$x = 6\frac{1}{2}$

It follows then that $(2x + 1)$ must be a factor of 7, i.e. $2x + 1 = 7n$ where n is any positive integer,

$$\text{therefore } x = \frac{7n - 1}{2} \text{ or } x = \frac{7n}{2} - \frac{1}{2}$$

6 (i) (*a*) and (*b*) See the Reference Section 2.A.9 and 10. To find the sector angle POR, consider the triangle POR and the lengths of its sides in terms of x.

(*c*) Percentage occupied by the sector

$$= \frac{\text{Area of sector}}{\text{Area of square}} \times 100\%$$

(ii) Volume of a cone $= \frac{1}{3}\pi r^2 h$ where r is the radius of the base. You must first calculate r, using the fact that: the circumference of the circular base = arc PQR, i.e.

$$2\pi r = \text{arc } PQR$$

$$\text{therefore:} \quad r = \frac{\text{arc } PQR}{2\pi}$$

7 (i) You need to draw a diagram as shown.

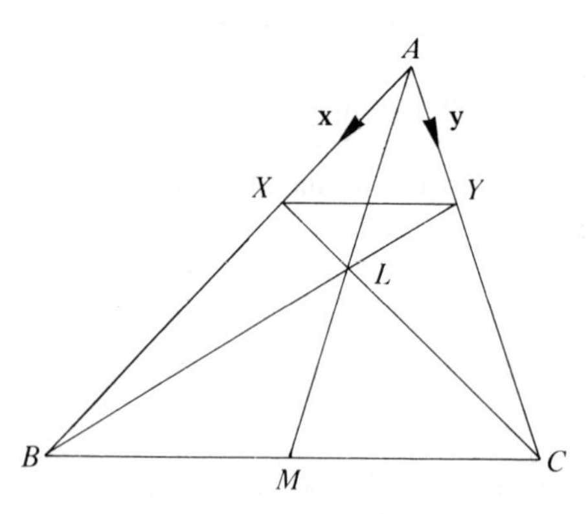

If $AX = \frac{1}{3}AB$ then $AB = 3AX$ and if $AY = \frac{1}{3}AC$ then $AC = 3AY$.

(ii) $\vec{XC} = \vec{XA} + \vec{AC}$.

(iii) Triangles XYL and CBL are similar and studying the vectors $\vec{XY}$ and $\vec{BC}$ will show that their corresponding sides are in the ratio 1 : 3, i.e. $XL = \frac{1}{3}CL$, thus $XL = \frac{1}{4}XC$.

(iv) Find $\vec{AL}$ using $\vec{AL} = \vec{AX} + \vec{XL}$ and then compare with $\vec{AM}$. If $\vec{AL}$ and $\vec{AM}$ have the same basic vector, then A, L and M must lie on the same straight line, i.e. they are collinear.

8 (i) (*b*) (i) Rewrite $x^2 - 2x - 6 = 0$ as $x^2 - 2x - 3 - 3 = 0$ and then as $x^2 - 2x - 3 = 3$
Now draw the straight line whose equation is $y = 3$ and read off the values of x at the points of intersection.

(ii) Rewrite $x^2 - 2x - 1 = 0$ as $x^2 - 2x - 3 + 2 = 0$ and then as $x^2 - 2x - 3 = -2$
Draw the straight line whose equation is $y = -2$ and read off the values of x at the points of intersection.

(ii) $3y = 4x - 9$ can be written as $y = \frac{4}{3}x - 3$ and the graph will have a gradient of $\frac{4}{3}$ with a y intercept of -3.

9 (i) The mean value $= \dfrac{\text{Total height of all children}}{100}$

The median value is really the middle height when the heights are arranged in order, i.e. between the 50th and 51st child. Looking at the table, you can see that it will be in the height group of 69 cm.

The Modal value is the most popular or most common height, i.e. 68 cm.

(ii) Take care!

$$\text{New mean height} = \frac{\text{Total height}}{\text{Total number of children}}$$

$$= \frac{\text{Total height of first sample} + (500 \times 72.03)}{600}$$

10 (i) See the Reference Section 6.F for combined transformations.

(ii) See the answer section for the detailed diagram.

Part 3
Answers

(Three-figure tables have been used.)

Section 1

Set 1

1 (i) 120 (ii) 9 (iii) 38 (iv) 69 (v) $33\frac{1}{3}\%$

2 (i) 17 (ii) $1\,000\,001_2$ (iii) 551_8 (iv) $1\,000\,000_2$ (v) 4

3 (i) £2.66 (ii) £1.28 (iii) £2.13 (iv) 512 cm^3 (v) 64 cubes

4 (i) 7 (ii) 6 (iii) 5.3 (iv) 2 (v) 20%

5 (i) 7 (ii) 11 (iii) −9 (iv) + or −9 (v) −3 or 4

6 (i) £30.00 (ii) 39 000 (iii) 3.48 (iv) 5.6×10^7 (v) 5.6×10^8

7 (i) 28 (ii) 14.95 (iii) $b = \frac{A}{\frac{1}{2}h}$ or $\frac{2A}{h}$ (iv) 5 (v) 7.5

8 (i) 0.5 (ii) 60° (iii) 30° (iv) 1.66* (v) 59° (vi) 31° (vii) 91°

9 (i) (*a*) 240 m^2 (*b*) 78.5 m^2 (*c*) 161.5 m^2

(ii) (*a*) 32 m (*b*) £12.80

10 (i) Intersection: (1,3) (ii) Intersection: (−2,−2)

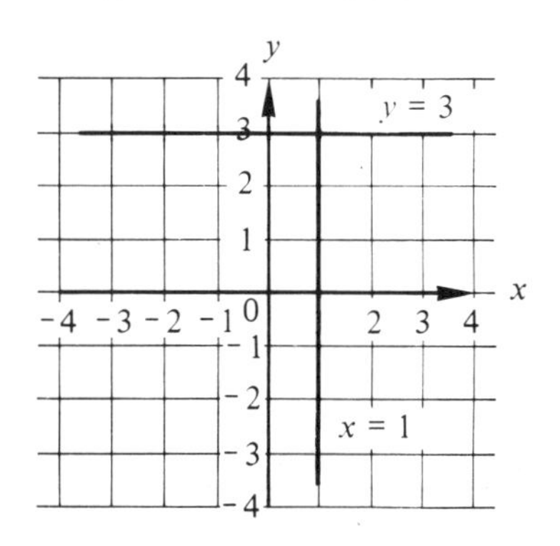

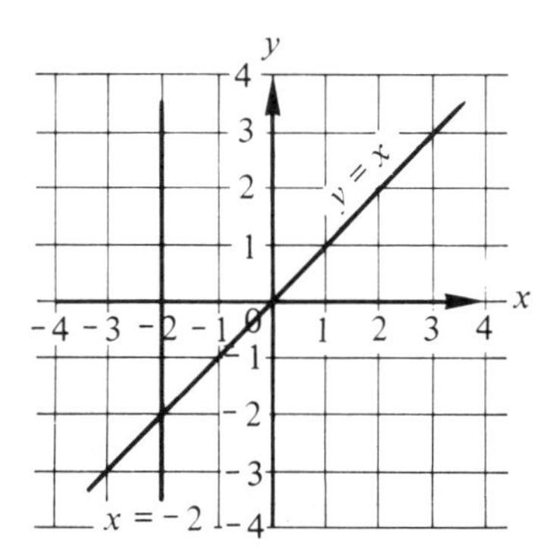

(iii)

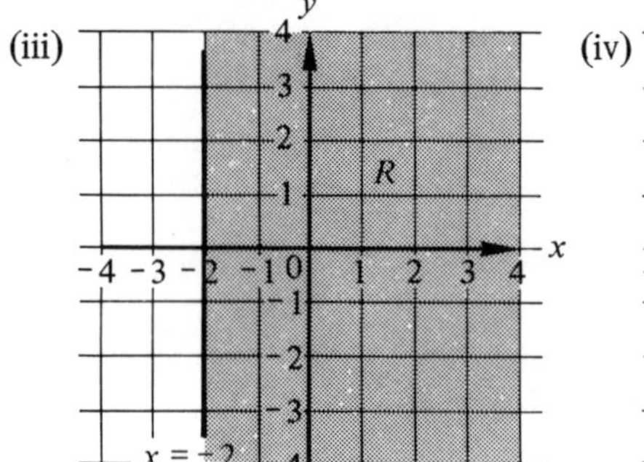

(iv)

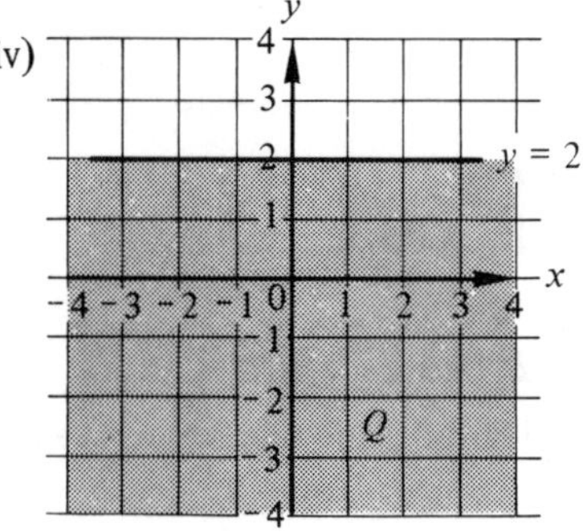

(v)

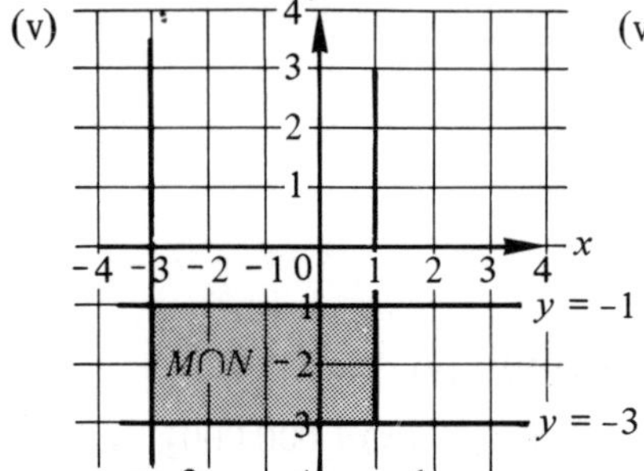

(vi)

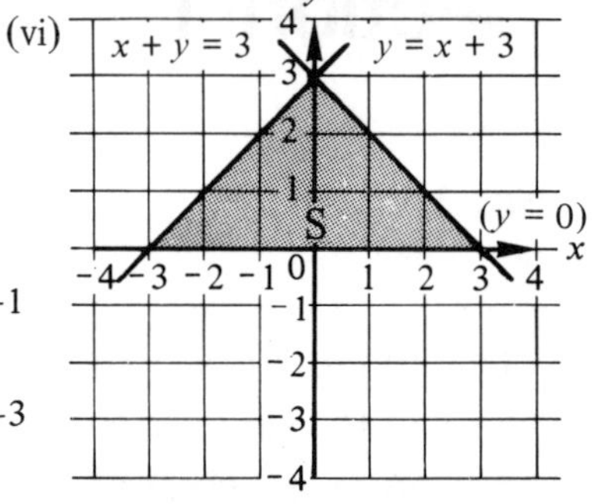

11 (i)

x	−4	−3	−2	−1	0	1	2	3	4
y	16	9	4	1	0	1	4	9	16

$y = x^2$

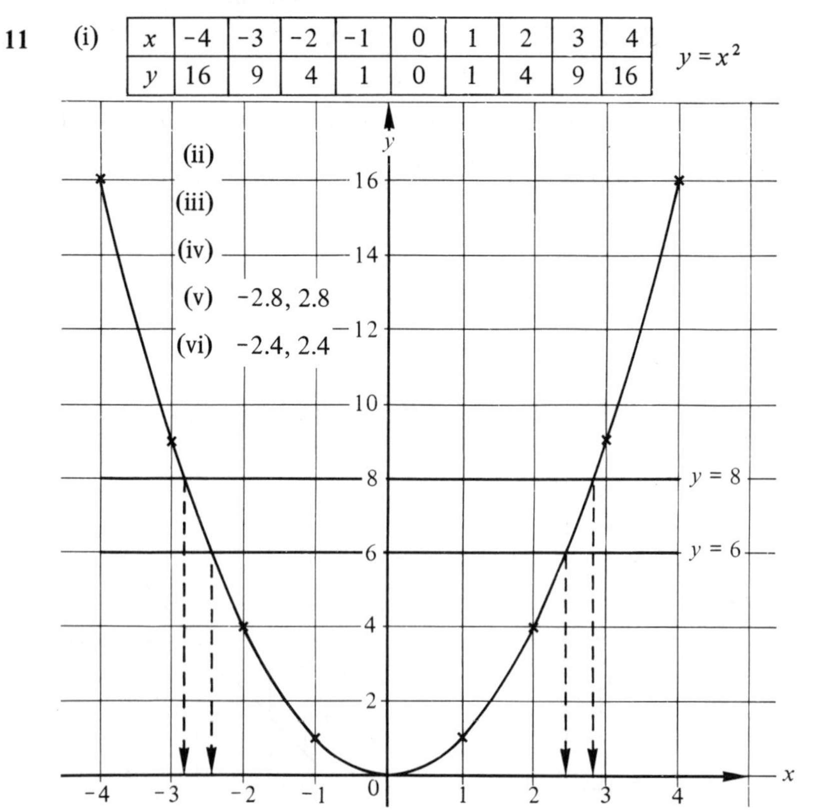

12 (i) $\begin{pmatrix} -4 & 8 & 2 \\ 6 & -2 & 12 \end{pmatrix}$ (ii) $\begin{pmatrix} 5 & 1 \\ 1 & 2 \end{pmatrix}$ (iii) $\begin{pmatrix} 1 & 3 \\ 7 & -2 \end{pmatrix}$

(iv) $\begin{pmatrix} 0 & 10 & 15 \\ -8 & 16 & 4 \end{pmatrix}$ (v) $\begin{pmatrix} 7 & -4 \\ -12 & 7 \end{pmatrix}$ (vi) C has 3 columns but B has only 2 rows.

13 (i) BC

(ii) D

(iii) $\begin{pmatrix} -4 \\ 2 \end{pmatrix}$

(iv) $\begin{pmatrix} -4 \\ -2 \end{pmatrix}$

(v) $\begin{pmatrix} 4 \\ -2 \end{pmatrix}$

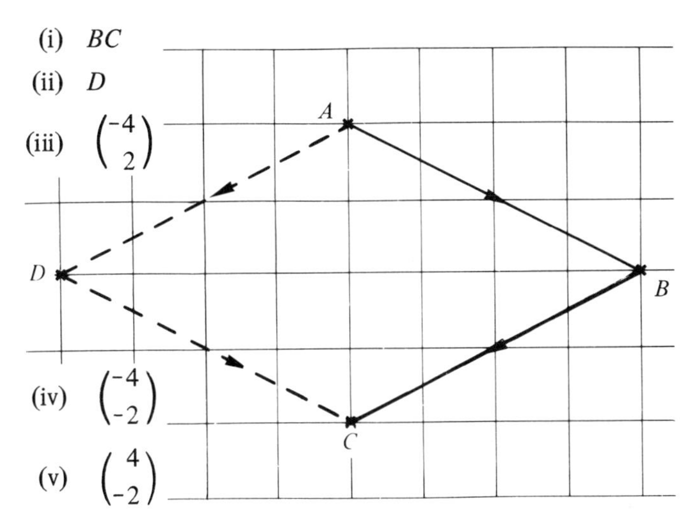

(vi) DC and AB are equal and parallel.

14 (i)

(i) A' (5,7), B' (7,5), C' (5,1), D' (3,5)

(ii) 3 cm^2

(iii) 12 cm^2

(iv) $\frac{1}{4}$

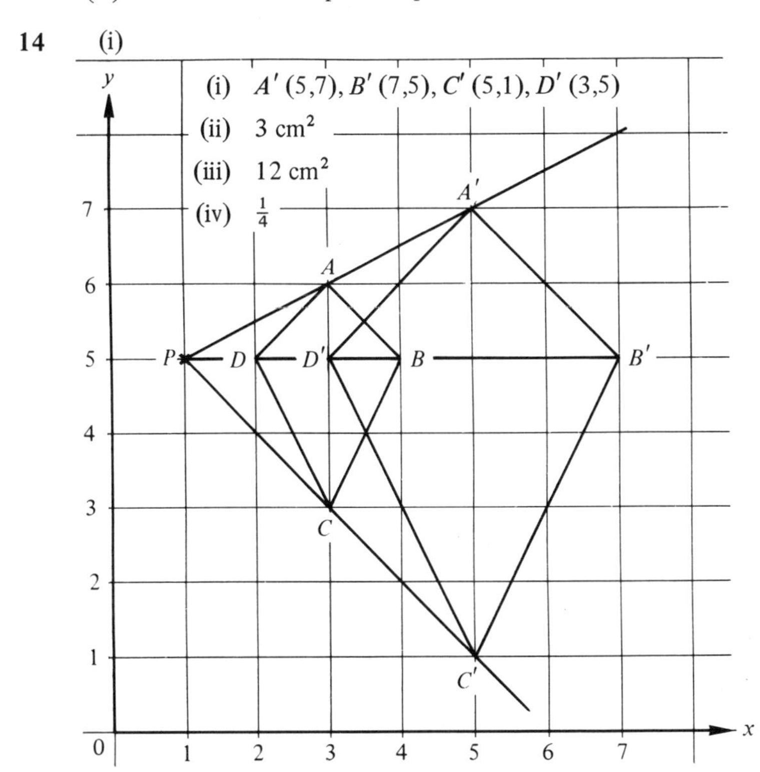

Set 2

1 (i) $10\,001_2$ (ii) 23_{10} (iii) 100_2 (iv) 10_{10} (v) $100\,001_2$ (vi) 1005_8

2 (i) £5.25 (ii) 24p (iii) 25.4 cm (iv) 0.81 m^2 (v) £9.80

3 (i) 5 (ii) -15 (iii) $2\frac{1}{2}$ (iv) 4 (v) ± 2

4 (i) 3,3,4,4,5,5,6,6,6,7,8,9 (ii) 6 (iii) 5.5 (iv) 5.5 (v) 6

5 (i) 38 000 (ii) 8 m (iii) 1.47 (iv) 2.3 (v) 0.2

6 (i) $2(a + 2b)$ (ii) $6(1 + 2x)$ (iii) $a(b + c + d)$ (iv) $a(a + 1)$ (v) $(a + b)(a - b)$

7 (i) 4.62 cm (ii) (*a*) 7.932 cm (*b*) 9 cm

8 (i) 15.7 (ii) 11 (iii) $r = \dfrac{C}{2\pi}$ (iv) $r = 2$ (v) $r = 7$

9 (i) Intersection: (−3,2)

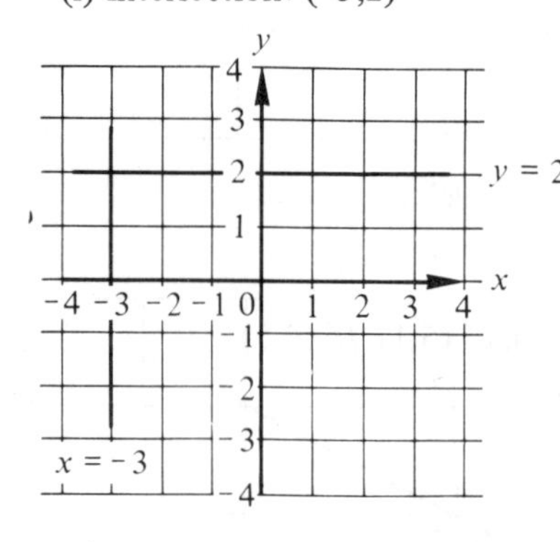

(ii) Intersection: (3,−3)

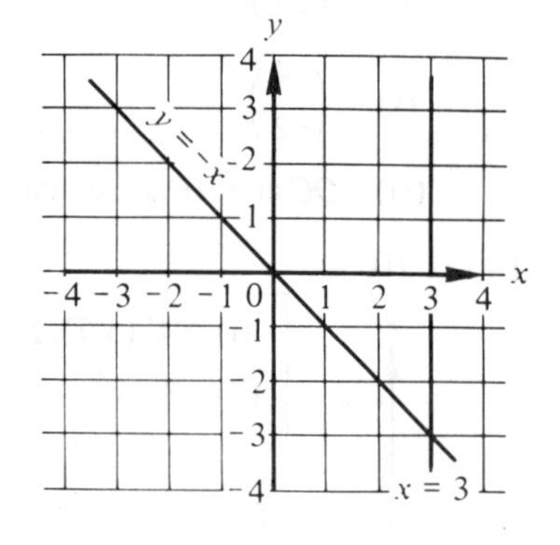

(iii)

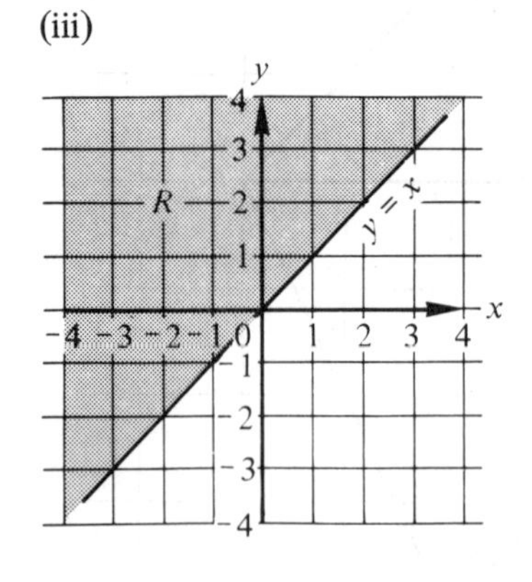

(iv)

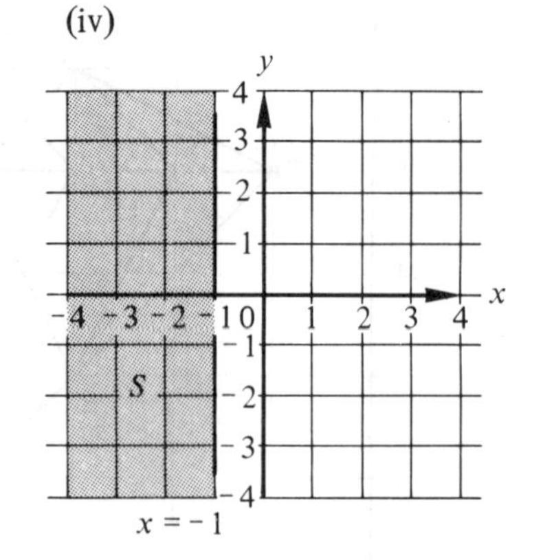

(v)

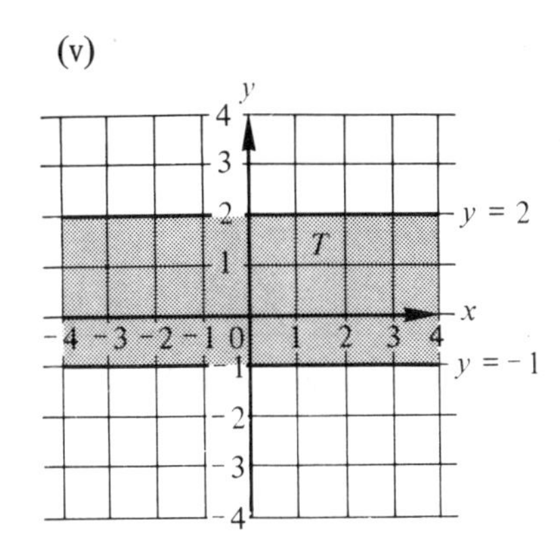

(vi)

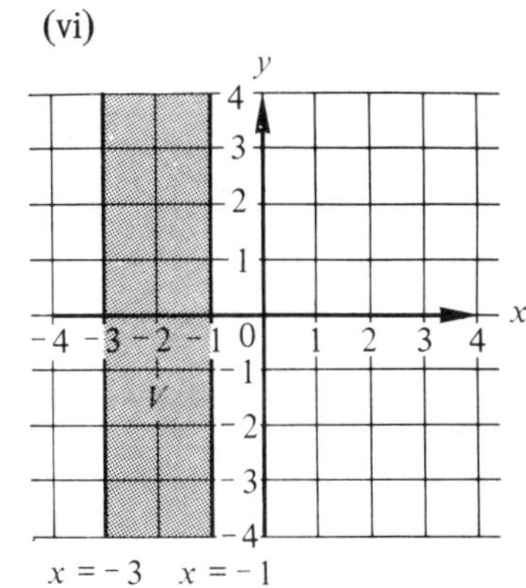

10 (i) $\begin{pmatrix} -9 & 12 & 6 \\ 6 & -3 & 12 \end{pmatrix}$ (ii) $\begin{pmatrix} 7 & -3 \\ 4 & 2 \end{pmatrix}$ (iii) $\begin{pmatrix} -1 & 1 \\ 0 & 4 \end{pmatrix}$ (iv) $\begin{pmatrix} 8 & -10 \\ 4 & -5 \end{pmatrix}$

(v) $\begin{pmatrix} 1 & -\frac{1}{2} \\ \frac{1}{2} & -\frac{1}{4} \end{pmatrix}$

11 (i) $\vec{QR}$

(ii) S

(iii) $\vec{RS} = \begin{pmatrix} 2 \\ -4 \end{pmatrix}$

(iv) $\vec{PS} = \begin{pmatrix} 8 \\ 0 \end{pmatrix}$

(v) $\vec{QR} = \frac{1}{2}\vec{PS}$ and $\vec{QR}$ is parallel to $\vec{PS}$

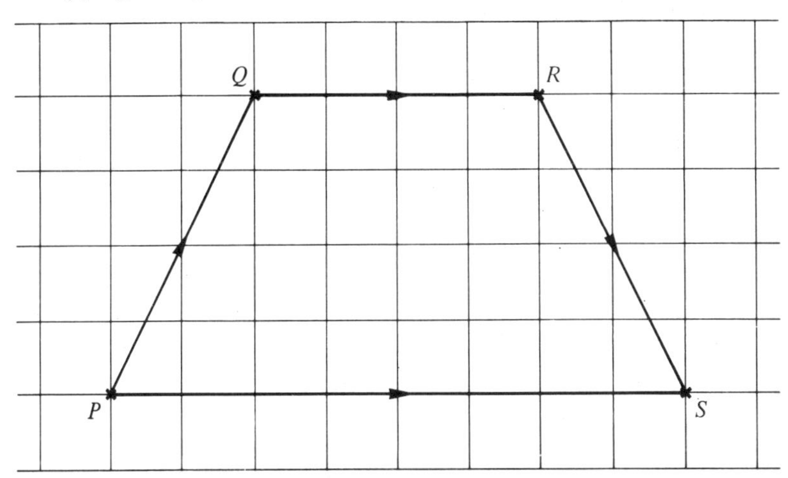

12 (i) 0.25 (ii) $\frac{25}{100}$ (iii) 25% (iv) $\frac{1}{64}$ (v) 2

13

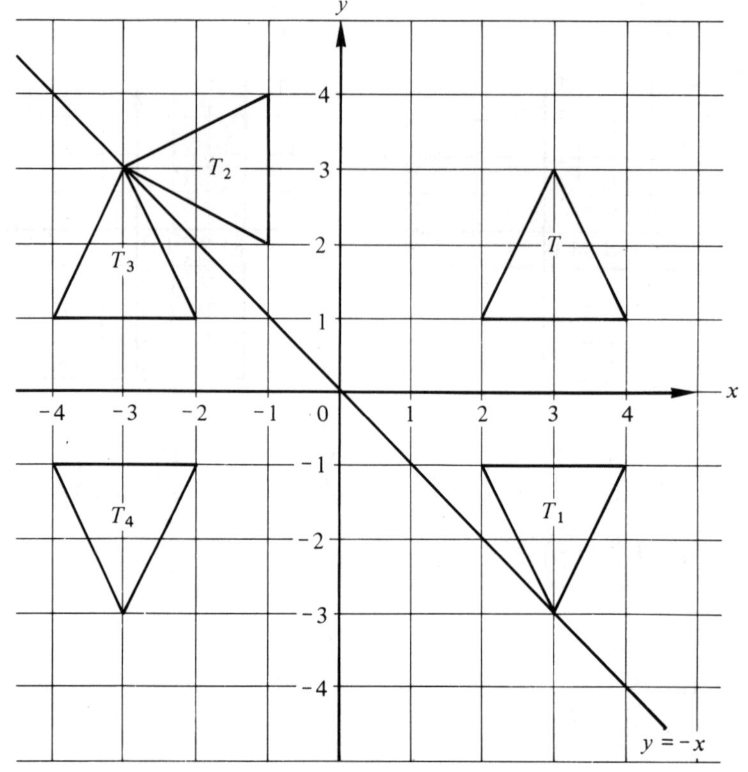

(vi) Rotation of 180°, centre O.

14 (i) (*a*) 20% (*b*) $12\frac{1}{2}$% (*c*) $6\frac{1}{4}$%

(ii) (*a*) $\frac{3}{20}$ (*b*) $\frac{4}{5}$ (*c*) $\frac{9}{20}$

(iii) 64%

(iv) 24

(v) (*a*) £1.40 (*b*) 25%

Set 3

1 (i) $1\,000\,000_2$ (ii) 513_{10} (iii) 320_{10} (iv) 347_8 (v) 6

2 (i) £1.34 (ii) £3.00 (iii) 1.3 m (iv) 45 km/h (v) £22.20

3 (i) $\frac{1}{2}$ (ii) -3 (iii) 24 (iv) 5 (v) + or -3

4 (i) 100 (ii) 15, 34, 54, 64, 95, 90 (iii) 352 (iv) 3.52

5 (i) £350 (ii) 535 000 (iii) 0.10 (iv) 1.0×10^6 (v) 1.0×10^{-6}

6 (i) $3(a + b)$ (ii) $5(x + 3y + 5z)$ (iii) $x(x^2 + x + 1)$

(iv) $(a + 3)(a - 3)$ (v) $(x + 2)(x + 3)$

7 (i) 6 cm (ii) 4.12 cm (iii) 0.8 (iv) 53.1° (v) 0.889 (vi) 27.3° (vii) 64.2°

8 (i) (*a*) 91 (*b*) 82.81 (ii) (*a*) 12 (*b*) 1.5 (iii) (*a*) 5 (*b*) 2.25

9 (i)

(ii)

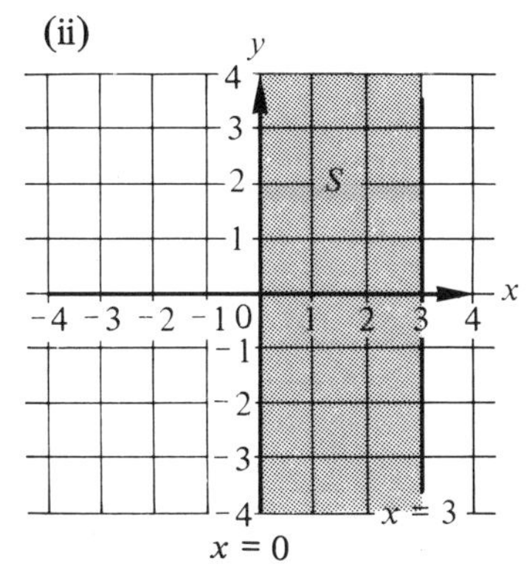

(iii)

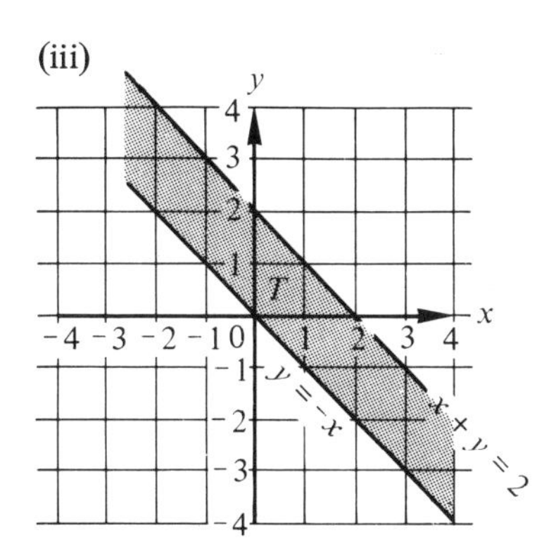

(iv)

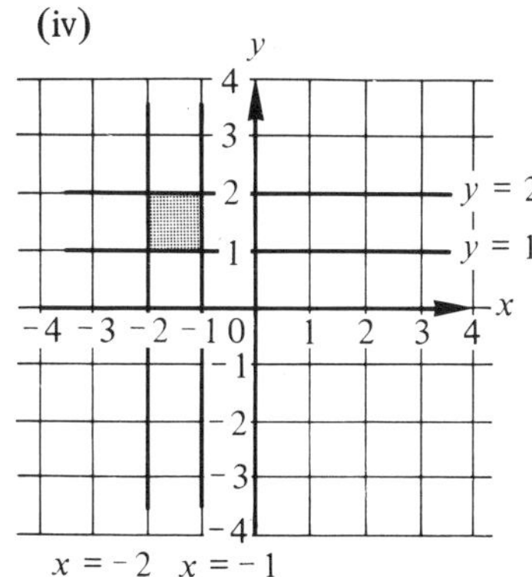

10 (i) $\begin{pmatrix} 0 & 4 \\ 8 & 0 \end{pmatrix}$ (ii) $\begin{pmatrix} 0 & -4 \\ -8 & 0 \end{pmatrix}$ (iii) $\begin{pmatrix} -17 & 10 & 22 \\ -24 & 14 & 31 \end{pmatrix}$ (iv) $\begin{pmatrix} 1 & 0 \\ 0 & 1 \end{pmatrix}$

(v) $\begin{pmatrix} 1 & 0 \\ 0 & 1 \end{pmatrix}$ (vi) *B* is the inverse of *A*.

11 (i) $\overrightarrow{BC}$

(ii) $\overrightarrow{AD}$

(iii) Parallelogram

(iv) $DC = \begin{pmatrix} -5 \\ -2 \end{pmatrix}$

(v) The opposite sides are equal and parallel.

12 (i) $\frac{7}{12}$ (ii) $\frac{1}{12}$ (iii) 2 (iv) $\frac{1}{12}$ (v) 5

13

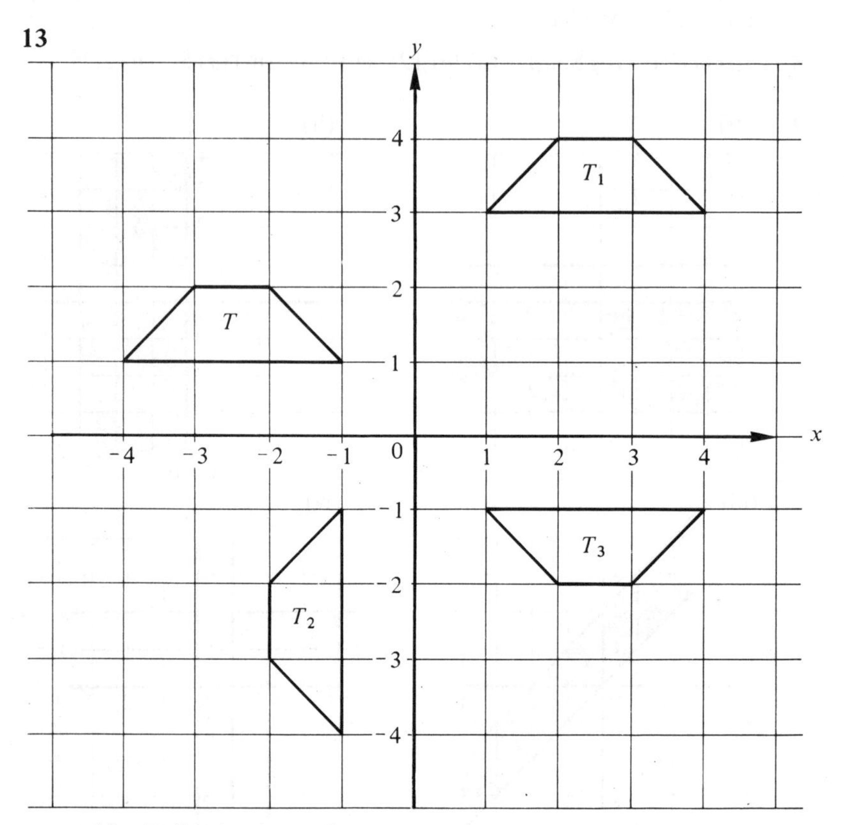

(v) Reflection in $y = 1$

14 (i) £8.00 (ii) £2.40 (iii) £5.60 (iv) 8.75% or $8\frac{3}{4}$% (v) 8.75%

Set 4

1 (i) 6 (ii) 7 (iii) 10 (iv) {12,5} (v) {5} (vi) 7

2 (i) 570 minutes (ii) 16 km/h (iii) 20°C (iv) multiply by 9, divide by 5 and add 32. (v) 8 km 47 m

3 (i) 9 (ii) −3 (iii) 24 (iv) 3 (v) + or −5

4

Score	Tally	Frequency	$S \times F$
4	11	2	8
5	111	3	15
6	~~1111~~	5	30
7	1111	4	28
8	11	2	16
9	111	3	27
10	1	1	10
	Totals:	20	134

(i) Mode: 6
(ii) Median: 6.5
(iii) Mean: 6.7

5 (i) $\frac{1}{4}$ (ii) 0.125 (iii) 0.75 (iv) $\frac{1}{50}$ (v) 25%

6 (i) $3(a+3)$ (ii) $7(2y+3z)$ (iii) $a(2a+1)$ (iv) $(a+b)(x+y)$
(v) $(x+2)(x+4)$

7 (i) (*a*) 3.84 cm (*b*) 7.016 cm (ii) (*a*) 4.5 cm (*b*) 6.72 or 6.73 cm

8 (i) (*a*) $x = \dfrac{b-a}{2}$ (*b*) 8.5 (*c*) −4.5

(ii) (*a*) $x = \dfrac{z}{y^2}$ (*b*) 7 (*c*) $y = \sqrt{\dfrac{z}{x}}$ (*d*) ± 2.5

9 (i)

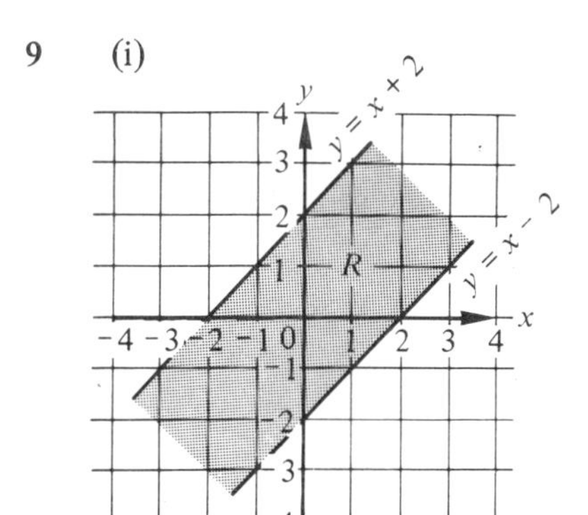

(ii)

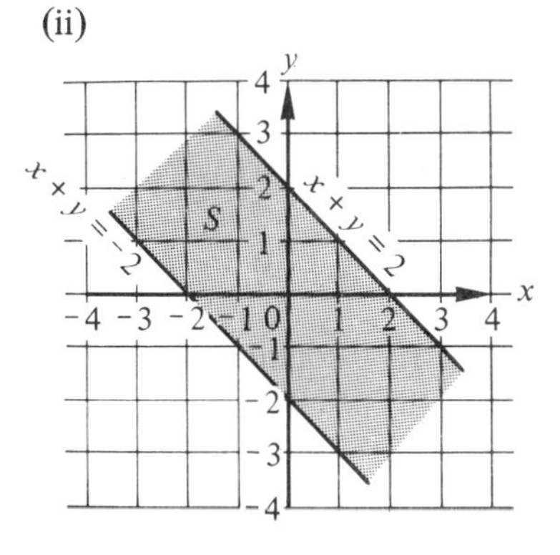

(iii)

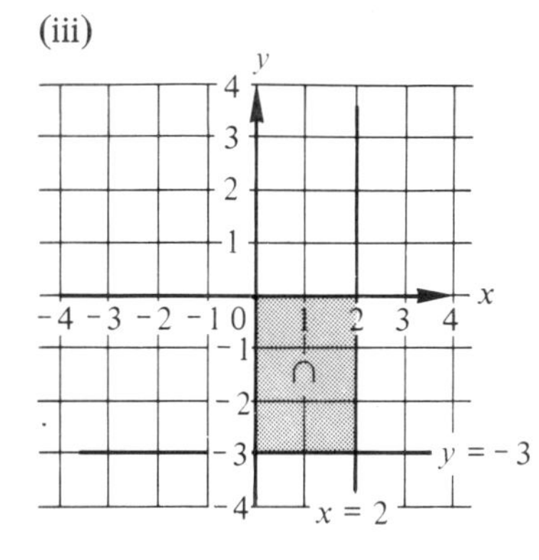

(iv)

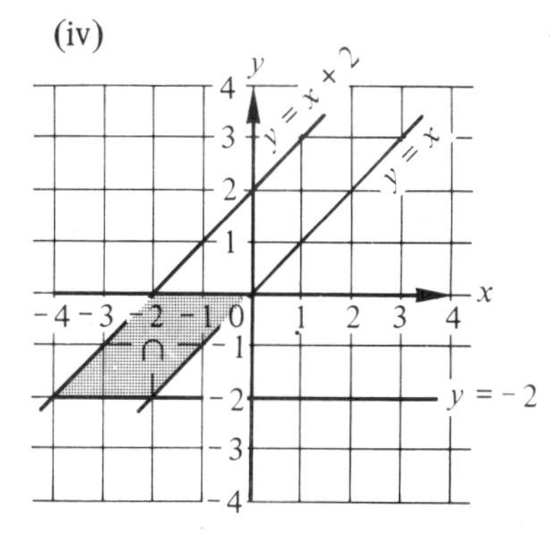

10 (i) $\begin{pmatrix} 1 & 5 \\ -3 & -4 \end{pmatrix}$ (ii) $\begin{pmatrix} -5 & 1 \\ 5 & -4 \end{pmatrix}$ (iii) $\begin{pmatrix} 7 & -18 \\ -6 & 19 \end{pmatrix}$

(iv) $\begin{pmatrix} 9 & -2 & 2 \\ -4 & 0 & 8 \end{pmatrix}$ (v) $\begin{pmatrix} -1 & -1 & 12 \\ -28 & 12 & -64 \end{pmatrix}$

11 (i) $\overrightarrow{BC}$

(ii) $\overrightarrow{CD}$

(iii) $\overrightarrow{DA} = \begin{pmatrix} -3 \\ 3 \end{pmatrix}$

(iv) Kite

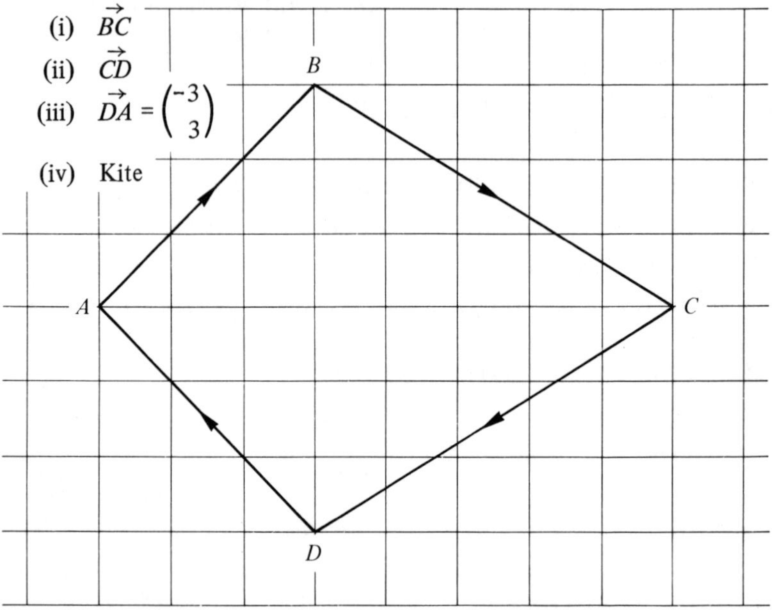

(v) *ABCD* has 2 pairs of equal sides but no parallel sides.

12 (i) $\frac{13}{15}$ (ii) $\frac{7}{15}$ (iii) $\frac{1}{5}$ (iv) $\frac{16}{81}$ (v) $\frac{2}{3}$

13 (i) Rotation of 180°, about 0, the origin

(ii) Reflection in $x = 0$ (or the y axis)

(iii) Rotation of 180°, about 0, the origin

(iv) Translation by the vector $\begin{pmatrix} 4 \\ 4 \end{pmatrix}$

(v) Rotation of 180°, about 0, the origin

(vi) Translation by the vector $\begin{pmatrix} -4 \\ -4 \end{pmatrix}$

14 (i) (*a*) $\frac{15}{100}$ (*b*) 0.15 (*c*) $\frac{3}{20}$

(ii) 81

(iii) (*a*) £6.90 (*b*) £39.10

(iv) (*a*) £1.86 (*b*) £14.26

Set 5

1 (i)

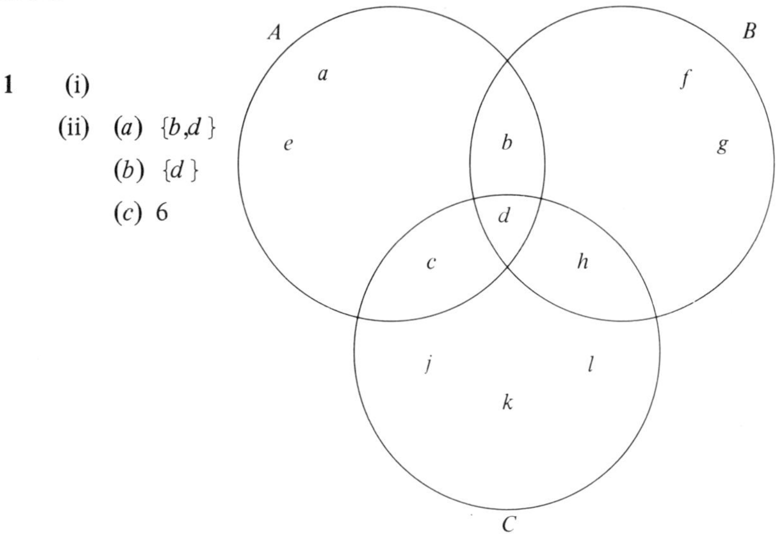

(ii) (*a*) $\{b, d\}$

(*b*) $\{d\}$

(*c*) 6

2 (i) £14.35 (ii) 12 km/litre (iii) 30 km/h (iv) 6 hours 40 minutes (v) 96.60 km

3 (i) 7 (ii) 4.5 (iii) -3 (iv) 26 (v) -4, 3

4 (i) 2,4,5,5,6,6,6,7,8,8,9 (ii) 6 (iii) 6 (iv) 6 (v) 12

5 (i) 210 (ii) 2.98 (iii) £36.50 (iv) 6.4×10^{-2} (v) 275.5

6 (i) $3(2a + 3b)$ (ii) $a^2(a + 1)$ (iii) $12(x + 2y + 5z)$ (iv) $\pi(A^2 + B^2)$ (v) $(a + 5)(a - 5)$

7 $x = 6.928, y = 4$

8 (i) (*a*) $4x$ (*b*) x^2 (*c*) $\sqrt{2x^2}$ or $x\sqrt{2}$

(ii) (*a*) 30.25 or 30.3 cm^3 (*b*) 7.78 cm

9 (i)

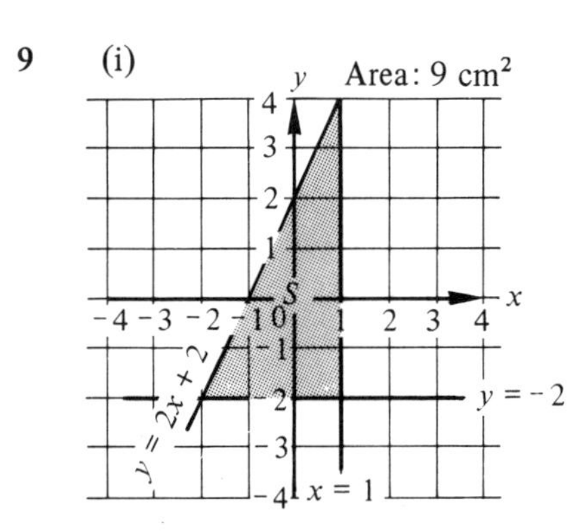

(ii)

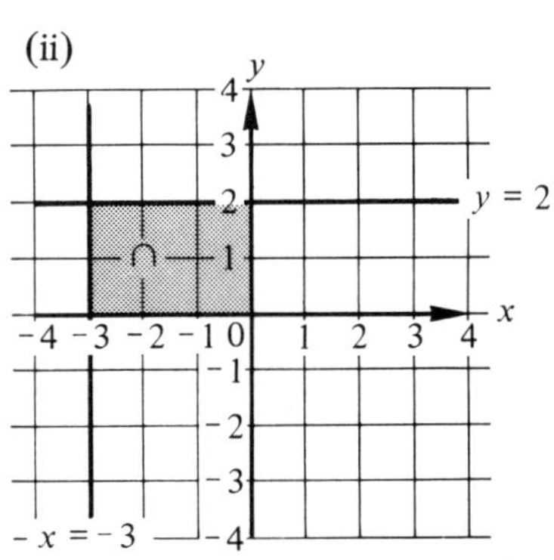

(iii)

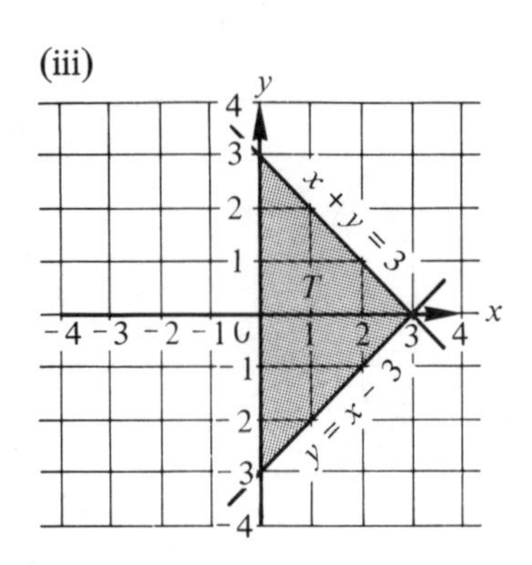

(iv)

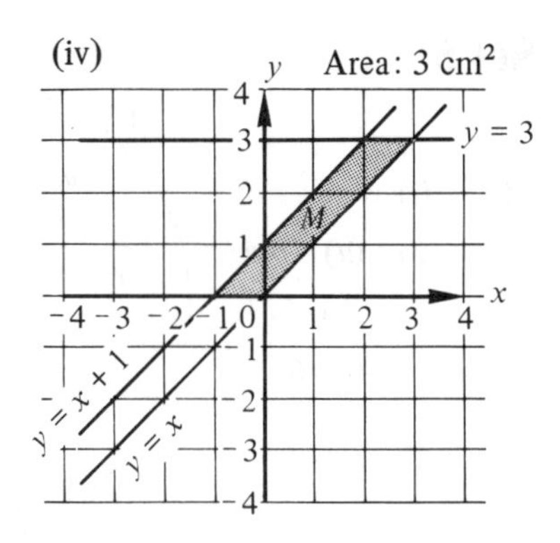

10 (i) 6 cm (ii) 10.4 cm (iii) 4 cm (iv) 6.93 cm
(v) 31.2 cm^2 (vi) 13.86 cm^2 (vii) 17.34 cm^2 (viii) $\frac{4}{9}$

11 (i) $\vec{BC}$
(ii) D
(iii) $\vec{DC} = \begin{pmatrix} 5 \\ 3 \end{pmatrix}$

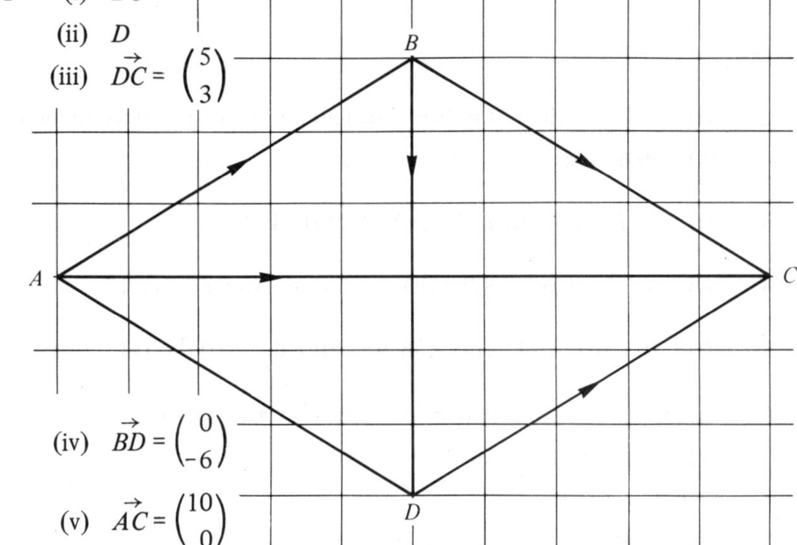

(iv) $\vec{BD} = \begin{pmatrix} 0 \\ -6 \end{pmatrix}$
(v) $\vec{AC} = \begin{pmatrix} 10 \\ 0 \end{pmatrix}$
(vi) The diagonals intersect at right angles.

12 (i) $\frac{19}{20}$ (ii) $\frac{11}{20}$ (iii) $\frac{1}{5}$ (iv) $59\frac{1}{2}$

13

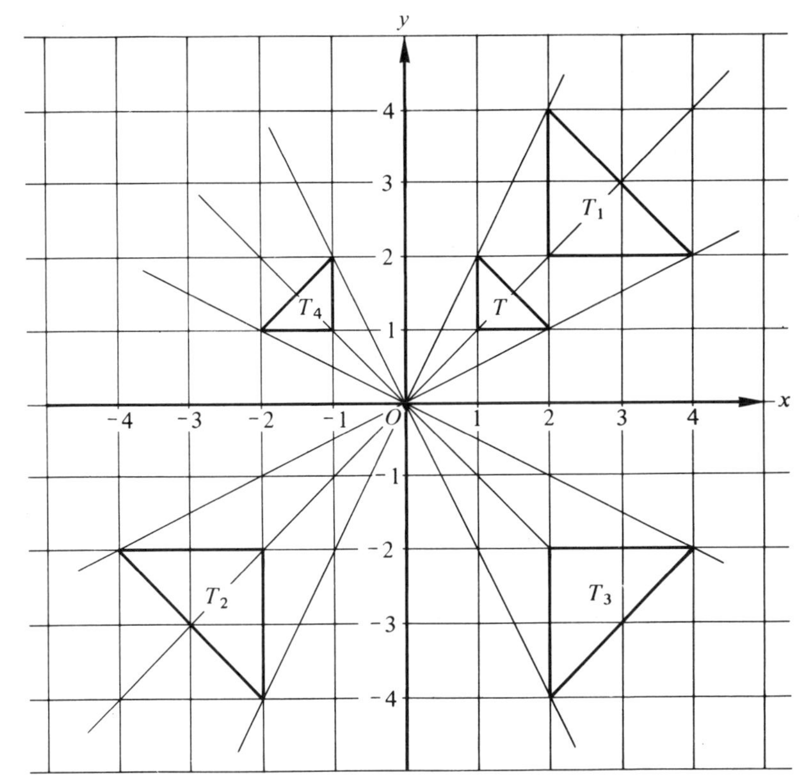

(vi) Rotation of 180° about 0, the origin.

14 (i) (*a*) $\frac{1}{10}$ (*b*) $\frac{1}{20}$ (*c*) $\frac{1}{40}$ (*d*) $\frac{7}{40}$

(ii) (*a*) 25% (*b*) $12\frac{1}{2}$% (*c*) $37\frac{1}{2}$% (*d*) $62\frac{1}{2}$%

(iii) 342

(iv) £2520

(v) £30.00

Set 6

1 (i)

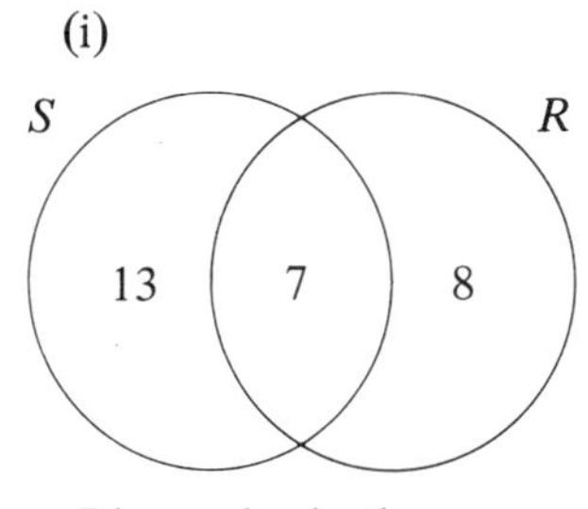

7 boys play both games

(ii)

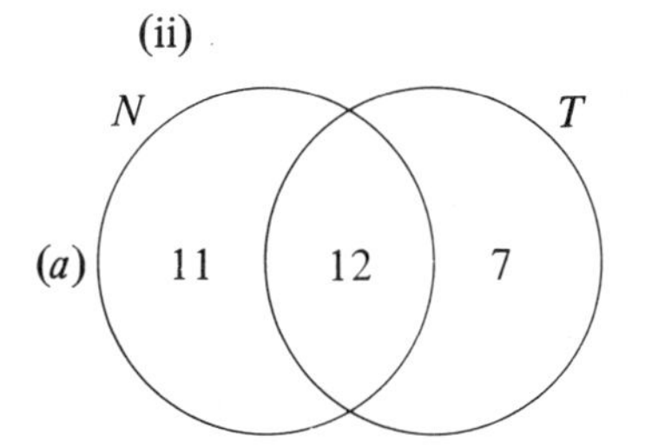

(*a*)

(*b*) 30 in the group.

2 (i) 7.5 cm (ii) 8.66 cm (iii) 56.3 or 56.25 m^2 (iv) 0.027 m^3 (v) 0.2 m

3 (i) $3\frac{1}{2}$ (ii) 50 (iii) $5\frac{1}{4}$ (iv) 8 (v) −4 or 3

4

Score	Frequency	$S \times F$
2	3	6
3	6	18
4	9	36
5	11	55
6	14	84
7	17	119
8	15	120
9	9	81
10	7	70
11	5	55
12	4	48
Totals:	100	692
	(i)	(ii) (iii)

(iv) Mean = 6.92
(v) $\frac{1}{6}$

5 (i) 4 (ii) x^2 (iii) 2.5×10^6 (iv) 4 000 400 (v) −6

6 (i) $b(y + 4z)$ (ii) $5(2a + 4b - 5c)$ (iii) $2x(x + 1)$
(iv) $(3a + 4b)(3a - 4b)$ (v) $(a + 3)(a + 5)$

7 (i) 10 m (ii) 7.5 m (iii) 0.75 (iv) 36.9° (v) 106.2° (vi) 75 m^2

8 (i) 125.6 (ii) 242 (iii) $r = \dfrac{A}{2\pi h}$ (iv) 3 (v) 4

(vi) The formula would be used to find the area of the curved surface of a cylinder.

9 (i)

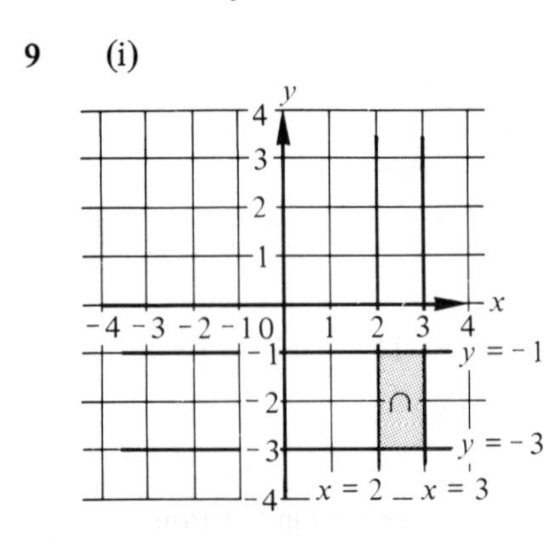

(ii)

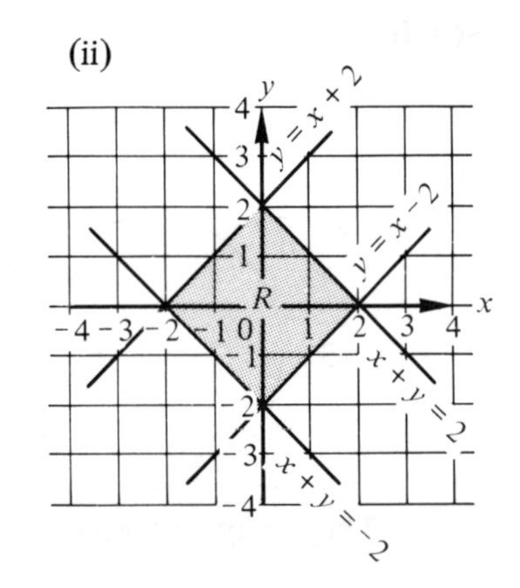

(iii)

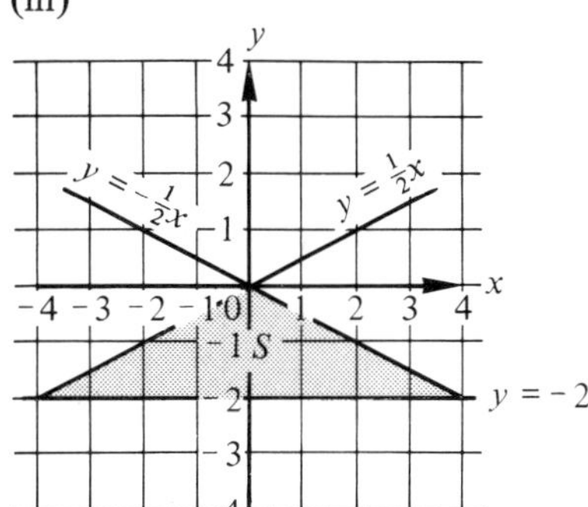

(c) Area: 8 cm^2

(iv)

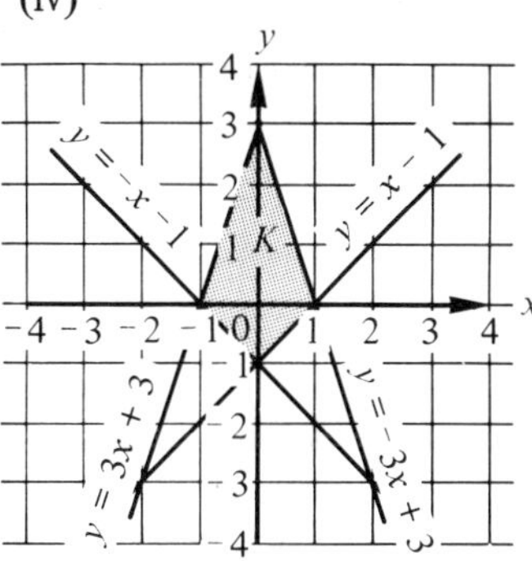

(c) K is a kite

10 (i) (a) $\begin{pmatrix} 6 & -5 \\ 3 & 3 \end{pmatrix}$ (b) $\begin{pmatrix} -2 & -1 \\ 5 & -3 \end{pmatrix}$ (c) $\begin{pmatrix} 2 & 1 \\ -5 & 3 \end{pmatrix}$ (d) $\begin{pmatrix} 2 & -3 \\ 4 & 0 \end{pmatrix}$

(e) $\begin{pmatrix} 11 & -13 \\ 16 & -8 \end{pmatrix}$ (ii) Matrix C is the identity matrix.

11

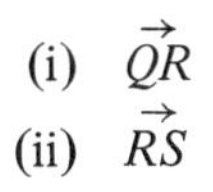

(i) $\overrightarrow{QR}$

(ii) $\overrightarrow{RS}$

(iii) V-bomber or arrowhead kite

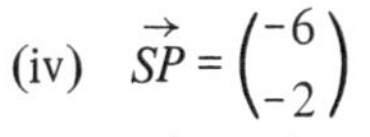

(iv) $\overrightarrow{SP} = \begin{pmatrix} -6 \\ -2 \end{pmatrix}$

(v) $\overrightarrow{QR} + \overrightarrow{RS} = \begin{pmatrix} 4 \\ 0 \end{pmatrix}$

(vi) $\overrightarrow{QS}$

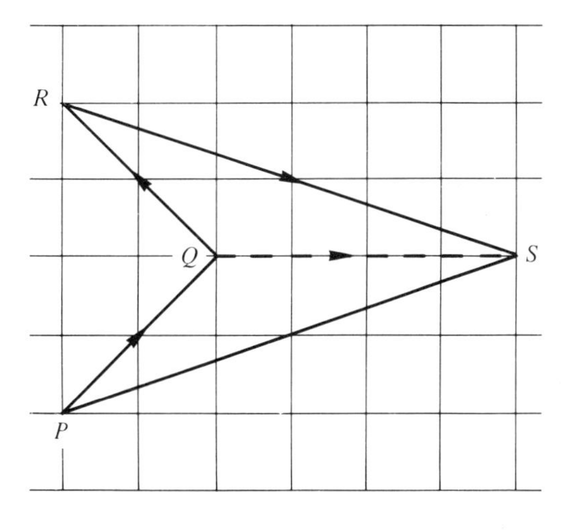

12 (i) $\frac{19}{24}$ (ii) $\frac{1}{24}$ (iii) $1\frac{1}{9}$ (iv) $\frac{9}{10}$ (v) 1

13 (i)

x	−3	−2	−1	0	1	2	3
y	18	8	2	0	2	8	18

$y = 2x^2$

(ii)

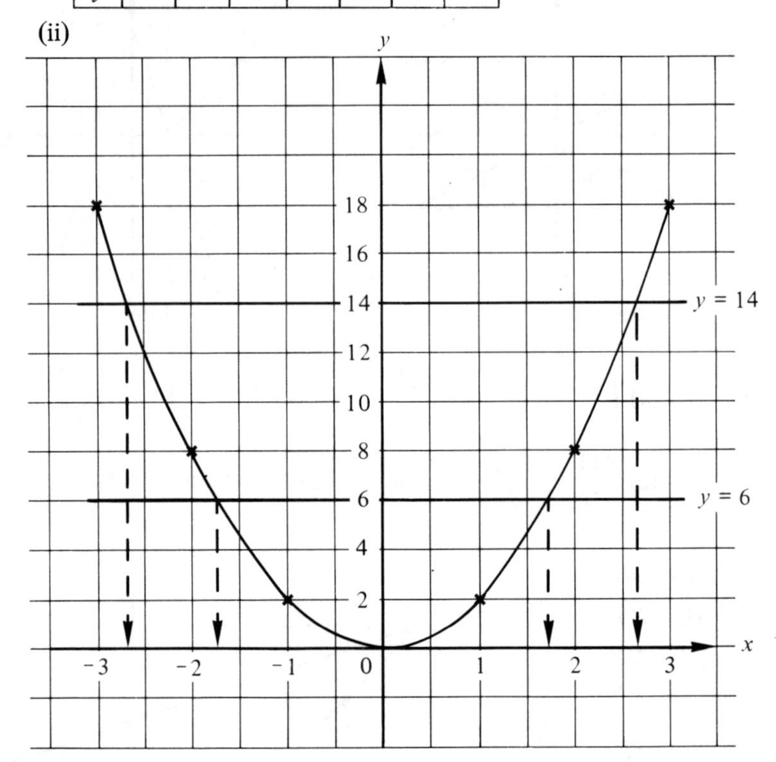

(iii) −1.7, + 1.7

(iv) −2.7, + 2.7

14 (i) £8.00 (ii) £88.00 (iii) £8.80 (iv) £96.80 (v) £9.68 (vi) £106.48

Set 7

1 (i) 39 (ii) 33 (iii) 15 (iv) 31 (v) 60 (vi) 20%

2 (i) 29.64 acres (ii) 404.7 hectares (iii) 44 a (iv) 105 000 m^2 (v) 100 hectares

3 (i) 7 (ii) 4.5 (iii) 5 (iv) + or −4 (v) + or −3

4 (i) 42.6* (ii) 66

5 (i) 0.05 (ii) 5% (iii) 0.67 (iv) $66\frac{2}{3}\%$ (v) 0.625

6 (i) $4(x + 2y + 6z)$ (ii) $6(2a - 3b - 4c)$ (iii) $2a(2a + 1)$
(iv) $(x + 1)(x - 1)$ (v) $(x + 2)(x - 5)$

7 (i) 48.6° (ii) 48.2° (iii) 68.2°

8 (i) (*a*) $3x$ cm (*b*) $P = 8x$ cm (*c*) $A = 3x^2$ cm^2
(*d*) $D = \sqrt{10x^2}$ or $x\sqrt{10}$ cm (ii) (*a*) 18.75 cm^2 (*b*) 7.91 cm

9 (i) 30° (ii) 4.6 cm (iii) 6.9 cm (iv) 15.9 cm^2
(v) 27.6 or 27.7 cm^2 (vi) 3.9 cm^2

10 (i) $\begin{pmatrix} 5 & 4 & 2 \\ 3 & 6 & 4 \\ 4 & 0 & 6 \\ 6 & 0 & 0 \\ 3 & 5 & 0 \end{pmatrix}$ (ii) $\begin{pmatrix} 9 \\ 11 \\ 8\frac{1}{2} \end{pmatrix}$ (iii) $\begin{pmatrix} 106 \\ 127 \\ 87 \\ 54 \\ 82 \end{pmatrix}$
£4.56

11 (i) $\vec{QR}, \vec{RS}, \vec{ST}, \vec{TU}$
(ii) Hexagon
(iii) $\vec{UP} = \begin{pmatrix} -4 \\ -2 \end{pmatrix}$
(iv) $\vec{PQ} + \vec{QR} = \begin{pmatrix} 7 \\ 4 \end{pmatrix}$
(v) $\vec{PR}$

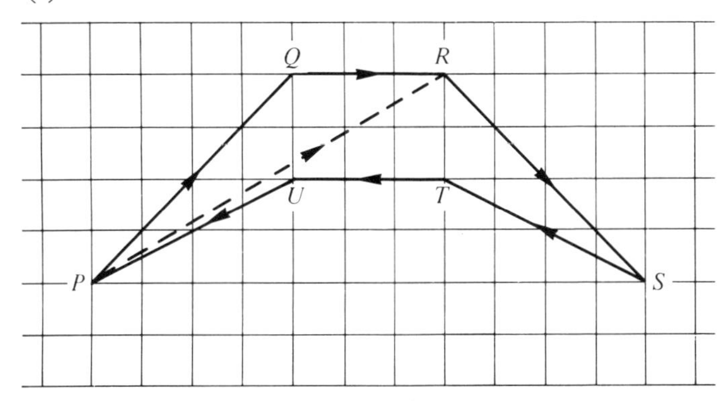

12 (i) $\frac{2}{3}$ (ii) (*a*) $1\frac{5}{24}$ (*b*) $-\frac{1}{24}$ (*c*) $\frac{14}{15}$ (*d*) 129

13

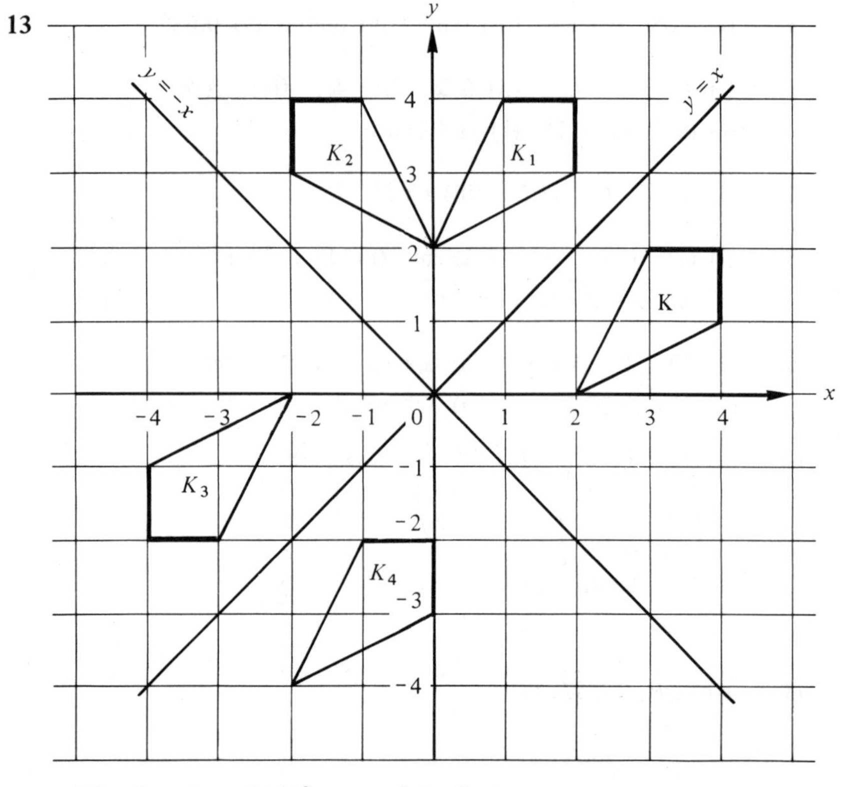

(vi) Rotation of 180°, centre (−2,−2)

14 (i) Mathematics: 76%, English: 70%, Science: 68%

(ii) 72.9%

(iii) 46 marks

Set 8

1 (i) 585_{10} (ii) 101_8 (iii) 29_{10} (iv) $110\,111_2$ (v) 7

2 (i) £7.75 (ii) 32p (iii) £20.80 (iv) £4.50 (v) 75p

3 (i) $3\frac{1}{2}$ (ii) 5 (iii) + or − 1 (iv) 2 (v) + or − 3

4 (i) I.N. Swinger: 30.1, L.A. Spinner: 15.4, M. Pace: 19.5

(ii) 15.4, 19.5, 30.1

5 (i) 10 010 (ii) 300 (iii) 2.9 (iv) −6 (v) 8.0×10^7

6 (i) $13(a + 2b + 4c)$ (ii) $2(x^2 + 2y^2 + 5z^2)$ (iii) $(a + b)(m + n)$
(iv) $\frac{1}{2}(a + b + 4c)$ (v) $3(x + y)(x - y)$

7 (i) 216.9° (ii) 16 km (iii) 12 km

8 (i) (*a*) 1.57 (*b*) $D = \dfrac{C}{\pi}$ (*c*) 21
(ii) (*a*) $u = v - ft$ (*b*) 37

9 (i) 22 cm² (ii) (*a*) 5 cm (*b*) 9.43 cm (*c*) 8.94 cm

10 (i) (*a*) $\begin{pmatrix} 22 \\ 8 \end{pmatrix}$ (*b*) $\begin{pmatrix} 26 \\ 10 \end{pmatrix}$ (*c*) $\begin{pmatrix} 52 & 12 \end{pmatrix}$ (*d*) $\begin{pmatrix} 90 \\ 81 \end{pmatrix}$ (ii) (3 by 3)

11 (i) $\vec{BC}, \vec{CD}, \vec{DE}$
(ii) $\vec{EA} = \begin{pmatrix} 4 \\ -3 \end{pmatrix}$
(iii) Pentagon
(iv) $\vec{BC} + \vec{CD} + \vec{DE} = \begin{pmatrix} -2 \\ 7 \end{pmatrix}$
(v) $\vec{BE}$

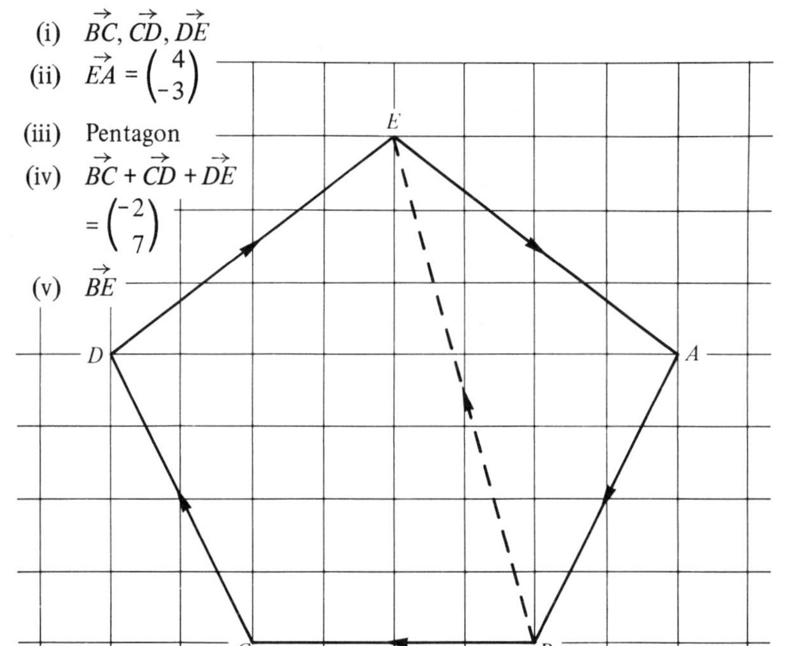

12 (i) $\{\frac{16}{24}, \frac{14}{24}, \frac{20}{24}, \frac{15}{24}, \frac{18}{24}\}$
(ii) $\{\frac{7}{12}, \frac{5}{8}, \frac{2}{3}, \frac{3}{4}, \frac{5}{6}\}$
(iii) $\frac{1}{4}$
(iv) $3\frac{11}{24}$
(v) $75\% - 62\frac{1}{2}\% = 12\frac{1}{2}\%$

13

x	-4	-3	-2	-1	0	1	2
y	9	4	1	0	1	4	9

$y = (x + 1)^2$

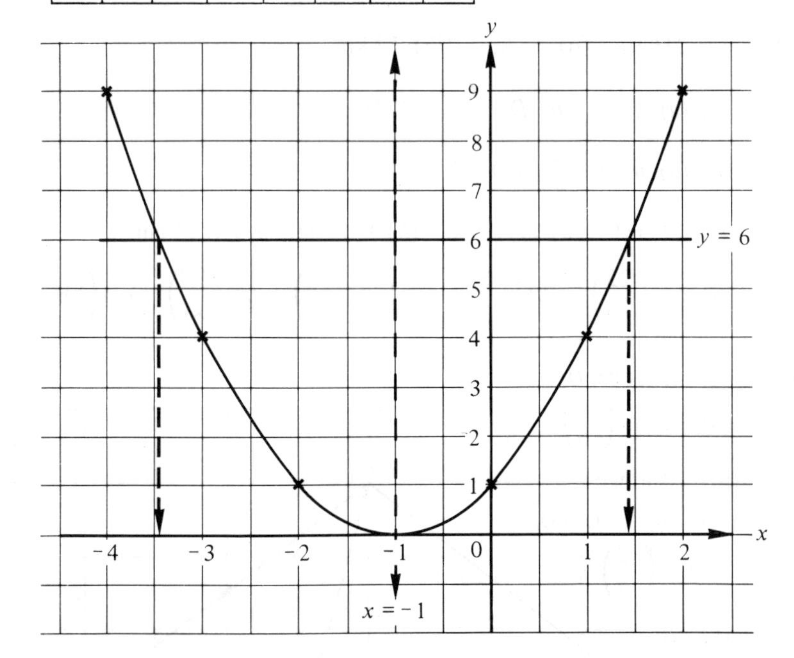

(iii) -3.45 and 1.45 (iv) $x = -1$

14 (i) (*a*) £5.75 (*b*) £23.00
(ii) (*a*) 8% (*b*) £81.00

Set 9

1 (i) 30 (ii) 80 (iii) 51 (iv) 11 (v) 5 (vi) $37\frac{1}{2}\%$

2 (i) 0.729 cm^3 (ii) 125 cubes (iii) (*a*) 36.125 cm^2 (*b*) 25.5 cm
(iv) 22.8 cm

3 (i) 2.5 (ii) -3 (iii) 32 (iv) 5 (v) + or -4

4 (i) 2,3,4,5,5,6,6,7,7,7,7,8,8,9,10
(ii) 7 (iii) 7 (iv) 6.27 (v) $66\frac{2}{3}\%$

5 (i) 0.33 (ii) $33\frac{1}{3}\%$ (iii) $\frac{1}{8}$ (iv) $12\frac{1}{2}\%$ (v) 0.73

6 (i) $7(2r + 3s - 5t)$ (ii) $d(d + 1)$ (iii) $4(1 - 2a - 3b + c)$
(iv) $(x + y)(a + 4)$ (v) $(x + 2)(x - 3)$

7 (i) (*a*)

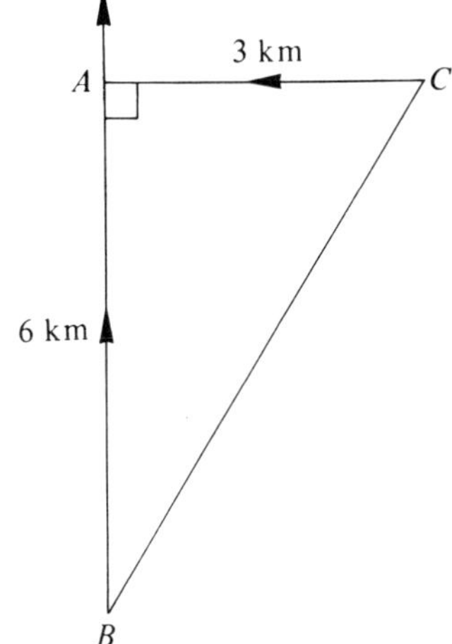

(*b*) 026.5° or 026.6° (*c*) 206.5°

(ii) The northerly component = the easterly component = 141.4 km

8 (i) (*a*) x^2 cm² (*b*) $4x^2$ cm² (*c*) $\sqrt{5x^2}$ or $x\sqrt{5}$ cm
(ii) (*a*) $AB = 11.2$ cm (*b*) perimeter = 44.8 cm

9 (i)

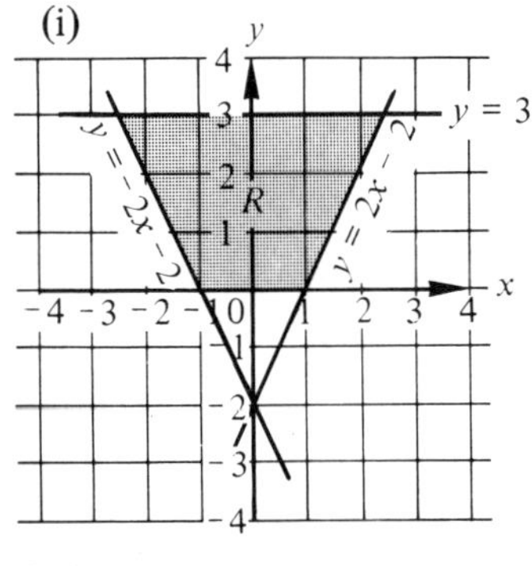

(ii)

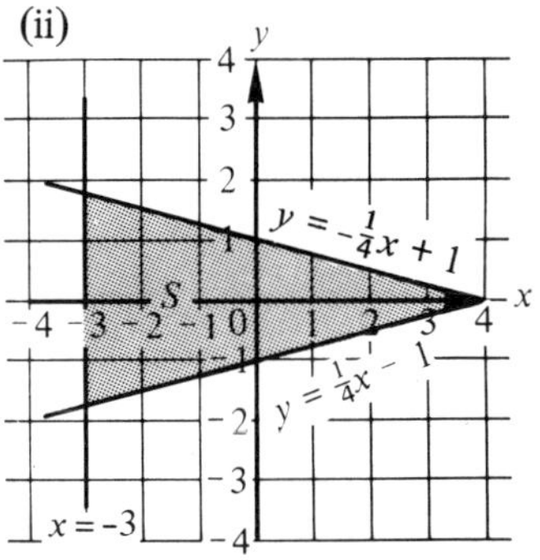

(iii)

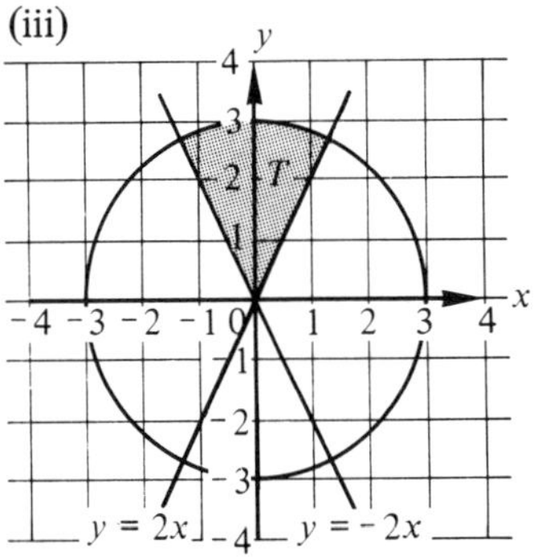

(iv) (*c*) (1,2)

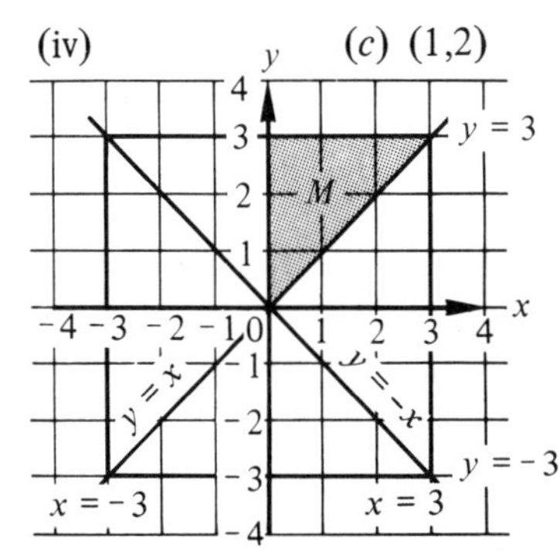

10 (i) $\begin{pmatrix} -1 & -1 \\ -2 & 6 \end{pmatrix}$ (ii) $\begin{pmatrix} 2 & 1 \\ 4 & -6 \end{pmatrix}$ (iii) $\begin{pmatrix} 9 & -5 \\ -6 & 3 \end{pmatrix}$ (iv) $\begin{pmatrix} 12 & 3 \\ -8 & -7 \end{pmatrix}$

(v) $\begin{pmatrix} 12 & -15 \\ 12 & -5 \end{pmatrix}$

11 (i) $\vec{BC}, \vec{CD}, \vec{DE}, \vec{EF}$.

(ii) Hexagon

(iii) $\vec{AF} = \begin{pmatrix} 2 \\ 3 \end{pmatrix}$

(iv) $\vec{BC} + \vec{CD} + \vec{DE} + \vec{EF}$

$= \begin{pmatrix} 0 \\ 6 \end{pmatrix}$

(v) $\vec{BF}$

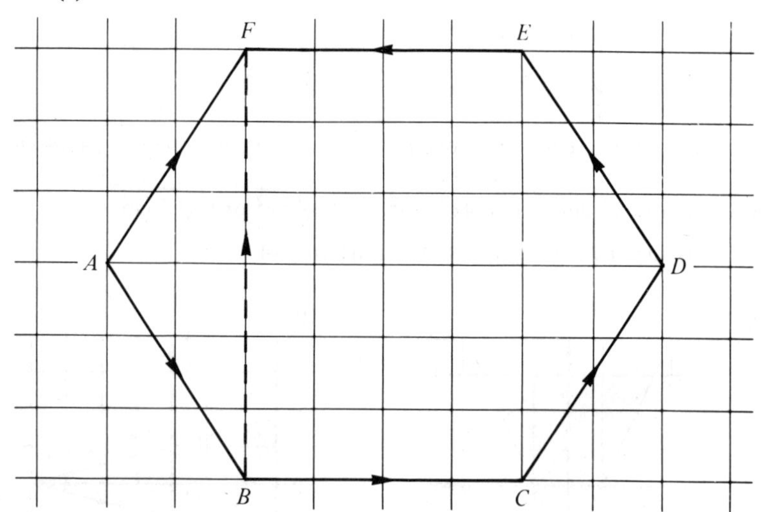

12 (i) $\{\frac{22}{30}, \frac{19}{30}, \frac{21}{30}, \frac{18}{30}\}$ (ii) $\{\frac{3}{5}, \frac{19}{30}, \frac{7}{10}, \frac{11}{15}\}$ (iii) $2\frac{2}{3}$

(iv) $\frac{2}{3}$ (v) 60%

13 (i) (*a*) Reflection in the line whose equation is $y = 6$

(*b*) Enlargement, Scale Factor + 2, centre of enlargement (0,3)

(*c*) Reflection in the line whose equation is $x = 5$

(*d*) Translation by the vector: $\begin{pmatrix} 7 \\ 0 \end{pmatrix}$

(*e*) Rotation of 180° about the point $(8\frac{1}{2}, 6)$

(ii) $\frac{1}{4}$

14 (i) £420 (ii) 1200 litres (iii) £300 (iv) £858 (v) £16.50

Set 10

1 (i) 80 (ii) 37 (iii) 52 (iv) 16 (v) 48 (vi) 60%

2 (i) £3.70 (ii) 0830 hours (iii) 1640 hours (iv) $11\frac{1}{2}$ hours
(ii) 2220 hours on Sunday

3 (i) $5\frac{1}{2}$ (ii) 7 (iii) 24 (iv) 0 or 5 (v) +3 or $-3\frac{1}{2}$

4 (i)

Children in family	Frequency	$C \times F$
1	7	7
2	9	18
3	7	21
4	4	16
5	2	10
6	1	6
Totals:	30	78

(ii)

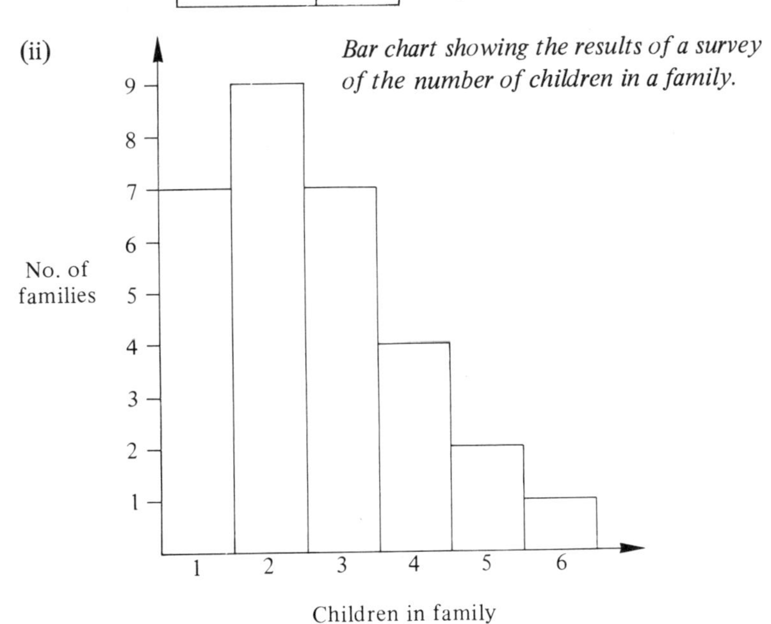

Bar chart showing the results of a survey of the number of children in a family.

(iii) Mean: 2.6

5 (i) 7.5×10^6 (ii) 8.7×10^{-7} (iii) 9.0×10^9
(iv) 2.0×10^{13} (v) 600

6 (i) $9(2a - 3b - 4c + 9)$ (ii) $4x(x + 3)$ (iii) $(x + 5)(a + b)$
(iv) $(a - 1)(a - 7)$ (v) $25(x + 1)(x - 1)$

7 (i) 105° (ii) 8.66 cm (iii) 5 cm (iv) 7.07 cm (v) 13.66 cm
(vi) 68.3 cm^2

8 (i) 31.4 (ii) $38\frac{1}{2}$ (iii) $\sqrt{\frac{A}{\pi}}$ (iv) 3 (v) 6.48

9 (i) $XB = 4$ cm, $YC = 12$ cm (ii) 160 cm^2 (iii) 6 cm^2
(iv) 30 cm^2 (v) 124 cm^2

10 (i) $\begin{pmatrix} 5 & 3 \\ -3 & -6 \end{pmatrix}$ (ii) $\begin{pmatrix} -1 & -1 \\ 0 & 3 \end{pmatrix}$ (iii) $\begin{pmatrix} -7 & 15 \\ 4 & -8 \end{pmatrix}$
(iv) $\begin{pmatrix} -58 & -45 \\ 32 & 24 \end{pmatrix}$

11

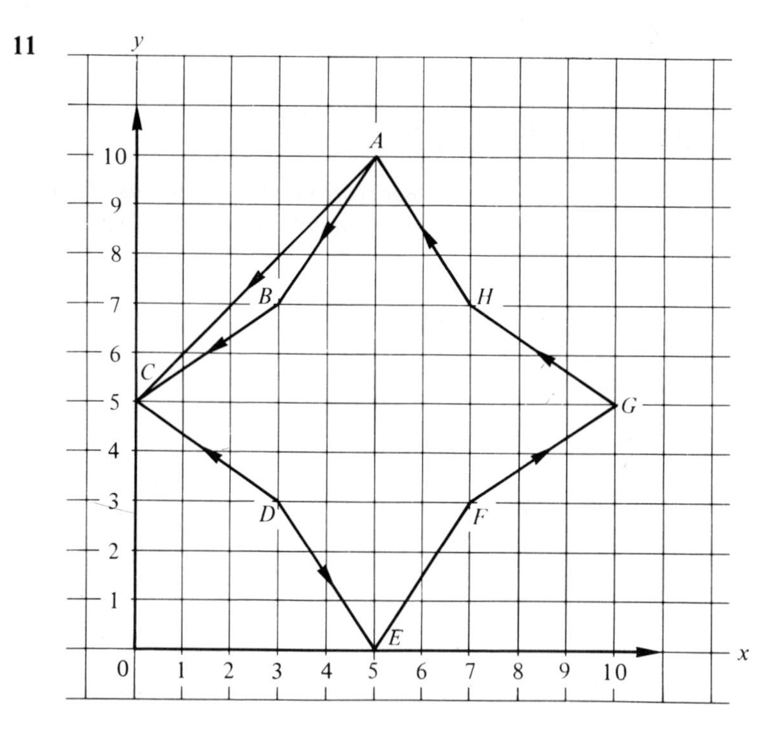

(i) Diagram above (ii) Octagon

(iii) $\vec{AB} = \begin{pmatrix} -2 \\ -3 \end{pmatrix}$, $\vec{BC} = \begin{pmatrix} -3 \\ -2 \end{pmatrix}$, $\vec{DE} = \begin{pmatrix} 2 \\ -3 \end{pmatrix}$,
$\vec{FG} = \begin{pmatrix} 3 \\ 2 \end{pmatrix}$, $\vec{HA} = \begin{pmatrix} -2 \\ 3 \end{pmatrix}$

(iv) (*a*) $\vec{BA} = \vec{EF} = \begin{pmatrix} 2 \\ 3 \end{pmatrix}$ (*b*) $\vec{GF} = \vec{BC} = \begin{pmatrix} -3 \\ -2 \end{pmatrix}$

(v) (*a*) $\vec{AB} + \vec{BC} = \begin{pmatrix} -5 \\ -5 \end{pmatrix}$ (*b*) $\vec{AC}$

12 (i) $1\frac{7}{18}$ (ii) $\frac{5}{18}$ (iii) $\frac{2}{3}$ (iv) $1\frac{1}{2}$ (v) $\frac{5}{6}$ (vi) 0

13 (i)

x	−2	−1	0	1	2	3	4
y	9	4	1	0	1	4	9

$y = (x - 1)^2$

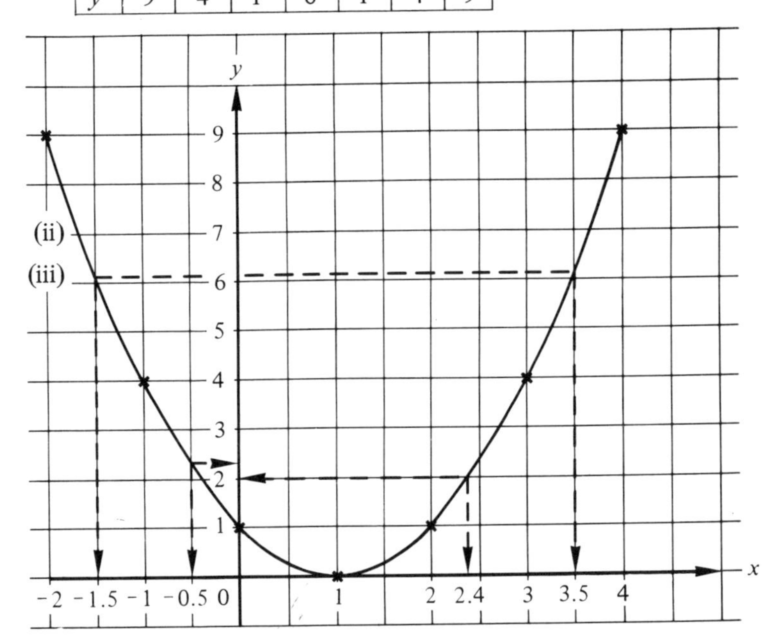

(iv) (*a*) $y = 2$ (*b*) $y = 2.2$ (v) $x = -1.5$ or $+3.5$

14 (i)

2nd die

+	1	2	3	4
1	2	3	4	5
2	3	4	5	6
3	4	5	6	7
4	5	6	7	8

1st die

(ii) (*a*) $\frac{1}{16}$ (*b*) $\frac{1}{4}$ (*c*) $\frac{1}{8}$

(*d*) $\frac{5}{8}$ (*e*) $\frac{3}{8}$ (*f*) $\frac{1}{4}$

(*g*) 1

Section 2

Set 1

1 (i) (*a*) Cuboid (*b*) 12 (*c*) 8 (ii) (*a*) 80 cm^2 (*b*) 42 cm^3
(iii) (*a*) G, Q (*b*) 6.95 cm

2 (i) 100 (ii) 9 (iii) 54 (iv) 64 (v) 15%

3 (i) (*a*) 0.000 002 56 (*b*) 2.56×10^{-6}
(ii) (*a*) 239 000 (*b*) 240 000
(iii) (*a*) 5.6×10^{-4} m^2 (*b*) 0.000 56 m^2 (*c*) 5.6 cm^2

4 (i) (*a*) $a = -4$ (ii) $x = 2\frac{1}{2}$ (iii) $x = -3$ or 0 (iv) $= \pm 6$
(v) $x = -1$ or 7

5 (i) £2400 (ii) £2040 (iii) £1785 (iv) 44%

6 (i) 100 (ii) 294 (iii) 2 (iv) 3 (v) 2.94 (vi) 36° (vii) 68%

7 (i) (*a*) $\begin{pmatrix} 9 & -4 \\ 0 & 8 \end{pmatrix}$ (*b*) $\begin{pmatrix} 0 & 11 \\ 3 & 2 \end{pmatrix}$ (*c*) $\begin{pmatrix} -1 & -2 \\ 1 & 2 \end{pmatrix}$

(ii) (*a*) B (*b*) $\begin{pmatrix} 2 & 5 \\ 1 & 3 \end{pmatrix}$ (*c*) $\begin{pmatrix} 1 & 0 \\ 0 & 1 \end{pmatrix}$

8 (i) 43.3 cm^2 (ii) 10.8 cm^2 (iii) 32.5 cm^2 (iv) 1300 cm^3
(v) $\frac{4}{1}$ (vi) 41.2 cm

9 (i) $x = -5$ or 3 (ii) $x = -2.35$ or 0.85

10 (i) 364 m (ii) 1060 m (iii) (*a*) 132.5 or 133 m/s (*b*) 478 km/h

11 (i) AB: $y = \frac{1}{2}x$, CD: $y = -\frac{1}{2}x$, EF: $x = 4$
(ii) $y \leqslant \frac{1}{2}x, y \geqslant -\frac{1}{2}x, x \leqslant 4$
(iii) 8 cm^2

12 (i)

x	-4	-3	-2	1	0	1	2	3	4
y	4	$2\frac{1}{4}$	1	$\frac{1}{4}$	0	$\frac{1}{4}$	1	$2\frac{1}{4}$	4

$y = \frac{1}{4}x^2$

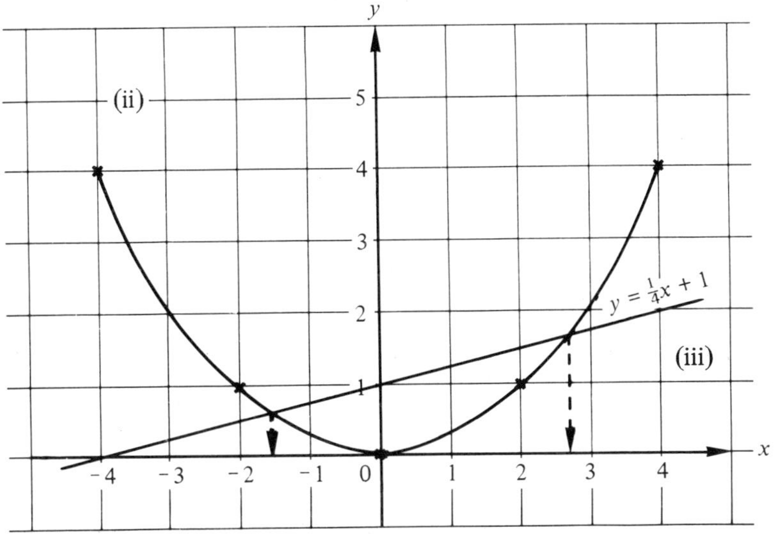

(iv) $x = -1.6, 2.6$

(v) $\frac{1}{4}x^2 - \frac{1}{4}x - 1 = 0$

or $x^2 - x - 4 = 0$

13 (i) 40° (ii) 5 m (iii) 9.4 m (iv) 4.925 m (v) 10 m

14

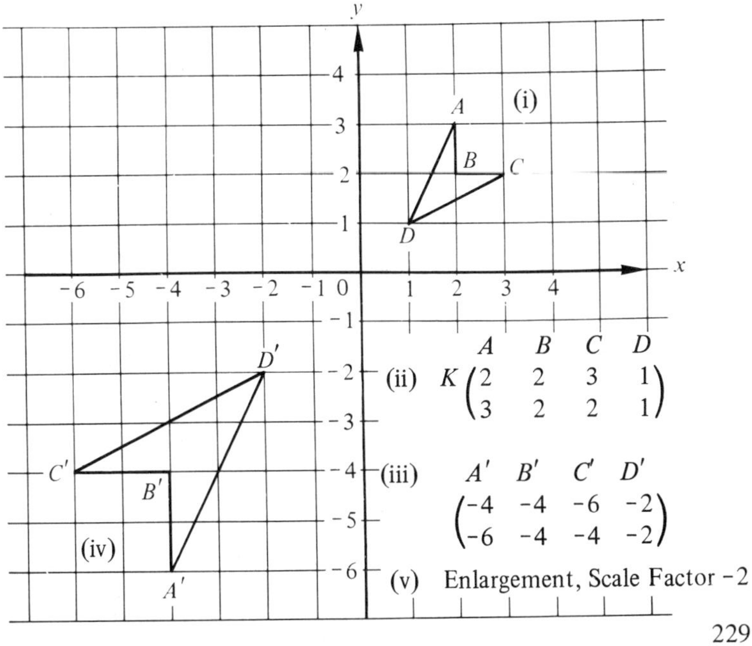

(ii) $K\begin{array}{c}\begin{array}{cccc}A & B & C & D\end{array}\\ \begin{pmatrix}2 & 2 & 3 & 1\\ 3 & 2 & 2 & 1\end{pmatrix}\end{array}$

(iii) $\begin{array}{c}\begin{array}{cccc}A' & B' & C' & D'\end{array}\\ \begin{pmatrix}-4 & -4 & -6 & -2\\ -6 & -4 & -4 & -2\end{pmatrix}\end{array}$

(v) Enlargement, Scale Factor -2

15 (i) H H H (*a*) $\frac{3}{8}$ (ii) (*a*) 13 31 51 71 91
H H T (*b*) $\frac{1}{8}$ 15 35 53 73 93
H T T 17 37 57 75 95
H T H 19 39 59 79 97
T H H (*b*) $\frac{3}{10}$
T H T
T T H
T T T

Set 2

1 (i) $4(x - 4)$ (ii) $a(a + b + c)$ (iii) $3(x + 2)(x - 2)$
(iv) $(p - 1)(p + 5)$ (v) $(1 + p)(1 - p)$

2 (i) (*a*) Rhombus (*b*) Isosceles trapezium (*c*) Tetrahedron
(ii) (*a*) 120° (*b*) 0.5 (*c*) 8.66 cm (*d*) 43.4 cm^2
(*e*) 173.2 or 173 cm^3

3 (i) (*a*) $\begin{pmatrix} 1 & 4 \\ 5 & -5 \end{pmatrix}$ (*b*) $\begin{pmatrix} 3 & -6 \\ 1 & 1 \end{pmatrix}$ (*c*) $\begin{pmatrix} -2 & 18 \\ -3 & 28 \end{pmatrix}$ (*d*) $\begin{pmatrix} 1 & -1 \\ -1\frac{1}{2} & 2 \end{pmatrix}$

4 (i) $x = 13$ (ii) $x = 2\frac{1}{2}$ (iii) $y = \pm 6$ (iv) $x = \pm 9$
(v) $x = -3$ or 5

5 (i) $1\,000\,001_2$ (ii) 55_8 (iii) 531_8 (iv) 54_{10} (v) $x = 5$

6

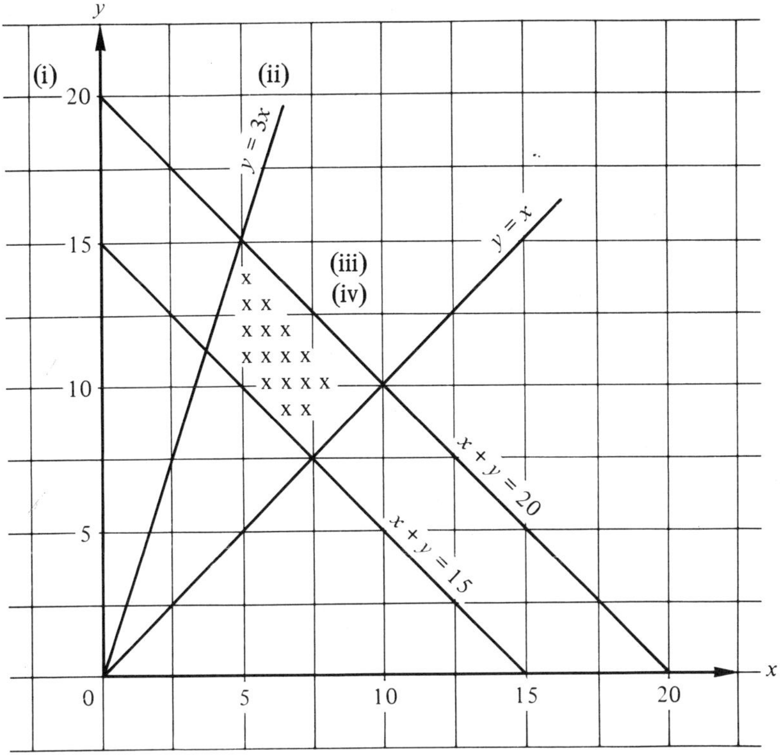

(v) 16 rectangles

7 (i) (*a*) 5.4×10^9 (*b*) 1.7×10^{-5} (ii) (*a*) 0.000 008 (*b*) 17 000

(iii) 1.5×10^4 (iv) (*a*) 6.0×10^{-3} by 4.0×10^{-3}

(*b*) (i) 2.4×10^{-5} km^2 (ii) 24 m^2

8 (i) 9 (ii) 12 (iii) 50 (iv) 5 (v) 21 (vi) 29

9 (i) (*a*) 68° (*b*) 56° (*c*) 34° (*d*) 34° (ii) (*a*) 2.48 (*b*) 12.4 cm

10

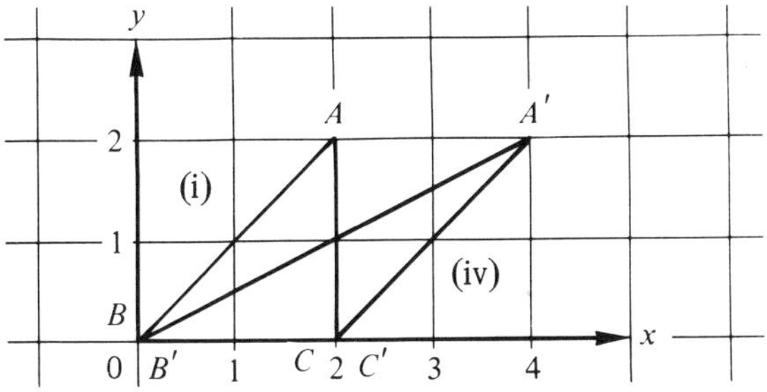

(ii) $K\begin{pmatrix} 2 & 0 & 2 \\ 2 & 0 & 0 \end{pmatrix}$ (iii) $\begin{pmatrix} 4 & 0 & 2 \\ 2 & 0 & 0 \end{pmatrix}$

(v) Shear, points on $y = 0$ invariable, Shear Factor 1.

11 (i) 0.5 (ii) 30° (iii) 10 m (iv) 1.33 (v) 53° (vi) 15.1 m

12 (i) $\begin{pmatrix} 90 \\ 81 \end{pmatrix}$ (ii) $\begin{pmatrix} 10 & -12 \\ -18 & 22 \end{pmatrix}$ (iii) $\begin{pmatrix} 8\frac{1}{4} & 2 \\ 17 & 0 \end{pmatrix}$ (iv) $a = 4, b = -6$

(v) $x = -2, y = -3, z = -6$

13 (i) 34.86 m^2 (ii) 3.64 m^2 (iii) 2.52 m^2 (iv) 6.16 m^2

(v) 28.7 m^2

14

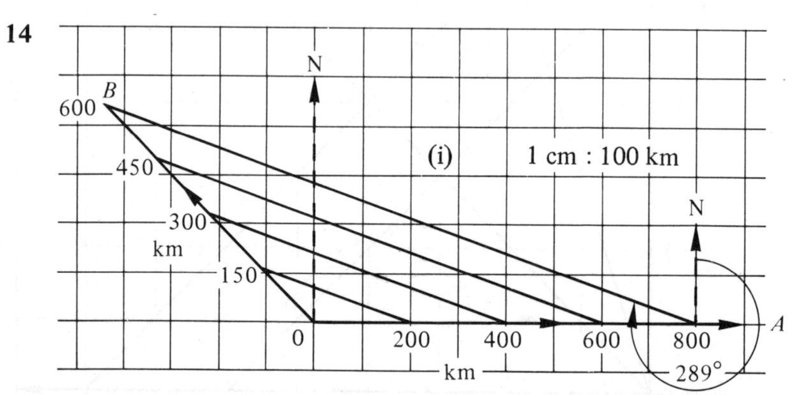

(ii) (*a*) 325 km (*b*) 650 km (*c*) 975 km (*d*) 1290 km

(iii) 289°

15 (i) **a** (ii) **b** (iii) 2**b** − 2**a** (iv) **b** (v) **a**

(vi) (*a*) *AC* is parallel to *DE*, *AC* = 2*DE* (vii) Parallelogram

(*b*) *AB* is parallel to *FE*, *AB* = 2*FE*

Set 3

1 (i) £2.01 (ii) 40 (iii) £25.00 (iv) $62\frac{1}{2}\%$ (v) £8.64

2 (i) (*a*) 4 cm (*b*) 2 cm (ii) (*a*) 2π cm (*b*) 2π cm

(*c*) 4π cm (*d*) 6π cm^2 (*e*) $24 - 6\pi$ cm^2

3 (i) $\begin{pmatrix} 3 & 4 & 10 & 3 \\ 4 & 6 & 9 & 4 \\ 6 & 9 & 4 & 2 \end{pmatrix}$ (ii) $\begin{pmatrix} 12 \\ 6 \\ 3 \\ -2 \end{pmatrix}$ (iii) $\begin{pmatrix} 84 \\ 103 \\ 134 \end{pmatrix}$

4 (i) $x = 1, y = 3$ (ii) $x = -4$ or $\frac{1}{2}$

5 (i) 2**b** (ii) (*a*) **a** + 2**b** (*b*) 2(**a** + 2**b**) or 2**a** + 4**b** (*c*) **a** + 4**b**

6 (i) (*a*) 51°N (*b*) 20°S (ii) (*a*) 3060 n.m. (*b*) 1200 n.m.
(*c*) 45°N (*d*) 37°S (*e*) 21 600 n.m.

7 (i) (*a*) $\frac{x-9}{2}$ (*b*) $\frac{1}{x(x+1)}$ (ii) $x = 1$ or 4

8 (i) (*a*) 64 (*b*) 64 (*c*) 144 (*d*) 1296 (ii) $k = \pm 5$ (iii) $k = \pm 3$

9 (i) (*a*) 138° (*b*) 69° (*c*) 42° (*d*) 69° (ii) (*a*) 2.61
(*b*) 15.66 cm

10

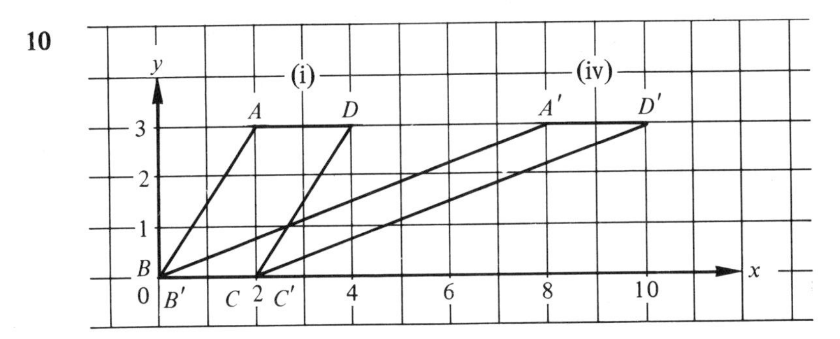

(ii) $K = \begin{pmatrix} 2 & 0 & 2 & 4 \\ 3 & 0 & 0 & 3 \end{pmatrix}$ (iii) $\begin{pmatrix} 8 & 0 & 2 & 10 \\ 3 & 0 & 0 & 3 \end{pmatrix}$

(v) Shear, points on $y = 0$ invariable, Shear Factor 2.

(vi) Parallel lines, areas, points on $y = 0$

11 125.1°

12 (i) $\begin{pmatrix} 16 & -10 & -1 \\ 6 & 11 & 12 \end{pmatrix}$ (ii) (*a*) $\begin{pmatrix} 0 & -6 \\ -4 & -2 \end{pmatrix}$ (*b*) $\begin{pmatrix} 1 & -3 \\ 1 & -3 \end{pmatrix}$

(*c*) $\begin{pmatrix} -2 & 8 \\ 7 & 1 \end{pmatrix}$

(iii) $b = -6, c = -8$

13

1 cm : 100 km

N

N

(i)

225°

720 km

074°

Base

180°

980 km

500 km

Plane

(ii) (*a*) 500 km

(*b*) 180°

14

$y = 2^x$

(i)

x	0	1	2	3	4	5	6
y	1	2	4	8	16	32	64

(ii) Graph

(iii) (*a*) 5 (*b*) 12 (*c*) 26

(iv) (*a*) $2^{3.3}$ (*b*) $2^{4.5}$ (*c*) $2^{5.5}$

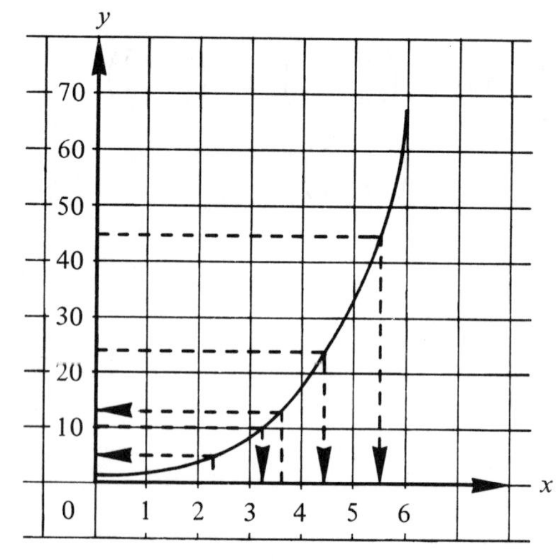

Powers of 2

15 (i) $\mathbf{a}+\mathbf{b}$ (ii) $\frac{3}{4}(\mathbf{a}+\mathbf{b})$ (iii) $\frac{1}{4}(\mathbf{a}+\mathbf{b})$ (iv) $\mathbf{b}-3\mathbf{a}$
(v) $\frac{3}{4}\mathbf{a}-\frac{1}{4}\mathbf{b}=\frac{1}{4}(3\mathbf{a}-\mathbf{b})$ (vi) $2\frac{1}{4}\mathbf{a}-\frac{3}{4}\mathbf{b}=\frac{3}{4}(3\mathbf{a}-\mathbf{b})$
(vii) (*a*) The points are colinear (*b*) $\frac{1}{3}$

Set 4

1 (i) $2(a+2b+3c)$ (ii) $x(x-1)$ (iii) $(a+b)(a-b)$
(iv) $(a+2)(a+13)$ (v) $(5k+2)(k-1)$

2 (i) 120° (ii) 150° (iii) 15° (iv) 43.3 cm^2 (v) 259.8 cm^2

3 (i) {4,8,12} (ii) {3,5,7,11} (iii) {3,4,6,9,12}
(iv) {3,4,5,6} (v) {7,8,9,10,11,12} (vi) {4,12}

4 (i) 6 (ii) 12 (iii) $\frac{1}{3}$

5 (i) 523 cm^3 (ii) 314 cm^2

6 (i) (*a*) $\dfrac{x-y}{6x-y}$ (*b*) $\dfrac{-2}{(x+1)(x-1)}$ or $\dfrac{-2}{(x^2-1)}$
(ii) (*a*) $x=3$ or 5 (*b*) $x=-2$ or 3

7 (i) 9 cm (ii) 120° (iii) 16.5 cm

8

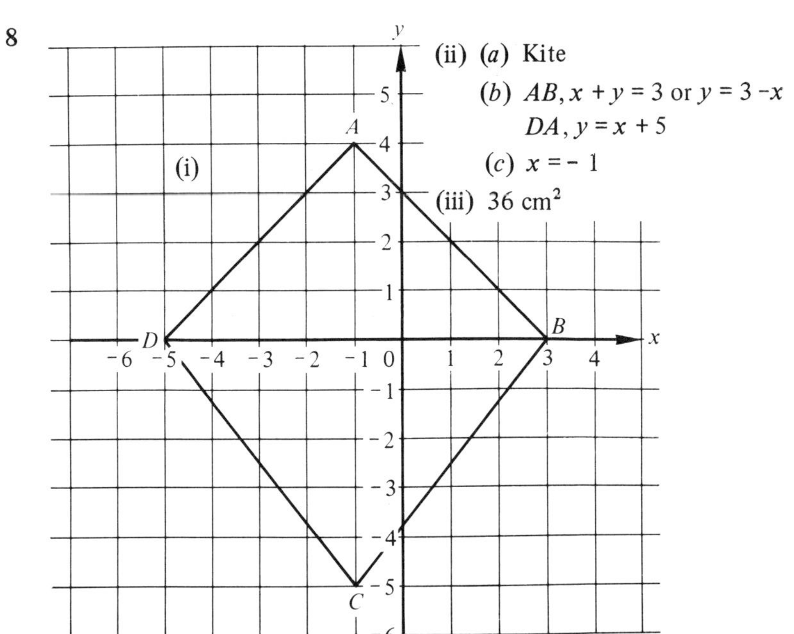

(ii) (*a*) Kite
(*b*) AB, $x+y=3$ or $y=3-x$
DA, $y=x+5$
(*c*) $x=-1$
(iii) 36 cm^2

9

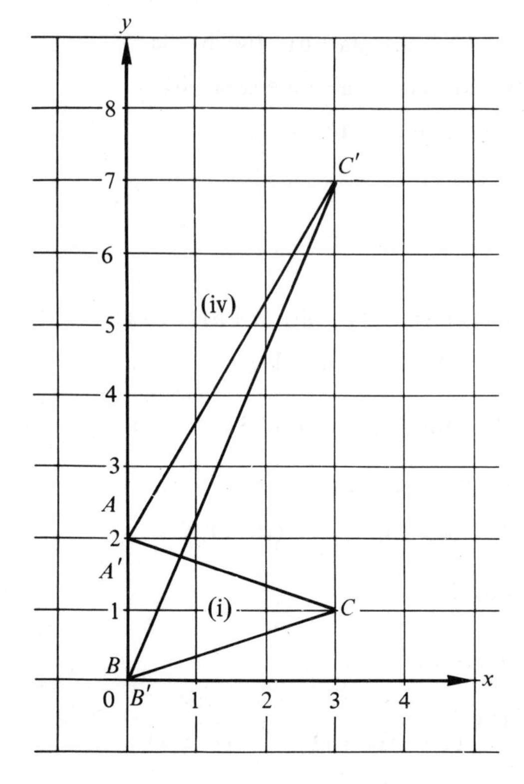

(ii) $M = \begin{pmatrix} 0 & 0 & 3 \\ 2 & 0 & 1 \end{pmatrix}$ (iii) $\begin{pmatrix} 0 & 0 & 3 \\ 2 & 0 & 7 \end{pmatrix}$

(v) Shear, points on $x = 0$ invariable, Shear Factor 2.

(vi) $ABC = A'B'C' = 3\text{ cm}^2$

10 (i) 9 cm (ii) 4.37 cm (iii) 42° (iv) 42° (v) 7.43 cm

11 (i) $a = 10, b = 2$ (ii) $A^{-1} = \begin{pmatrix} -2 & 3 \\ 5 & -7 \end{pmatrix}$ (iii) $x = 2, y = -1$

12 (i) 45 (ii) 75 (iii) 150 (iv) 45% (v) 60 (vi) 34 (vii) 26

13 (i) 25 cm (ii) 70 m (iii) 4 cm (iv) 10 cm (v) 500 m^2

14 (i) $\mathbf{a}$ (ii) $\frac{2}{3}\mathbf{a}$ (iii) $\mathbf{b}$ (iv) $-\frac{2}{3}\mathbf{b}$ (v) $\mathbf{a} - \mathbf{b}$ (vi) $\frac{2}{3}\mathbf{a} - \frac{2}{3}\mathbf{b}$
(vii) $\frac{3}{2}$

Set 5

1 (i) £2.38 (ii) 47% (iii) £2.40 (iv) 50% (v) £168

2 (i) 24 cm (ii) 2.4 (iii) 134.8° (iv) 120 cm^2 (v) 6000 cm^3

3 (i) $\begin{pmatrix} 2 & 2 \\ -1 & 5 \end{pmatrix}$ (ii) $\begin{pmatrix} -6 & 4 \\ 3 & 3 \end{pmatrix}$ (iii) $\begin{pmatrix} 6 & 6 \\ -3 & 15 \end{pmatrix}$ (iv) $\begin{pmatrix} \frac{1}{2} & \frac{1}{2} \\ 1 & 2 \end{pmatrix}$

(v) $\begin{pmatrix} 4 & -1 \\ -2 & 1 \end{pmatrix}$

4 (i) $a = 3, b = -1$ (ii) $x = -\frac{1}{3}$ or 2

5 (i) (*a*) 30° (*b*) 60° (*c*) 17.3 cm (ii) (*a*) 10.5 cm (*b*) 52.3 cm^2

6 (i) 6000 cm^3 (ii) 4710 cm^3 (iii) 1290 cm^3 (iv) 21.5%

7 (i) 16 hours (ii) 2200 hours (iii) 56 litres (iv) £14
(v) 1 hour 30 minutes

8 (i) 25% (ii) $\frac{19}{180}$ (iii) 4 (iv) 100_2 (v) 11

9 (i) (*a*) 74° (*b*) 148° (*c*) 74° (ii) 3.49 (iii) 13.96 cm
(iv) 27.92 cm^2 (v) 55.84 cm^2

10

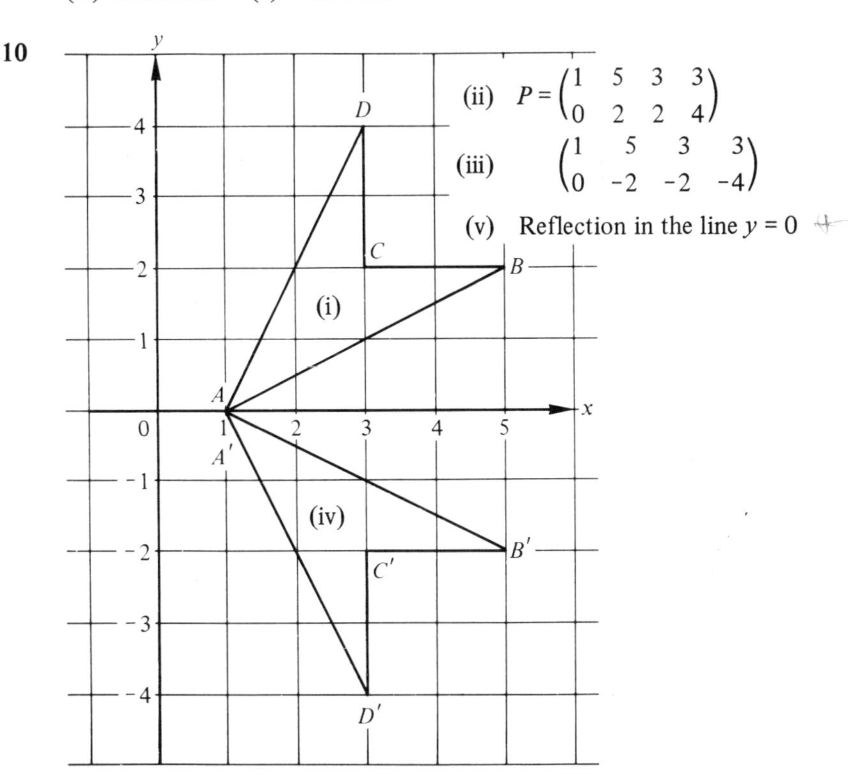

(ii) $P = \begin{pmatrix} 1 & 5 & 3 & 3 \\ 0 & 2 & 2 & 4 \end{pmatrix}$

(iii) $\begin{pmatrix} 1 & 5 & 3 & 3 \\ 0 & -2 & -2 & -4 \end{pmatrix}$

(v) Reflection in the line $y = 0$

11 (i) 36.9° (ii) 12 cm^2 (iii) 96.9° (iv) 9.17 cm

12 (i) (*a*) £36 (*b*) £198 (*c*) £324 (ii) 2.4 km (iii) 1000 litres

13 (i) 1512 hours (ii) 50 km/hr (iii) 1330 hours (iv) 120 km/h
(v) 20 km/h (vi) 1348 hours (vii) 24 km (viii) 1417 or 1418 hours
(ix) 1430 hours (x) 75 mph

14 (i) $\mathbf{b}$ (ii) $\frac{1}{4}\mathbf{b}$ (iii) $-\mathbf{a}$ (iv) $-\frac{1}{4}\mathbf{a}$ (v) $\mathbf{b} - \mathbf{a}$ (vi) $\frac{1}{4}\mathbf{b} - \frac{1}{4}\mathbf{a}$
(vii) $\frac{1}{4}$

15 (i)

x	1	2	3	4	6	8	12	24
y	24	12	8	6	4	3	2	1

$xy = 24$

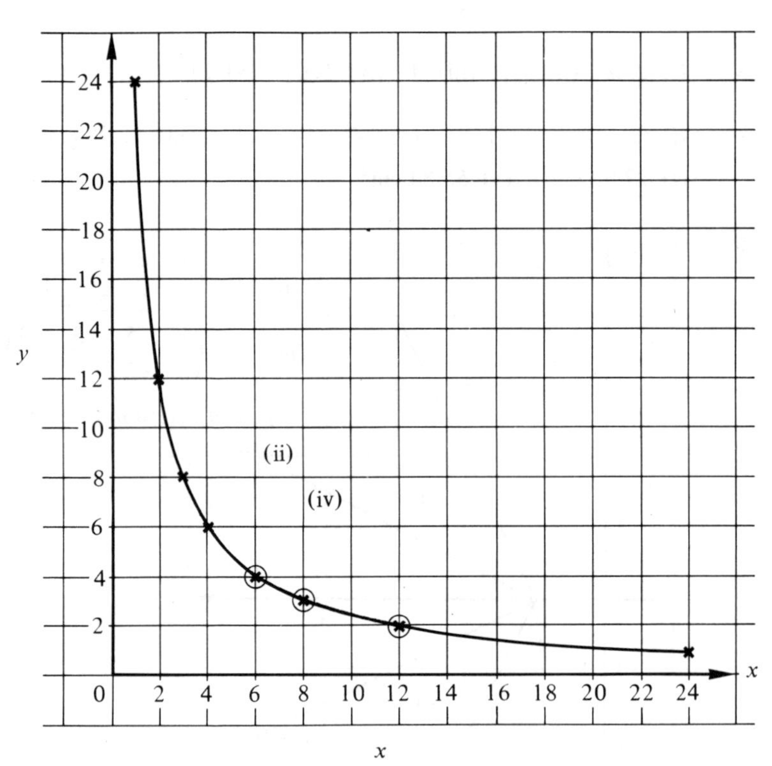

(iii) $y = \dfrac{24}{x}$ or $x = \dfrac{24}{y}$

(v) (6,4), (8,3), (12,2)

Set 6

1 (i) $5(a + 3b + 2)$ (ii) $(3a + 5b)(3a - 5b)$ (iii) $3(x + 4)(x - 4)$
(iv) $(y + 3)(y + 5)$ (v) $(3a + b)(3a - 2b)$

2 (i) (*a*) Triangular prism (*b*) 5 cm
(ii) (*a*) 4 cm (*b*) 6 cm^2 (*c*) 108 cm^2 (*d*) 48 cm^3 (*e*) 36.9°

3

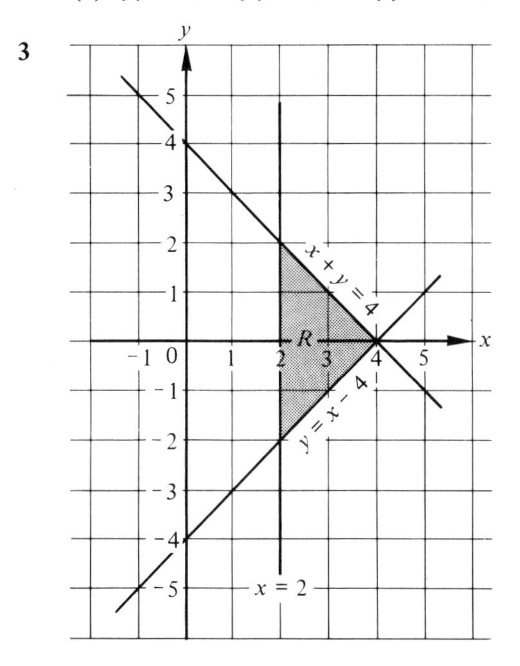

4 (i) $x = 8$ (ii) $x = 24$ (iii) $a = 2$ (iv) $x = -2$ or 10
(v) $x = -\frac{1}{2}$ or 3

5 (i) 65° (ii) $57\frac{1}{2}$° (iii) $122\frac{1}{2}$°

6 (i) 314 m^2 (ii) 628 m^3 (iii) 785 m^3 (iv) 1413 m^3 (v) 12.5 m
(vi) (*a*) 392.5 m^2 (*b*) 125.6 m^2

7 (i) 10^{13} (ii) 10^{11} (iii) 10^1 (iv) 10^{15} (v) 10^4 (vi) 10^6

8 (i) £1.20 (ii) £30 (iii) (*a*) $\frac{4}{25}$ (*b*) $\frac{8}{125}$

9 (i) Regular pentagon (ii) (*a*) 72° (*b*) 54° (*c*) 108°
(iii) 6 right angles (iv) 2.94 cm (v) 4.045 cm (vi) 29.4 cm
(vii) 59.4 cm^2

10

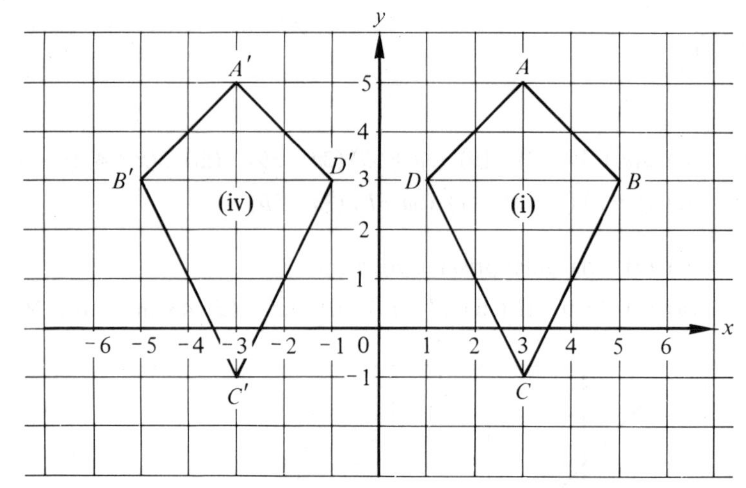

(ii) $P = \begin{pmatrix} 3 & 5 & 3 & 1 \\ 5 & 3 & -1 & 3 \end{pmatrix}$ (iii) $\begin{pmatrix} -3 & -5 & -3 & -1 \\ 5 & 3 & -1 & 3 \end{pmatrix}$

(v) Reflection in the line $x = 0$

11 (i) 15 cm (ii) 62° (iii) 31° (iv) 4.808 or 4.81 cm

12 (i) $20\frac{1}{4}$ (ii) 5 (iii) $(x + 4)(x + 4)$ (iv) 1,2 (v) −1,5

13 (i)

x	−2	−1	0	1	2	3	4
y	6	1	−2	−3	−2	1	6

$y = x^2 - 2x - 2$

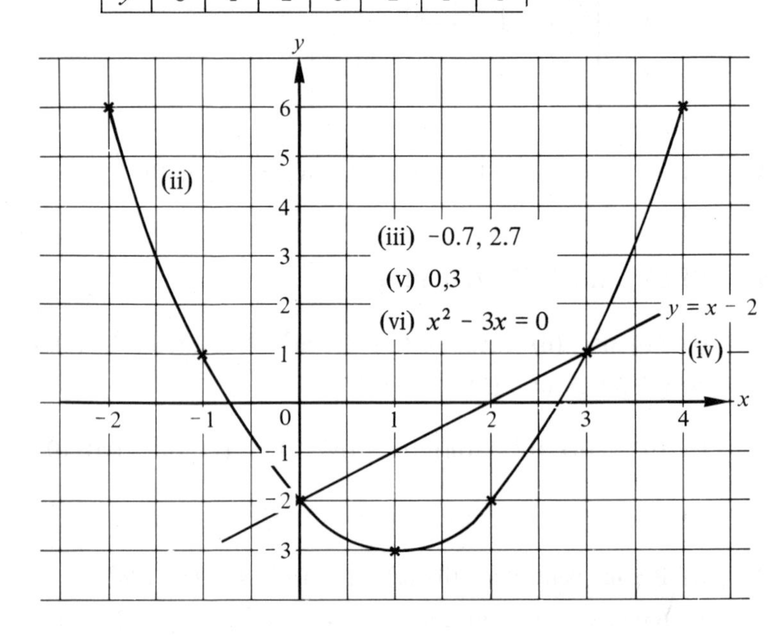

14 (i) (*a*) $\mathbf{b} - 2\mathbf{a}$ (*b*) $\mathbf{a} + \mathbf{b}$ (*c*) $\frac{1}{2}(\mathbf{a} + \mathbf{b})$ (*d*) $\frac{1}{2}(\mathbf{b} - 2\mathbf{a})$
(*e*) $\frac{1}{2}\mathbf{a}$
(ii) XY is parallel to CD, $XY = \frac{1}{2}CD$

15 (i) (*a*) 10π cm^2 (*b*) 18π cm (*c*) 2500π cm^3 (*d*) 250π cm^3
(ii) £1.55
(iii) (*a*) 1 : 8 (*b*) £124

Set 7

1 (i) £5.52 (ii) 64% (iii) 192 (iv) £1600 (v) £41.00

2 (i) 1728 cm^3 (ii) 36 cm^3 (iii) $\frac{1}{48}$

3 (i) (*a*) $\begin{pmatrix} 21 & -21 \\ -28 & 28 \end{pmatrix}$ (*b*) The determinant = 0 (ii) $\begin{pmatrix} 14 & 3 \\ -10 & 5 \end{pmatrix}$

4 (i) $x^2 + 12 + x^2 + 2 + x^2 + 12 + x^2 + 2 = 92$
or $4x^2 + 28 = 92$
(ii) (*a*) 28 cm (*b*) 18 cm (*c*) 504 cm^2

5 (i) 30° (ii) 60° (iii) 4.8 cm (iv) 3.4 cm

6 (i) 16.74 cm^3 (ii) 12.56 cm^3 (iii) 29.3 cm^3 (iv) 322.3 g

7 (i) $4x^2$ m^2 (ii) πx^2 m^2 (iii) $4x^2 + \pi x^2$m^2 (iv) $x^2(4 + \pi)$
(v) 714 m^2

8 (i) A: £4000, B: £6000, C: £8000 (ii) 80°, 120°, 160°.

9 (i) (*a*) $(-4,1)$ (*b*) $(0,-3)$
(ii) (*a*) $y = \frac{1}{2}x + 3$ (*b*) $y = \frac{1}{2}x - 3$ (*c*) $y = -x$ (*d*) $y = -x - 3$
or $x + y = -3$
(iii) $x \geqslant 0, y \leqslant -x, y \geqslant \frac{1}{2}x - 3$
(iv) $y = \frac{1}{2}x$ (v) $y = -\frac{1}{2}x - 1$

10

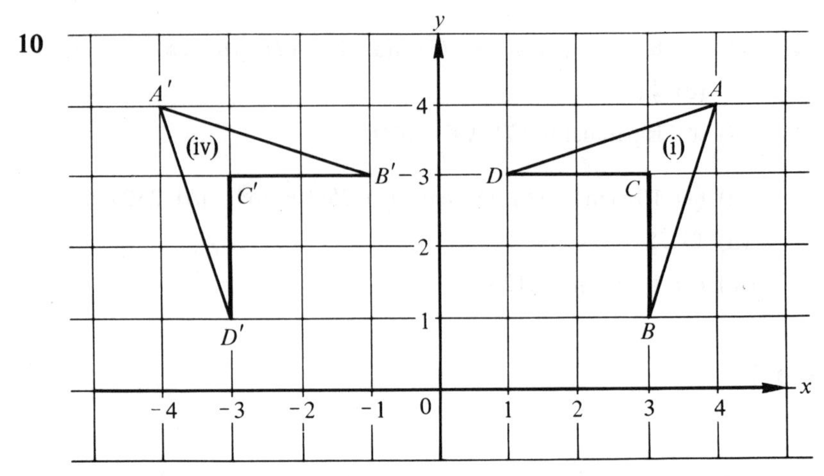

(ii) $Q = \begin{pmatrix} 4 & 3 & 3 & 1 \\ 4 & 1 & 3 & 3 \end{pmatrix}$

(iii) $\begin{pmatrix} -4 & -1 & -3 & -3 \\ 4 & 3 & 3 & 1 \end{pmatrix}$

(v) Rotation of +90° about 0.

11 (i) 132.8° (ii) 18.35 cm

12 (i) (*a*) 053° (*b*) 143° (*c*) 225° (*d*) 117° (*e*) 260°
(ii) (*a*) 66 or 67 km (*b*) 91 km (*c*) 110 km
(iii) 3 hours 40 minutes (iv) 40 km/h

13 (i) $x = 2, y = 5$ (ii) $x = -\frac{1}{3}$ or 2

14 (i) (*a*) **b** (*b*) 2**b** (*c*) −**b** (*d*) 2**b** (ii) (*a*) **b** − 3**a** (*b*) **b** − 3**a**
(iii) (*a*) *BX* and *YD* are equal and parallel (*b*) *BY* = *XD*
(iv) Parallelogram

15 (i) (*a*) $12x$ cm (*b*) x^2 cm² (*c*) $6x^2$ cm² (*d*) x^3 cm³
(*e*) $\sqrt{2x^2}$ or $x\sqrt{2}$ cm (*f*) $\sqrt{x^2 + 2x^2}$ or $\sqrt{3x^2}$ or $x\sqrt{3}$ cm (ii) (*a*) 9 cm (*b*) 729 cm³

Set 8

1 (i) $6(x + 2y + 1)$ (ii) $x^2(x + 1)$ (iii) $(x + 12)(x - 12)$
(iv) $(a - 1)(a - 11)$ (v) $6(x + 2)(x + 2)$ or $6(x + 2)^2$

2 (i) 150° (ii) 2.6 cm (iii) 3.9 cm^2 (iv) 24.6 cm^2
(v) 9 or 9.04 cm^2 (vi) 2.25 or 2.26 cm^2 (vii) 2.12 cm

3 (i) (*a*) 29 640 (*b*) 4446 (ii) 40%

4 (i) $x = 0, y = 4$ (ii) $x = 0.47$ or 2.14

5 (i) (*a*) $\mathbf{z} - \mathbf{y}$ (*b*) $\mathbf{z} - \mathbf{y} - \mathbf{x}$ (*c*) $\mathbf{x} + \mathbf{y} - \mathbf{z}$
(ii) $\frac{1}{3}\mathbf{p} - \frac{1}{3}\mathbf{q}$ or $\frac{1}{3}(\mathbf{p} - \mathbf{q})$

6 (i) $2r$ (ii) $V = \pi r^2 \times 2r$ (iii) $2\pi r^3$ (iv) $\frac{2}{3}$ (v) Equal areas

7 (i) 15.7 (ii) 47.1 (iii) $r = \dfrac{C}{2\pi}$ (iv) $r = 10$ (v) $r = 14$

8 (i) (*a*) 36 (*b*) 21 (*c*) 58 (ii) (*a*) 64 (*b*) $\frac{1}{64}$ (*c*) 64

9 (i) $R = 1.3$ cm (ii) πR^2 (iii) $r = 1$ cm (iv) πr^2
(v) $\pi R^2 - \pi r^2$ (vi) (*a*) $\pi(R^2 - r^2)$ (*b*) $\pi(R + r)(R - r)$
(vii) 2.17 cm^2

10

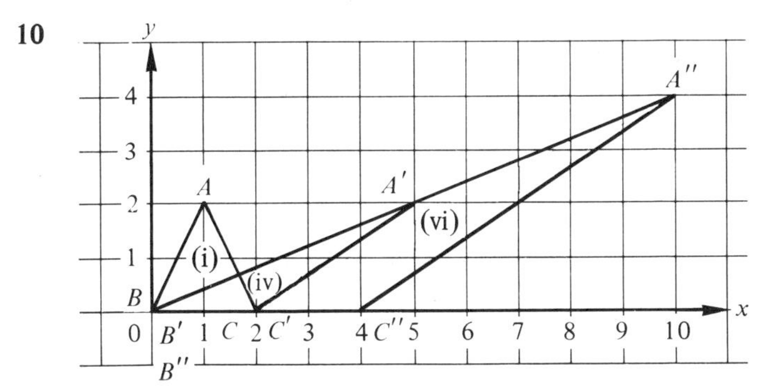

(ii) $M = \begin{pmatrix} 1 & 0 & 2 \\ 2 & 0 & 0 \end{pmatrix}$ (iii) $\begin{pmatrix} 5 & 0 & 2 \\ 2 & 0 & 0 \end{pmatrix}$

(v) $\begin{matrix} A'' & B'' & C'' \end{matrix}$ $\begin{pmatrix} 10 & 0 & 4 \\ 4 & 0 & 0 \end{pmatrix}$ (vii) $\begin{pmatrix} 2 & 4 \\ 0 & 2 \end{pmatrix}$

11 (i) 13 cm (ii) 120°

12 (i) (*a*) $\mathbf{a} + \frac{1}{2}\mathbf{b}$ (*b*) $2\mathbf{a} + \mathbf{b}$ (*c*) $\mathbf{b} + \frac{1}{2}\mathbf{a}$ (*d*) $2\mathbf{b} + \mathbf{a}$ (*e*) $\mathbf{b} - \mathbf{a}$
(*f*) $\mathbf{b} - \mathbf{a}$
(ii) *CA* and *GH* are equal and parallel.

13

14 (i) $x^2 = (x - 9)^2 + 15^2$ (ii) $x = 17$ (iii) $AB = 17, AC = 8$ cm

15 (i)

$y = \frac{1}{2}x^2 - 2$

x	-4	-3	-2	-1	0	1	2	3	4
y	6	$2\frac{1}{2}$	0	$-1\frac{1}{2}$	-2	$-1\frac{1}{2}$	0	$2\frac{1}{2}$	6

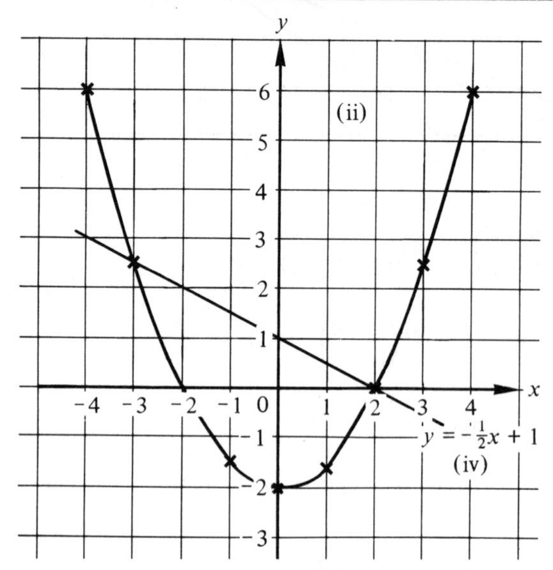

(iii) $x = -2, 2$ (v) $x = -3, 2$

(vi) $\frac{1}{2}x^2 + \frac{1}{2}x - 3 = 0$ or $x^2 + x - 6 = 0$

Set 9

1 (i) £3.75 (ii) $233\frac{1}{3}\%$ (iii) £20.00 (iv) £14.85 (v) 16%

2 (i) $A = \pi R^2 - \pi r^2$ (ii) $\pi(R + r)(R - r)$ (iii) 62.8 (iv) 942 cm^3

3 (i) 3.74, 6.48 cm

4 (i) 9 (ii) ± 2 (iii) $2\frac{1}{2}$ (iv) $-2\frac{1}{2}$ or 6 (v) 2 or 9

5 (i) $\frac{1}{2}$ (ii) $x = 11$ (iii) 15

6 (i) 15 cm (ii) 12 cm (iii) 1152 cm^3 (iv) 1.33 (v) 53°

7 (i) (*a*) $x = b - a$ (*b*) $x = \dfrac{b}{a}$ (*c*) $x = \dfrac{ab}{c}$ (*d*) $x = \dfrac{4h}{3}$

(ii) (*a*) $v = 8$ (*b*) $s = \dfrac{v^2 - u^2}{2f}$ (*c*) $s = 4$

8 (i)

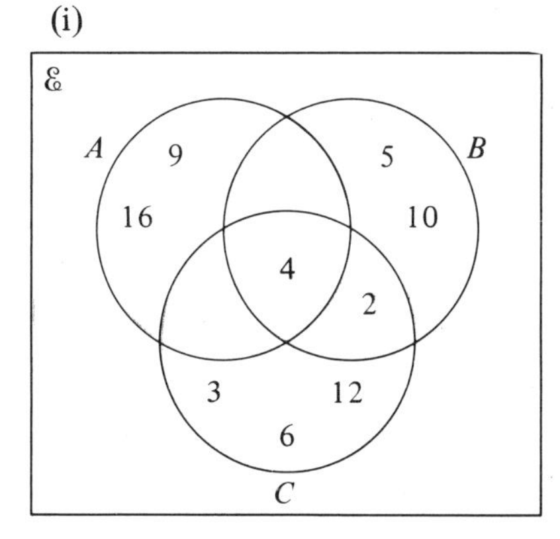

(ii) 1

(iii) $\frac{2}{5}$

9 (i) (*a*) 30° (*b*) 120° (*c*) 150°

(ii) (*a*) 0.866 (*b*) 0.866 (*c*) 0.5 (*d*) -0.5

(iii) 7.85 cm (iv) 25.7 cm (v) 39.25 cm^2 (vi) 13.08 cm^2

10

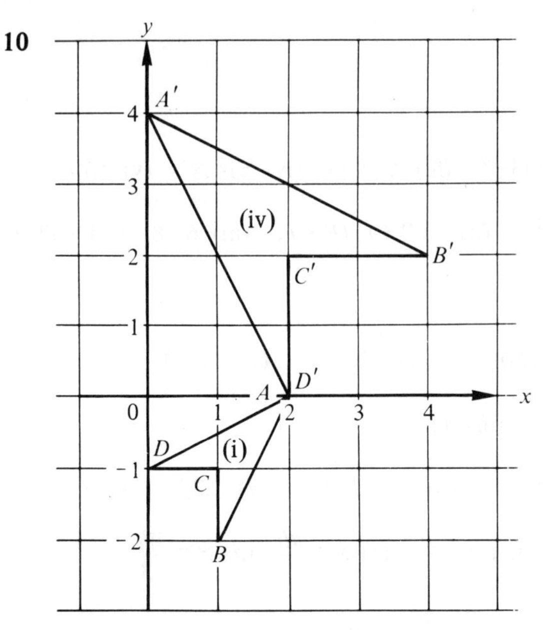

(ii) $K = \begin{pmatrix} 2 & 1 & 1 & 0 \\ 0 & -2 & -1 & -1 \end{pmatrix}$

(iii) $\begin{pmatrix} 0 & 4 & 2 & 2 \\ 4 & 2 & 2 & 0 \end{pmatrix}$

(v) Rotation +90° about 0.
Enlargement, Scale
Factor 2, centre 0.

(vi) 1 : 4 or $\frac{1}{4}$

11 (i) 31.4 cm² (ii) 29.4 cm² (iii) 2 cm²

12 (i) (*a*) $24x$ (*b*) $3x^2$ (*c*) $2x^2$ (*d*) $6x^2$ (*e*) $22x^2$ (*f*) $6x^3$
(*g*) $\sqrt{13x^2}$ or $x\sqrt{13}$ (*h*) $\sqrt{14x^2}$ or $x\sqrt{14}$
(ii) (*a*) $x = 4$ cm (*b*) 352 cm²

13 (i) (*a*) $6\mathbf{a} - 2\mathbf{b}$ (*b*) $\frac{1}{4}(6\mathbf{a} - 2\mathbf{b}) = 1\frac{1}{2}\mathbf{a} - \frac{1}{2}\mathbf{b}$
(*c*) $1\frac{1}{2}(\mathbf{a} + \mathbf{b})$ or $1\frac{1}{2}\mathbf{a} + 1\frac{1}{2}\mathbf{b}$ (*d*) $2(\mathbf{a} + \mathbf{b})$ or $2\mathbf{a} + 2\mathbf{b}$
(ii) (*a*) *PT* is parallel to *QR* (*b*) $\dfrac{PT}{QR} = \dfrac{3}{4}$

14

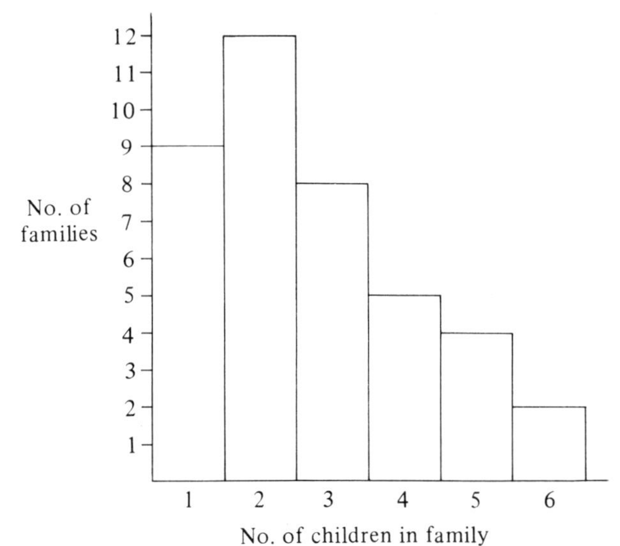

(i) 40

(ii) 109

(iii) 2

(iv) 2.7

(v) (*a*) $12\frac{1}{2}\%$ (*b*) $22\frac{1}{2}\%$

(vi) The mode would probably have been 6, as large families were commonplace in those days.

15 (i) (*a*) $7.9 < CX < 9.6$ cm (*b*) 8,9 cm

(ii) (*a*) $3x + 1 = 2(x + 2)$ (*b*) $x = 3$ (*c*) $AB = 5, AC = 10$ cm

(*d*) 8.66 cm

Set 10

1 (i) $x(x^2 + x + 1)$ (ii) $(3 + x)(3 - x)$ (iii) $3(2x + 3y)(2x - 3y)$
(iv) $(a - 3)(a + 7)$ (v) $(2x - 3)(x + 2)$

2 (i) 44 cm (ii) 7 cm (iii) 154 cm^2 (iv) 836 cm^2 (v) 1848 cm^3
(vi) 616 cm^3

3 (i) (*a*) £360 (*b*) £216 (*c*) £153 (ii) £1071 (iii) (*a*) £90
(*b*) 5%

4 (i) $y = 2x - 3$

x	4	0	−1
y	5	−3	−5

$y = -\frac{1}{2}x + 2$

x	4	0	−2
y	0	2	3

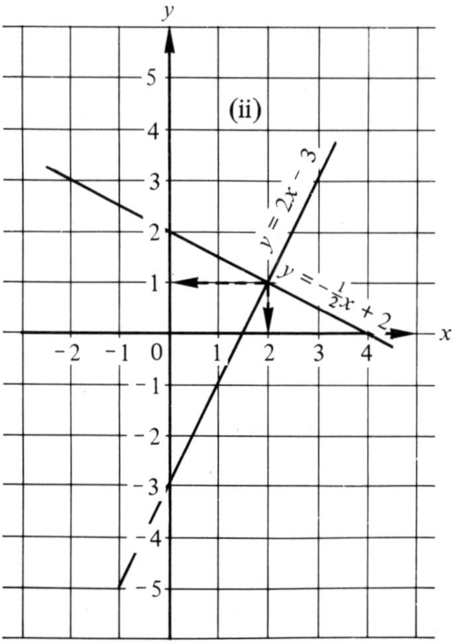

(ii)

(iii) $x = 2, y = 1$

5 (i) 513_{10} (ii) 101_8 (iii) 452_{10} (iv) $1\,000\,000_2$ cm^2 (v) 3
(vi) $x = 8, y = 2$

6 (i) (*a*) Cuboid (*b*) 90 cm^3 (ii) (*a*) Triangular prism
(*b*) 51 cm^3
(iii) (*a*) Hemisphere (*b*) 2093.3 cm^3 (iv) (*a*) Cone (*b*) 314 cm^3

7 (i) (*a*) $x = 4p$ (*b*) $x = \dfrac{b+c}{2}$ (*c*) $x = \dfrac{c - ab}{a}$ (*d*) $x = \dfrac{c}{a+b}$

(ii) (*a*) 1760 (*b*) $r = \dfrac{A}{2\pi h}$ (*c*) $r = 7$

8 (i) £9000, £15 000, £21 000 (ii) 72°, 120°, 168° (iii) 3 : 5

9

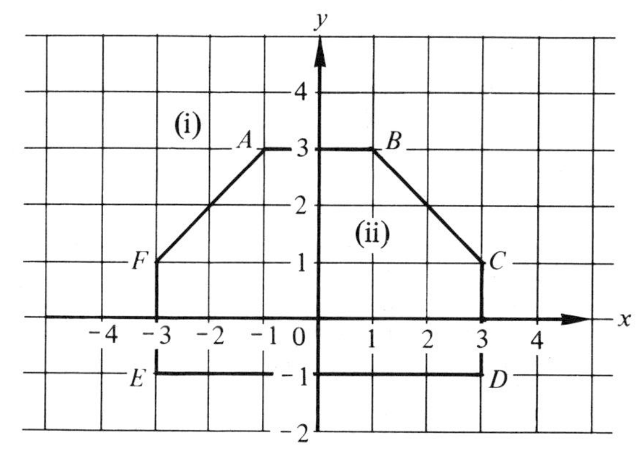

(iii) Reflection in $y = -x$

(iv) (*a*) $x + y = 4$
(*b*) $y = -1$
(*c*) $y = x + 4$

(v) $y = \frac{1}{3}x$

(vi) $-\frac{1}{3}$

(vii) Isosceles trapezium

10

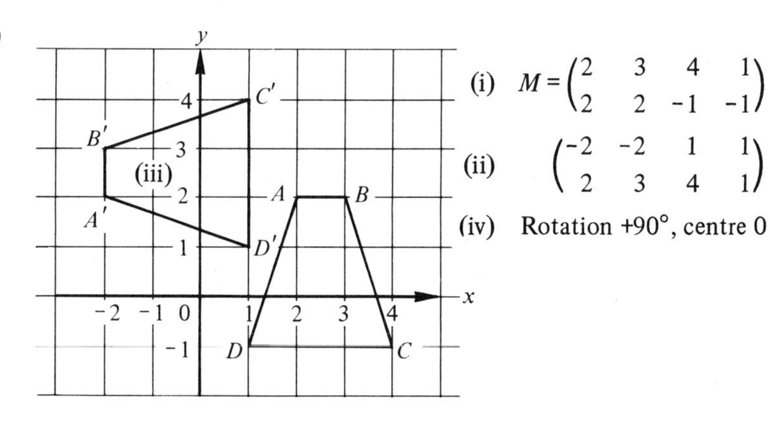

(i) $M = \begin{pmatrix} 2 & 3 & 4 & 1 \\ 2 & 2 & -1 & -1 \end{pmatrix}$

(ii) $\begin{pmatrix} -2 & -2 & 1 & 1 \\ 2 & 3 & 4 & 1 \end{pmatrix}$

(iv) Rotation +90°, centre 0

11 (i) 73.7° or 73.8° (ii) 36.9° or 36.85°

12 (i) $(4x + 1)^2 = (x + 2)^2 + (4x)^2$ (ii) (3,4,5) and (5,12,13)

13 (i) (*a*) 204° (*b*) 314° (*c*) 081° (ii) (*a*) (i) 201° (ii) 109° (iii) 289° (*b*) 065°

14 (i) 13 m (ii) 22.6° (iii) 16.76 or 16.8 m (iv) 17.3 or 17.4° (v) 20 m (vi) 14° (vii) 30 m^2 (viii) 40 m^2 (ix) 50 m^2 (x) 104 m^2 (xi) 39.1°

15 $y = -x^2 + 3$

(i)

x	−3	−2	−1	0	1	2	3
y	−6	−1	2	3	2	−1	−6

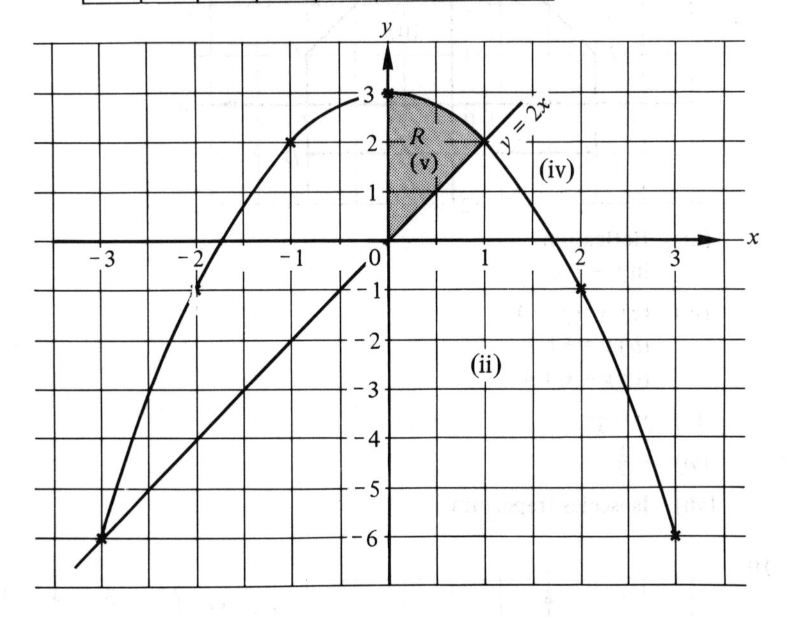

(iii) ± 1.7

(vi) The graph of $y = -x^2 + 3$ is symmetrical about the y axis and the y intercept is +3. The graph is a parabola.

Section 3

Set 1

1 (i) £4340 (ii) £4067.20 (iii) £3660.48 (iv) 15.7%

2 (i) (*a*) $x = 1$ (*b*) $x = -4$ or 9 (ii) (*a*) $(3x + 2)(x + 5)$
(*b*) 302×105

3 (i) 95 cm^3 (ii) 270 cups.

4 (i) (*a*) 2 (*b*) $-1\frac{1}{4}$ (*c*) -4 (*d*) 2 (ii) $-5.7, 0.7$
(iii) $s = \sqrt{r - pt}$

5 (i) 81.5 m (ii) (*a*) 16.3 m/s (*b*) 58.7 km/h

6 (i) (*a*) $\frac{4}{7}$ (*b*) $\frac{2}{5}$ (ii) (*a*) $\frac{5}{36}$ (*b*) $\frac{5}{18}$ (iii) $\frac{13}{425}$

7 (ii) $\begin{pmatrix} -2 & 0 \\ 0 & -2 \end{pmatrix}$

(iv) $\begin{pmatrix} -\frac{1}{2} & 0 \\ 0 & \frac{1}{2} \end{pmatrix}$

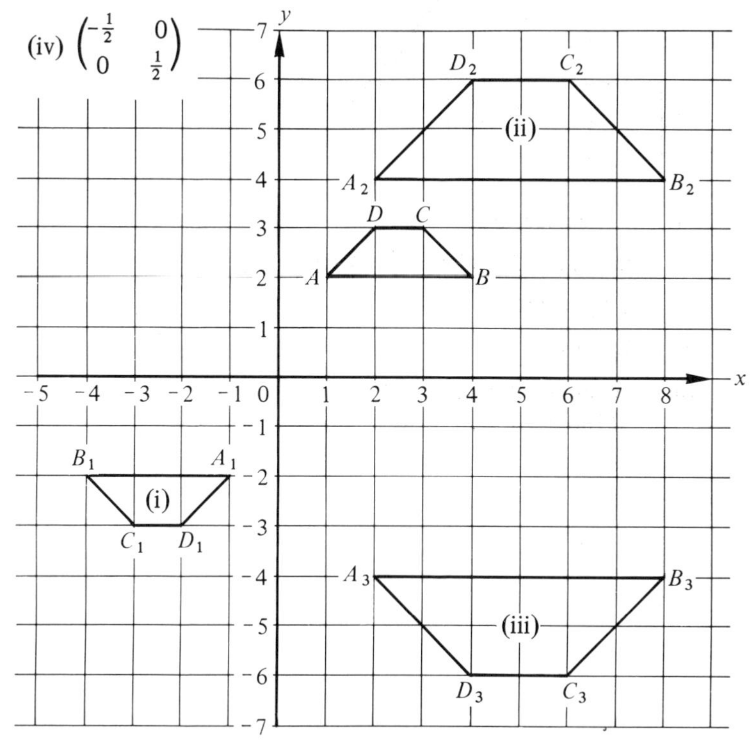

8 $$y = \frac{10}{x} - x$$

x	1	2	3	4	5	6	7	8	9
y	9	3	0.33	−1.5	−3	−4.33	−5.57	−6.75	−7.89

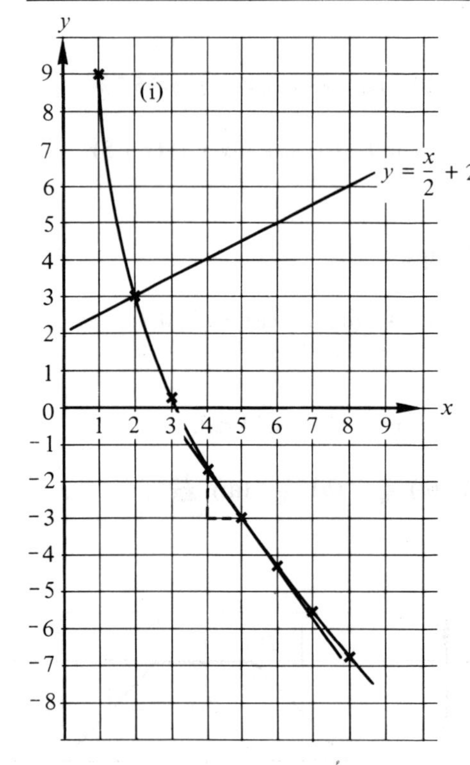

(ii) $x = 2$

(iii) -1.4

9 (i) $y \leqslant 3x, x \leqslant 3y$

$2x + 3y \leqslant 360$ $(5x + 7\frac{1}{2}y \leqslant 900)$

$x + y \leqslant 150$

(ii) Graphs

(iii) 60 lambs, 80 piglets
£1160 profit.

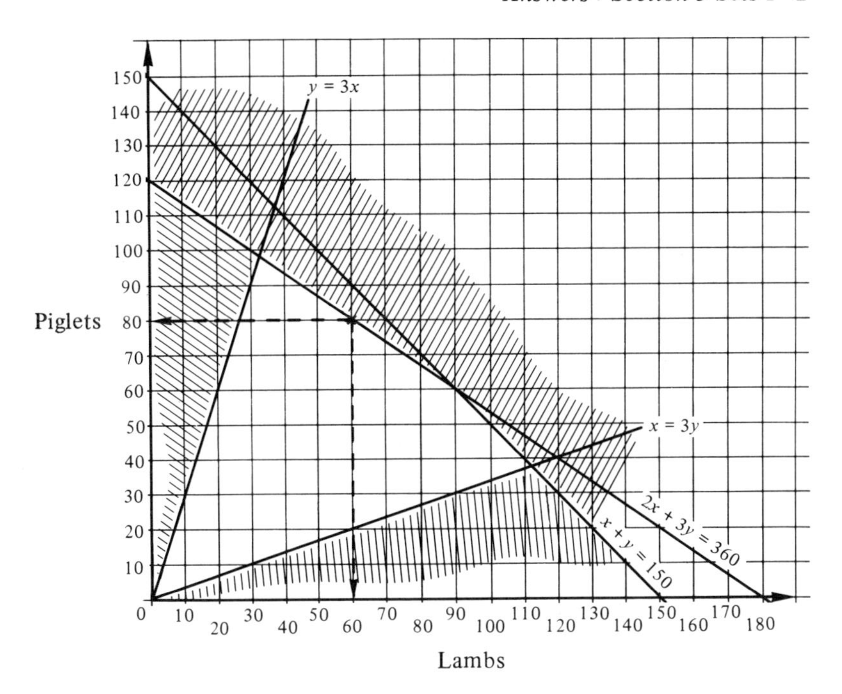

10 (i) $12\mathbf{a}$ (ii) (*a*) $3\mathbf{a} - 3\mathbf{b}$ (*b*) $2\mathbf{a} - 2\mathbf{b}$ (*c*) $2\mathbf{a} + \mathbf{b}$ (*d*) $4\mathbf{a} + 2\mathbf{b}$ (*e*) $4\mathbf{a} - \mathbf{b}$ (*f*) $12\mathbf{a} - 3\mathbf{b}$

(iii) $\frac{x}{y} = \frac{1}{3}$

Set 2

1 (i) 32.25% (ii) £8.06 (iii) £1177

2 (i) $b = -10, c = 25$ (ii) $\frac{2}{x+2}$ (iii) $x = 1$ or 4

3 (i) (4,0) (ii) $y = -\frac{1}{2}x + 2$ (iii) 12 cm^2

(iv) $x \geqslant 0, y \geqslant 0, y \leqslant -\frac{1}{2}x + 2$

4 (i) 4 (ii) (*a*) (*b*) 6.92

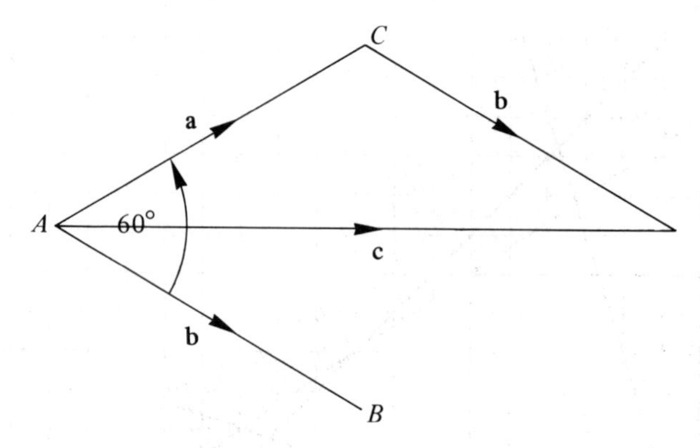

5 (i) (*a*) $\frac{1}{6}$ (*b*) $\frac{1}{36}$ (ii) (*a*) 216 (*b*) $\frac{1}{36}$ (*c*) $\frac{1}{24}$

6 (i) $n = 19$

(ii) (*a*) $0.049\ \text{cm}^2$
(*b*) $4.93 \times 10^{-2}\ \text{cm}^2$

7 $y = 2x^2 + 4x + 7$

(i)

x	−4	−3	−2	−1	0	1	2
y	9	−1	−7	−9	−7	−1	9

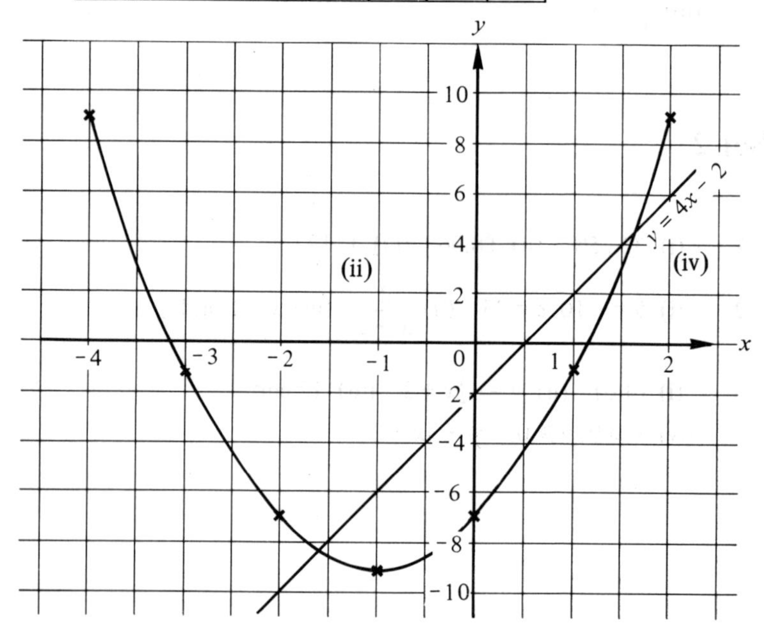

(iii) (*a*) $-3.1, 1.1$ (iv) ± 1.6

(*b*) $-2, 0$ (v) $2x^2 - 5 = 0$

(*c*) $-3.55, 1.55$ or $2x^2 = 5$

8 (i) 21.8 cm (ii) $XT = 2$, $YV = 4$ cm (iii) 252π cm^3

(iv) $\frac{1}{27}$, $9\frac{1}{3}\pi$ cm^3 (v) $\frac{8}{1}$, $74\frac{2}{3}\pi$ cm^3 (vi) $65\frac{1}{3}\pi$ cm^3

9 (i) $x \geqslant 16, y \geqslant 26$

$x + y \leqslant 100$

$3x + 5y \leqslant 360$

$45x + 40y \leqslant 3600$

$(9x + 8y \leqslant 720)$

(iii) 86 hectares

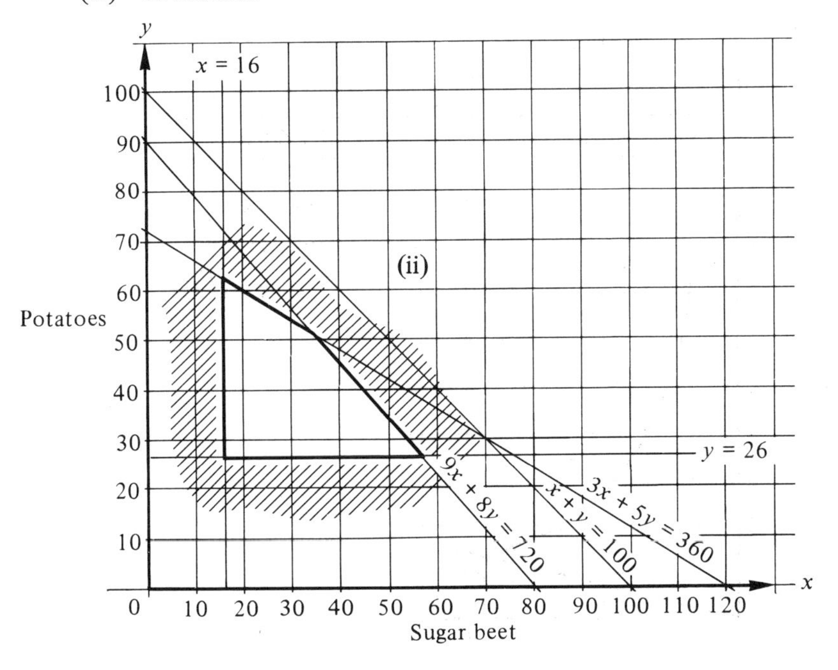

10 (i) 29.7° (ii) 50 m (iii) 35.4 m

Set 3

1 (i) (*a*) 21% (*b*) 33.1% (ii) (*a*) £332.50 (*b*) £47.50

(*c*) £190.00 (iii) 40%

2 (i) $b = -1, c = -12$ (ii) 10 (iii) $x = -2$ or 5

3 (i) $D = 8x^3 - \frac{4}{3}\pi x^3$ (ii) 3810 cm³

$D = x^3(8 - \frac{4}{3}\pi)$

4 (i) 5000 m (ii) 400 m (iii) 0.08 (iv) 4.6°

5 (i) (*a*) Kite (*b*) 40 cm²

(ii) 67.4° (iii) 58.8 cm² (iv) 9.4 cm² (v) 4.84 cm²

6 (i) (*a*) $(a - 5)(x + 4)$ (*b*) $a = 5$ for any value of x

$x = -4$ for any value of a

(ii) (*a*) $\pi R^2 - 100\pi r^2$ (*b*) $\pi(R + 10r)(R - 10r)$ (*c*) 5652 cm²

7 $y = \frac{8}{x}$

(i)

x	1	2	3	4	5	6
y	8	4	2.67	2	1.60	1.33

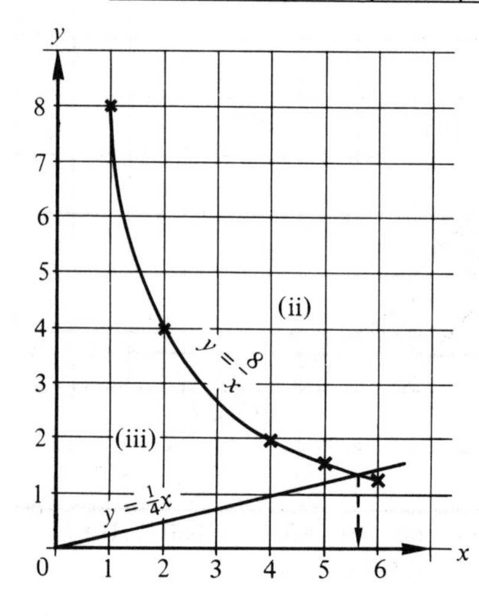

(iv) 5.65

(v) $x^2 = 32$

$x = \sqrt{32}$

8 (i) (*a*) **a** − **b** (*b*) $\frac{1}{2}$(**a** − **b**) (*c*) $\frac{1}{2}$**b** (*d*) $\frac{1}{2}$**a** (*e*) **a**

(ii) (*a*) *GE* and *AC* are parallel and equal. (*b*) *ACEG* is a parallelogram.

9 (i) (*a*) 10.4 m (*b*) 15.6 m² (*c*) 37.7 m² (*d*) 22.1 m²

(ii) (*a*) 188.4 m² (*b*) £848

10 (i) 61.9° (62°) (ii) 17 cm (iii) 28.1° (28°) (iv) 14.8 cm

(v) 35.8° (vi) 84.2°

Set 4

1 (i) (*a*) £2448 (*b*) £2142 (ii) (*a*) £214 (*b*) 10%

2 (i) 25 (ii) $x = \frac{1}{2}$ or 2 (iii) $n = 5$

3 (i) $V = 64x^3 - 16\pi x^3 = 16x^3(4 - \pi)$ (ii) 4704 cm³

4 (i) $\frac{1}{2}\begin{pmatrix} 7 & -1 \\ -5 & 1 \end{pmatrix}$ (ii) $\begin{pmatrix} 1 & 1 \\ 5 & 7 \end{pmatrix}\begin{pmatrix} x \\ y \end{pmatrix} = \begin{pmatrix} 2 \\ 4 \end{pmatrix}$

(iii) $\begin{pmatrix} 1 & 0 \\ 0 & 1 \end{pmatrix}\begin{pmatrix} x \\ y \end{pmatrix} = \frac{1}{2}\begin{pmatrix} 10 \\ -6 \end{pmatrix}$ (iv) $x = 5, y = -3$

5 (i) 204.1 cm (ii) 12.56 cm (iii) 1766.25 cm³

6 $y = x^2 + 5x$

(i)

x	−6	−5	−4	−3	−2	−1	0	1
y	6	0	−4	−6	−6	−4	0	6

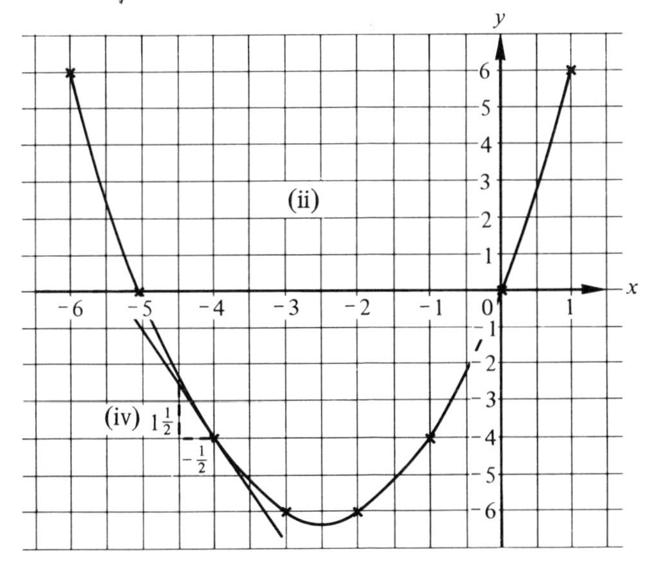

(iii) (*a*) − 5,0
(*b*) − 1.4, − 3.6
(*c*) − 5.4, 0.4

(iv) − 3

(v) $x = -2.5$

7 (i) 8 cm (ii) 6 by 4 by 3 cm (iii) 512 cm^3

8

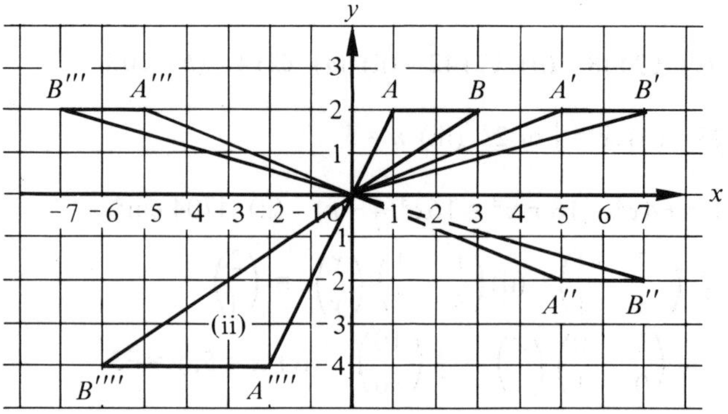

(i) (*a*) $\begin{pmatrix}1 & 2\\0 & 1\end{pmatrix}$ Shear, Scale Factor 2, $y = 0$ invariant

(*b*) $\begin{pmatrix}1 & 0\\0 & -1\end{pmatrix}$ Reflection in $y = 0$ (x axis)

(*c*) $\begin{pmatrix}-1 & 0\\0 & -1\end{pmatrix}$ Rotation of 180°, centre O

(ii) Enlargement, Scale Factor −2, centre of enlargement O.

9 (i) 90° (ii) 100 m (iii) 18 m (iv) 10.2° (v) 263°

10 (i) (*a*) 62.8 cm^2 (*b*) 47.6 cm^2 (*c*) 15.2 cm^2 (*d*) 24.32 cm

(ii) The answers would be multiplied by 4

Set 5

1 (i) £13 840 (ii) £424 (iii) 7%

2 (i) $\{x : x > 7\}$ or $\{x : x = 8, 9, 10 \ldots\}$ (ii) $x = 4, y = -7$

(iii) $x = -1\frac{1}{4}$

3 (i) $P = 8x + 4\pi x$: $P = 4x(2 + \pi)$

(ii) $A = 16x^2 + 4\pi x^2$: $A = 4x^2(4 + \pi)$

(iii) (*a*) 7.2 m (*b*) 3.5 m^2

4 $y = x^2 + 3x - 2$

(i)

x	-5	-4	-3	-2	-1	0	1	2
y	8	2	-2	-4	-4	-2	2	8

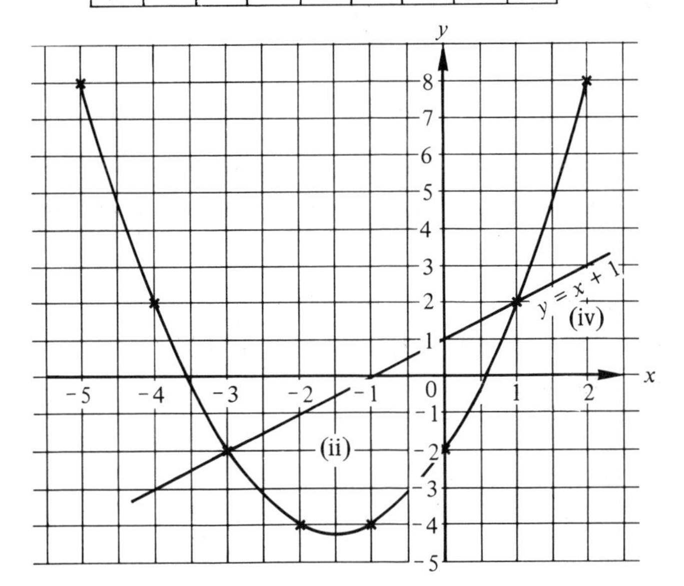

(iii) (*a*) $-3.6, 0.6$
(*b*) $-4.2, 1.2$

(iv) (*a*) $1, -3$
(*b*) $x^2 + 2x - 3 = 0$

5 (i) (*a*) $\dfrac{x}{16}$ (*b*) $\dfrac{x+8}{24} = \dfrac{x}{16} + \dfrac{1}{4}$, $x = 4$

(ii) (*a*) 13, 15, 17, 19, 31, 35, 37, 39, 57, 59, 51, 53, 79, 71, 73, 75, 91, 93, 95, 97

(*b*) $\frac{11}{20}$

6 (i) 39° (ii) 19.5° (iii) 25.5° (iv) 109.5° (v) 70.5°

7 (i) $(x + 17)(x + 11) = 432$ and $x = 7$

(ii) (*a*) 12 (*b*) Dodecagon

8 (i) (*a*) $(2x + 3y)(3a - 4b)$ (*b*) $(3t - 1)(t + 2)$

(*c*) $3(b + 4)(b - 4)$ (*d*) $\left(\dfrac{a^2}{9} + \dfrac{b^2}{4}\right)\left(\dfrac{a}{3} + \dfrac{b}{2}\right)\left(\dfrac{a}{3} - \dfrac{b}{2}\right)$

(ii) $\dfrac{100 \times 6.96}{10 \times 6.96} = \dfrac{696}{69.6} = 10$

9 (i) 12.6 cm (ii) 477 cm^2 (iii) 360 cm^3 (iv) 18.4°
(v) 19.2 cm (vi) 19.6 cm (vii) 11.6° or 11.7°

10 (i) 7.86 km (ii) 317.9° (318°)

Set 6

1 (iii) $r = 1.9$, $h = 4.9$ cm (iv) (*a*) 69.9 cm^3 (*b*) 79.55 cm^3 (v) (*a*) 9.65 cm^3 (*b*) £40.53

2 (i) (*a*) {9,10,11,12,13,14,15,16,17,18,19}
(*b*) {11,12,13,14,15} (*c*) {13,14,15,16,17,18}
(*d*) {13,14,15} (*e*) {9,10,19}

(ii)

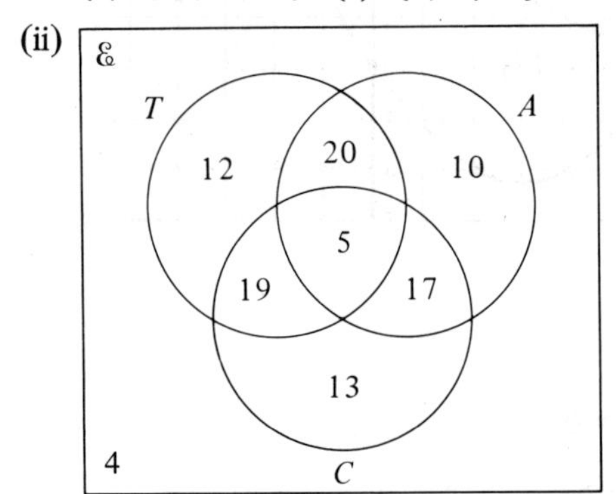

5 students took part in all three activities.

3 (i)

Cumulative frequency
5
16
38
85
157
260
366
522
582
600

(ii)

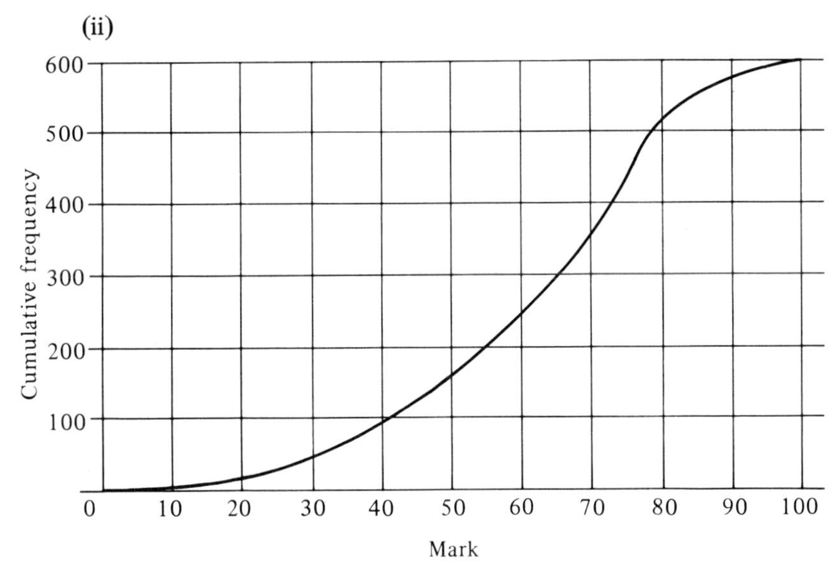

(iii) (*a*) 65 (*b*) 75 (*c*) 49 (*d*) 26
(iv) (*a*) 58% (*b*) 26.2%

4 (i) £589 (ii) 0.679%

5 (i) $7(2x + 1)$ (if 7 is one of the factors of the expression, it must be divisible by 7)

(ii) $x = 3, x = 6\frac{1}{2}$ (not unique solutions)

(iii) $x = 3\frac{1}{2}n - \frac{1}{2}$ or $\frac{7}{2}n - \frac{1}{2}$

6 (i) (*a*) $\frac{2\pi x}{3}$ (*b*) $\frac{2}{3}\pi x^2$ (*c*) $\frac{100\pi}{6}$ % (ii) $\frac{\pi x^2}{27} \times h$

7 (i) $\mathbf{y} - \mathbf{x}$, $3\mathbf{x}$, $3\mathbf{y}$, $3\mathbf{y} - 3\mathbf{x}$, $\frac{3}{2}\mathbf{x} + \frac{3}{2}\mathbf{y}$ or $1\frac{1}{2}(\mathbf{x} + \mathbf{y})$

(ii) $3\mathbf{y} - \mathbf{x}$

(iii) Triangles *XYL* and *CBL* are similar: the vectors $\overrightarrow{XY}$ and $\overrightarrow{BC}$ show that their corresponding sides are in the ratio 1 : 3; therefore $XL = \frac{1}{3}\ CL$ and $\frac{1}{4}XC$.

(iv) *A*, *L* and *M* are colinear.

8 $y = x^2 - 2x - 3$

(i)

x	−2	−1	0	1	2	3	4
y	5	0	−3	−4	−3	0	5

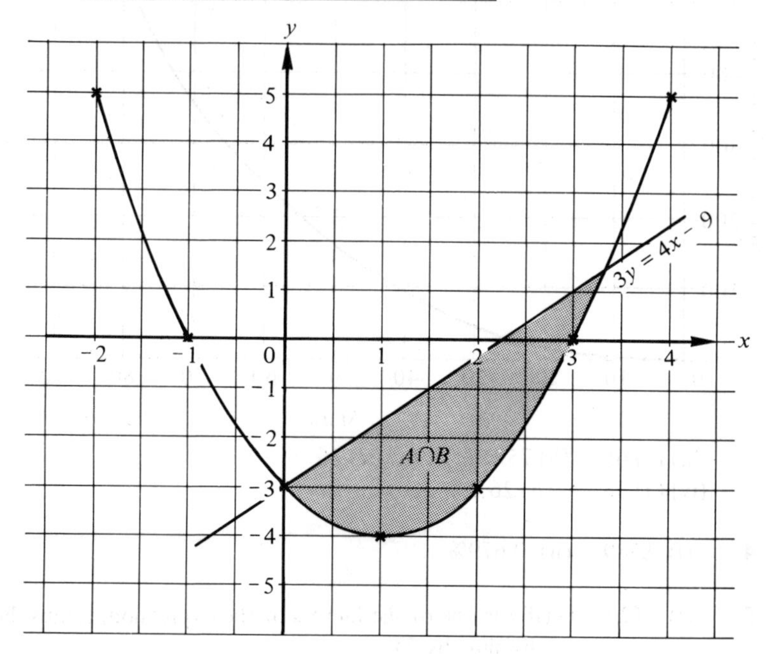

(*a*) −4

(*b*) (i) −1.7, 3.65

(ii) −0.4, 2.4

(iii) (*a*) 0, 3.35

(*b*) Shaded region

9 (i) Mode: 68, mean: 69.51, median 69, therefore the formula is valid.

(ii) 71.6

10 (i) $T_3 = \begin{pmatrix} -1 & 0 \\ 0 & 1 \end{pmatrix}$

T_1 = Rotation about (0,0), + 270° or − 90°

T_2 = Reflection in $y = x$ or $x = y$

T_3 = Reflection in $x = 0$ (y axis)

(ii)

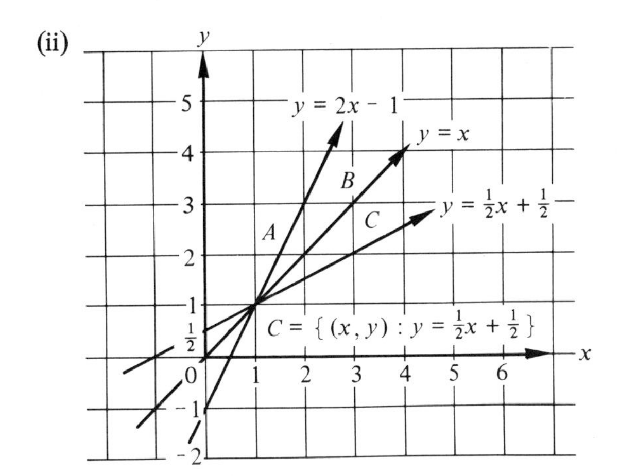

Part 4
Reference section

Contents

1. Units

Length

unit: metre (m)

1 m	=	1000 millimetres (mm)
1 m	=	100 centimetres (cm)
1000 m	=	1 kilometre (km)

Mass (weight)

unit: gram (g)

1000 g	=	1 kilogram (kg)
1000 kg	=	1 tonne (t)

Cubic capacity

unit: litre (l)

1 litre	=	1000 millilitres (ml)
1 litre	=	1000 cm^3
100 litres	=	1 hectolitre (hl)

Area

unit: 1 m^2

1 cm^2	=	100 mm^2
1 m^2	=	10 000 cm^2
100 m^2	=	1 are (a)
10 000 m^2	=	1 hectare (ha)
100 ha	=	1 km^2

Speed

1 m/s	=	3.6 km/h
5 m/s	=	18 km/h

Conversion of metric units to common British units

100 m	$\simeq$	109 yards
8 km	$\simeq$	5 miles
1 litre	$\simeq$	1.76 pints
1 kg	$\simeq$	2.2 lb
1 ha	$\simeq$	2.47 acres
48 km/h	$\simeq$	30 mph
80 km/h	$\simeq$	50 mph

2. Areas and volumes

A. Area: Plane figures

1 Rectangle

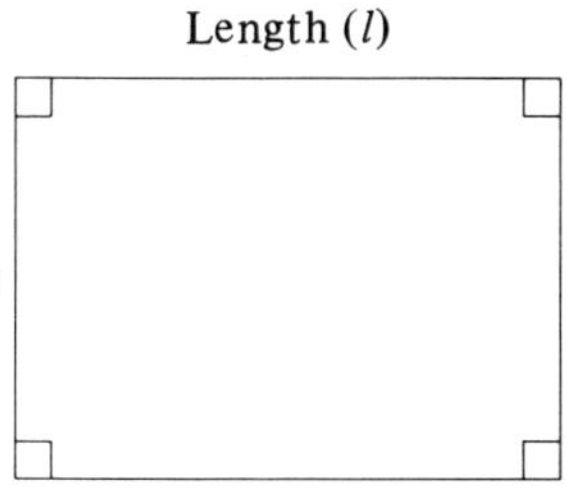

(i) Area = Length × Breadth,
i.e. $A = lb$

(ii) $l = \frac{A}{b}$

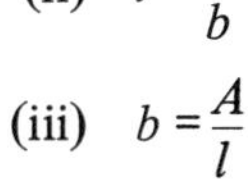

(iii) $b = \frac{A}{l}$

2 Square

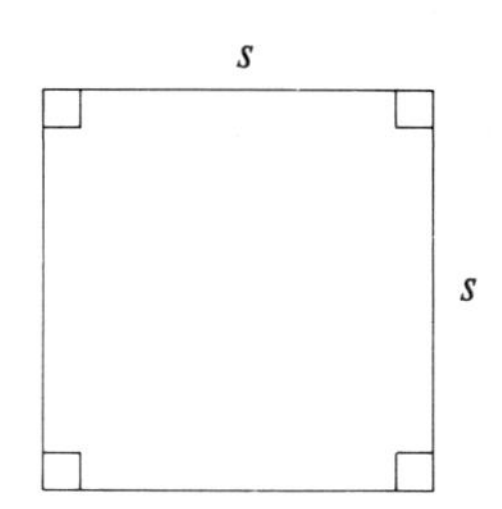

(i) Area = $s \times s$,
i.e. $A = s^2$

(ii) $s = \sqrt{A}$

3 Triangle

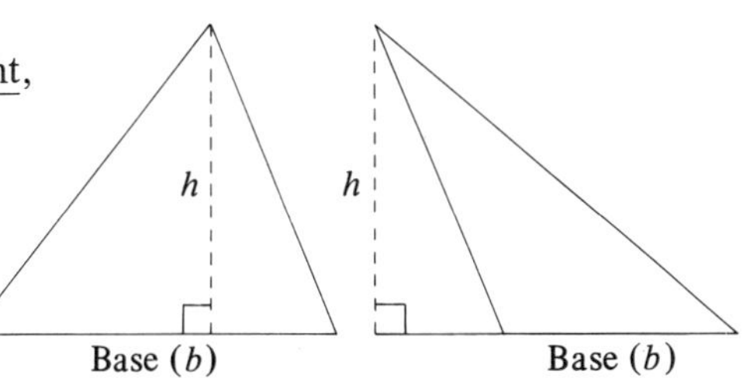

(i) Area = $\frac{\text{Base} \times \text{Height}}{2}$,

or: $A = \frac{1}{2}bh$

(ii) $b = \frac{2A}{h}$

(iii) $h = \frac{2A}{b}$

4 Parallelogram

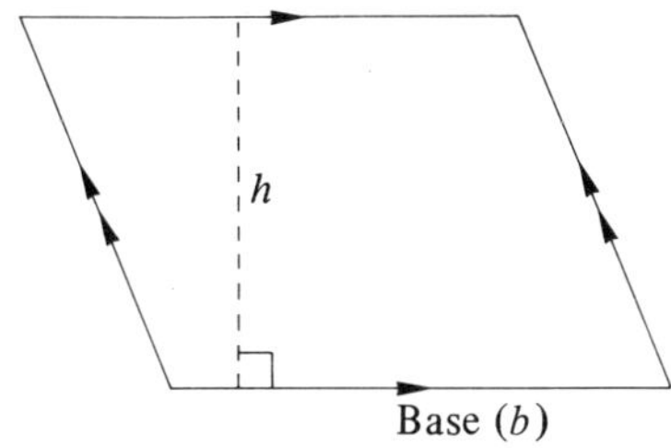

(i) Area = Base × Height,
i.e. $A = bh$

(ii) $b = \frac{A}{h}$

(iii) $h = \frac{A}{b}$

5 Rhombus

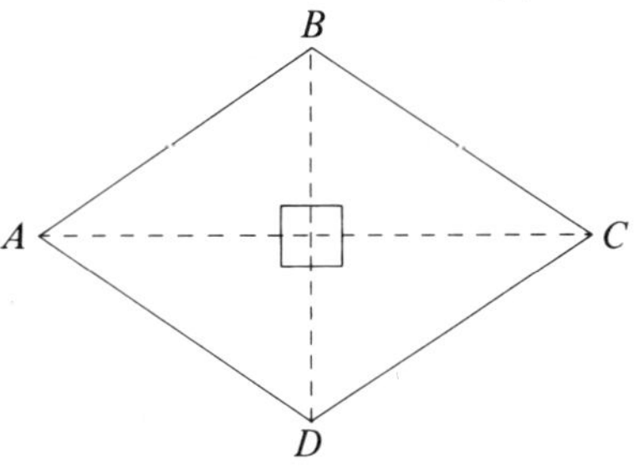

AC is the long diagonal
BD is the short diagonal

Area = $\frac{AC \times BD}{2}$

or $\frac{1}{2}(AC \times BD)$

6 Kite

AC is the long diagonal
BD is the short diagonal

Area = $\dfrac{AC \times BD}{2}$

or $\frac{1}{2}(AC \times BD)$

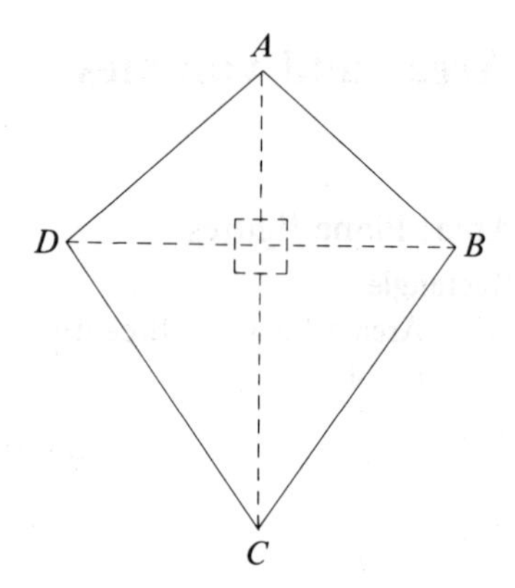

7 Trapezium

a and b are the parallel sides
h is the perpendicular height

(i) Area = $\left(\dfrac{a+b}{2}\right) \times h$

or $\frac{1}{2}(a+b) \cdot h$

(ii) $h = \dfrac{2A}{a+b}$

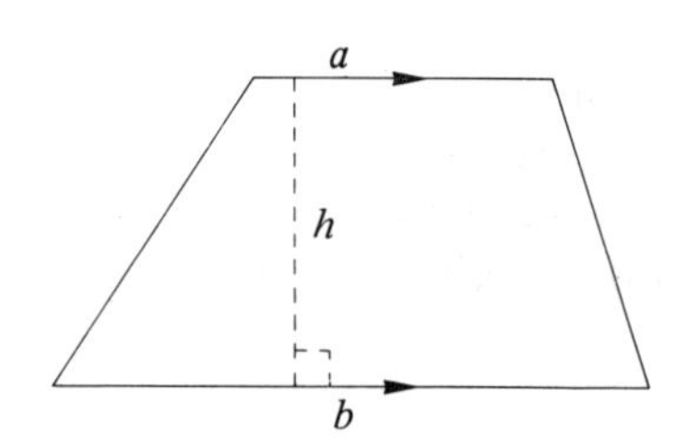

8 Circle

(i) Circumference (C)

(a) $C = 2\pi r$ or πD

(b) $r = \dfrac{C}{2\pi}$

(c) $D = \dfrac{C}{\pi}$

(ii) Area = $\pi \times r \times r$

(a) $A = \pi r^2$

(b) $r = \sqrt{\dfrac{A}{\pi}}$

($\pi = 3.14$ unless otherwise stated)

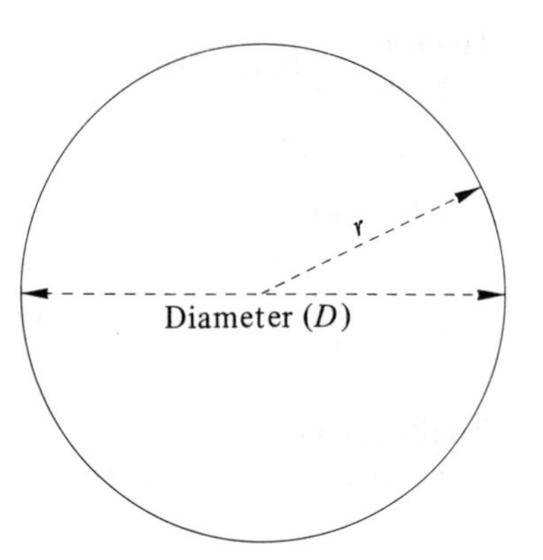

9 Length of an arc

AB is the arc subtended by angle $\alpha°$
r is the radius of the circle

Length of $AB = \dfrac{\alpha°}{360°} \times 2\pi r$

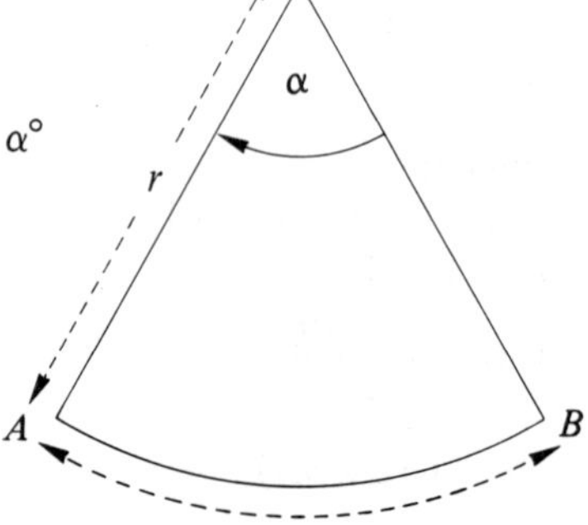

10 Area of a sector

OAB is the sector of a circle, radius r and centre O

Angle α is the sector angle

Area of sector = $\dfrac{\alpha^\circ}{360^\circ} \times \pi r^2$

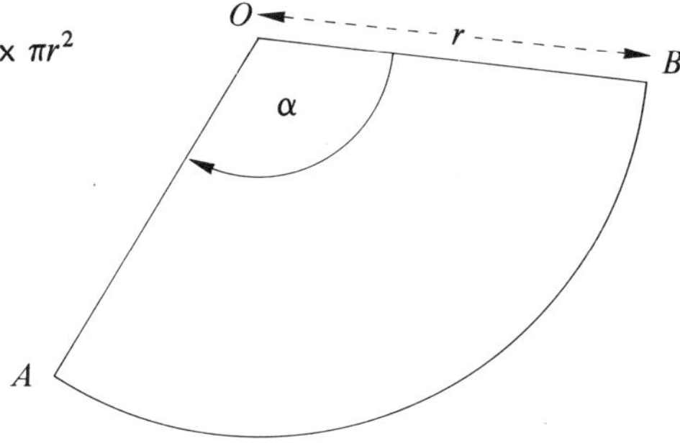

B. Surface area and volume: Solids

1 Cube

s is the length of the side

(i) Total surface area
$= 6 \times s \times s$
$= 6s^2$

(ii) Volume
$= s \times s \times s$
$= s^3$

(iii) $s = \sqrt[3]{V}$

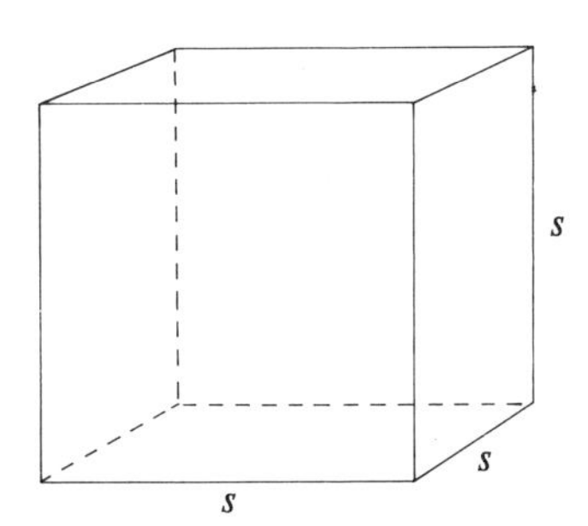

2 Cuboid

(i) Total surface area
$= 2(lb + lh + bh)$

(ii)

(*a*) Volume $= l \times b \times h$
$= lbh$

(*b*) $l = \dfrac{V}{bh}$

(*c*) $b = \dfrac{V}{lh}$

(*d*) $h = \dfrac{V}{lb}$

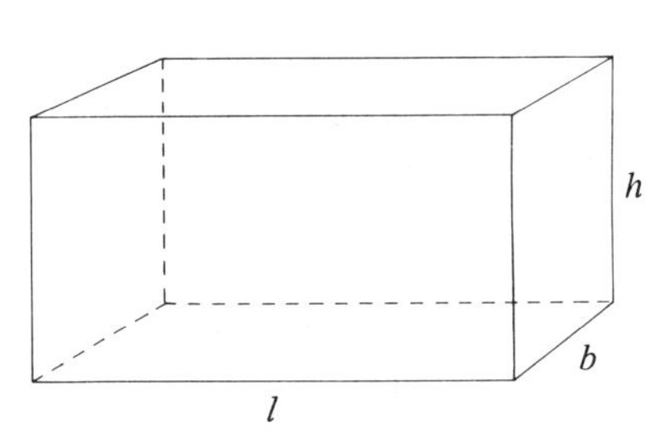

3 Prisms

(i) Volume: general formula
V = Area of end × Length (or Height)

(ii) h or l = $\dfrac{\text{Volume}}{\text{Area of end}}$

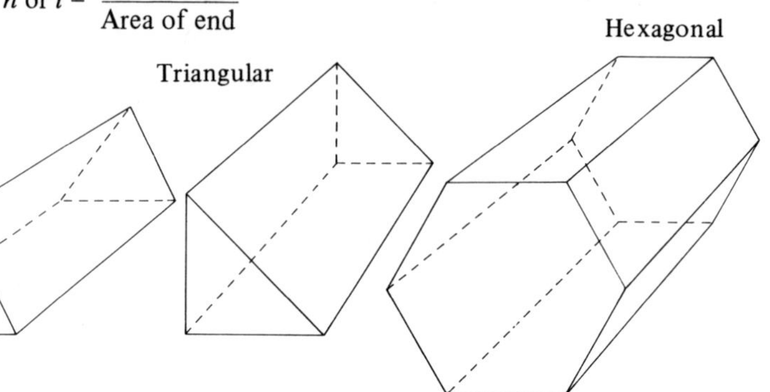

4 Cylinder

(i) Area:

(*a*) Area of curved surface
$= 2\,\pi rh$

(*b*) Area of ends
$= 2\,\pi r^2$

(*c*) Total surface area
$= 2\,\pi rh + 2\,\pi r^2$
$= 2\pi r\,(h + r)$

(ii) Volume:

(*a*) $V = \pi r^2 h$

(*b*) $h = \dfrac{V}{\pi r^2}$

(*c*) $r = \sqrt{\dfrac{V}{\pi h}}$

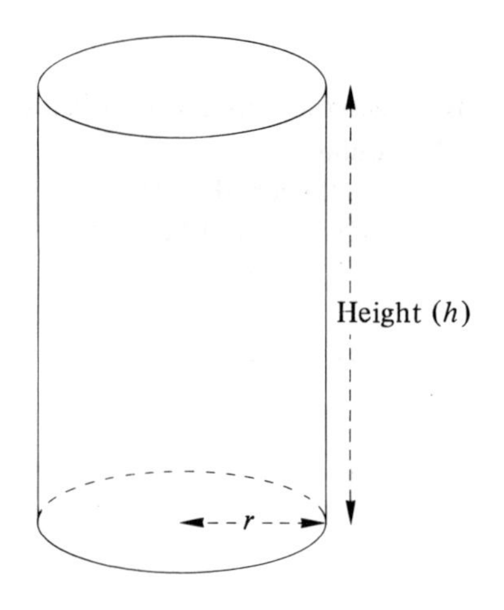

5 Pyramid (base may be any shape)

(i) Volume: general formula
$V = \frac{1}{3}$ × Base area × Height

(ii) Height = $\dfrac{\text{Volume}}{\frac{1}{3}\text{ Base area}}$

or $\dfrac{3V}{\text{Base area}}$

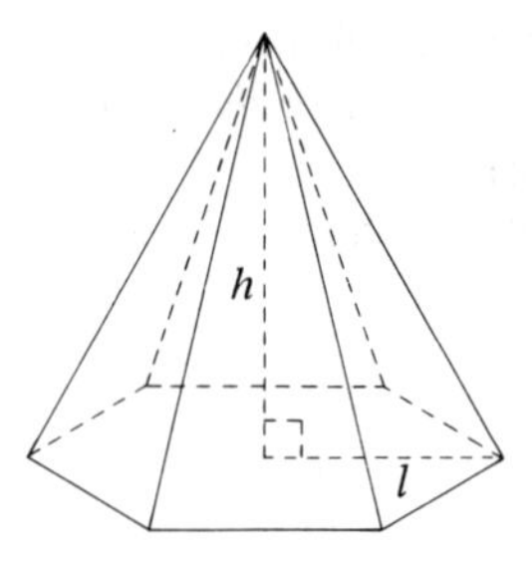

6 Cone (Right circular)

r = radius of base
h = perpendicular height
l = slant height

(i) Area:

(*a*) Curved surface $= \pi r l$

(*b*) Base $= \pi r^2$

(*c*) Total surface area
$= \pi r l + \pi r^2$
$= \pi r (l + r)$

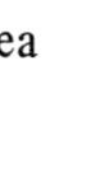

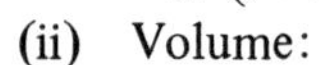

(ii) Volume:

(*a*) $V = \frac{1}{3} \times$ Base area $\times$ Perpendicular height

$= \frac{1}{3} \times \pi r^2 \times h$

$= \frac{1}{3} \pi r^2 h$

(*b*) $h = \dfrac{V}{\frac{1}{3} \pi r^2}$ or $\dfrac{3V}{\pi r^2}$

(*c*) $r = \sqrt{\dfrac{V}{\frac{1}{3} \pi h}}$ or $\sqrt{\dfrac{3V}{\pi h}}$

7 Sphere

(i) Surface area:
$A = 4 \times \pi \times r^2$
$= 4 \pi r^2$

(ii) Volume:

$V = \frac{4}{3} \pi r^3$

(iii) $r = \sqrt[3]{\dfrac{V}{\frac{4}{3}\pi}}$

or $r = \sqrt[3]{\dfrac{3V}{4\pi}}$

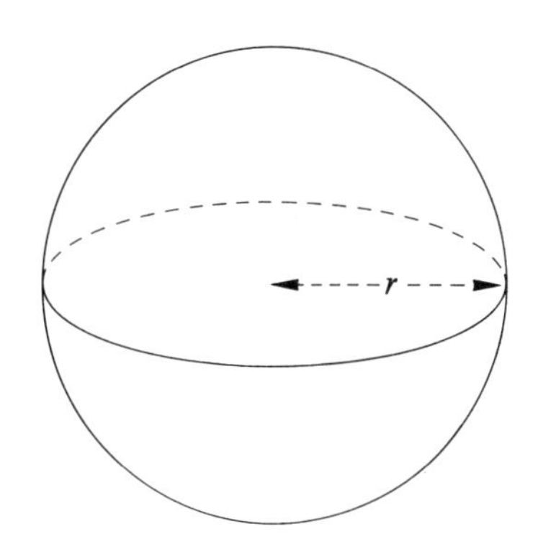

3. Trigonometry

A. ABC abc notation

Side a is opposite angle A
Side b is opposite angle B
Side c is opposite angle C

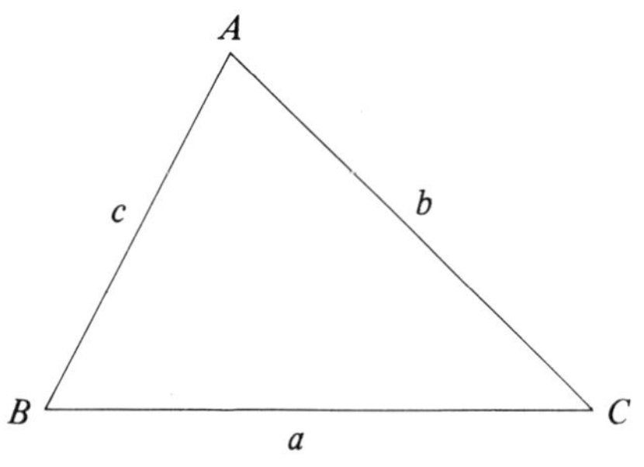

B. The right-angled triangle

The ratios which follow use the notation of this diagram

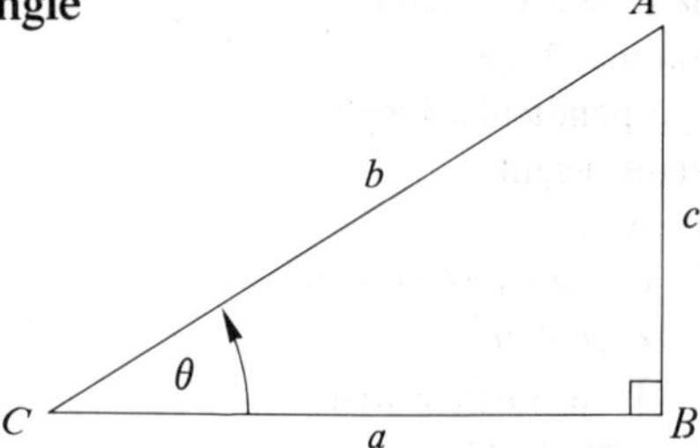

Trigonometrical ratios and formulae

(i) *Sine ratio*	(ii) *Cosine ratio*	(iii) *Tangent ratio*
$\sin\theta = \dfrac{\text{Side opposite}}{\text{Hypotenuse}}$	$\cos\theta = \dfrac{\text{Side adjacent}}{\text{Hypotenuse}}$	$\tan\theta = \dfrac{\text{Side opposite}}{\text{Side adjacent}}$
(*a*) $\sin\theta = \dfrac{c}{b}$	(*a*) $\cos\theta = \dfrac{a}{b}$	(*a*) $\tan\theta = \dfrac{c}{a}$
(*b*) $c = b\sin\theta$	(*b*) $a = b\cos\theta$	(*b*) $c = a\tan\theta$
(*c*) $b = \dfrac{c}{\sin\theta}$	(*c*) $b = \dfrac{a}{\cos\theta}$	(*c*) $a = \dfrac{c}{\tan\theta}$
		[or $a = c \times \tan(90° - \theta)$]

C. Acute and obtuse-angled triangles

(i) *Area* $= (\tfrac{1}{2}ab\sin C)$ or $(\tfrac{1}{2}bc\sin A)$ or $(\tfrac{1}{2}ac\sin B)$

(ii) *Sine formula* $\dfrac{a}{\sin A} = \dfrac{b}{\sin B} = \dfrac{c}{\sin C}$

Use this formula when you are given 2 angles and 1 side (AAS) or 2 sides and a non-included angle (SSA), but remember the 'ambiguous' case which may occur when given (SSA).

(iii) *Cosine formula*

(*a*) $a^2 = b^2 + c^2 - 2bc\cos A$
or $b^2 = a^2 + c^2 - 2ac\cos B$
or $c^2 = a^2 + b^2 - 2ab\cos C$

Use this formula when given 2 sides and the included angle (SAS).

(*b*) $\cos A = \dfrac{b^2 + c^2 - a^2}{2bc}$

or $\cos B = \dfrac{a^2 + c^2 - b^2}{2ac}$

or $\cos C = \dfrac{a^2 + b^2 - c^2}{2ab}$

Use this formula when given 3 sides (SSS)

D. Trigonometrical ratios of angles from 0° to 360°

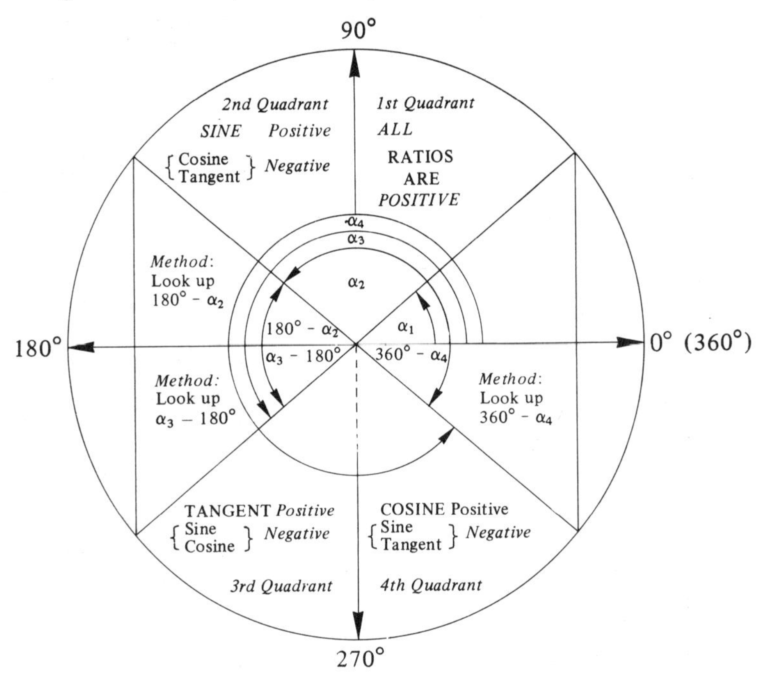

4. Sets

1	$\in$	:	'is a member of'
2	$\subset$	:	'is a subset of'
3	$\cap$	:	'intersection of sets'
4	$\{\}$ or $\emptyset$	;	'the empty set'
5	$\cup$	:	'the union of sets'
6	$\mathcal{E}$	:	'the universal set'
7	$n(A)$	:	'the number of members in set A'
8	A'	:	'the complement of A', i.e. the members of the universal set which are *not* members of A
9	$\{x : x \geqslant 2\}$	:	means that the value of x is such that x is greater than or equal to 2

10 If set $A = \{x : x \leqslant 12\}$ on the set of natural numbers, then:
set $A = \{1, 2, 3...12\}$

11 Closure of sets: If a set is closed under an operation, it means that the operation results in another member of the same set.

12 Number sets:

The set of natural numbers $\{1, 2, 3 \ldots .\}$
The set of counting numbers (or whole numbers) $\{0, 1, 2, \ldots\}$
The set of integers (negative and positive) $\{\ldots -2, -1, 0, 1, 2 \ldots\}$
The set of prime numbers $\{2, 3, 5, 7, 11 \ldots\}$

5. Venn diagrams

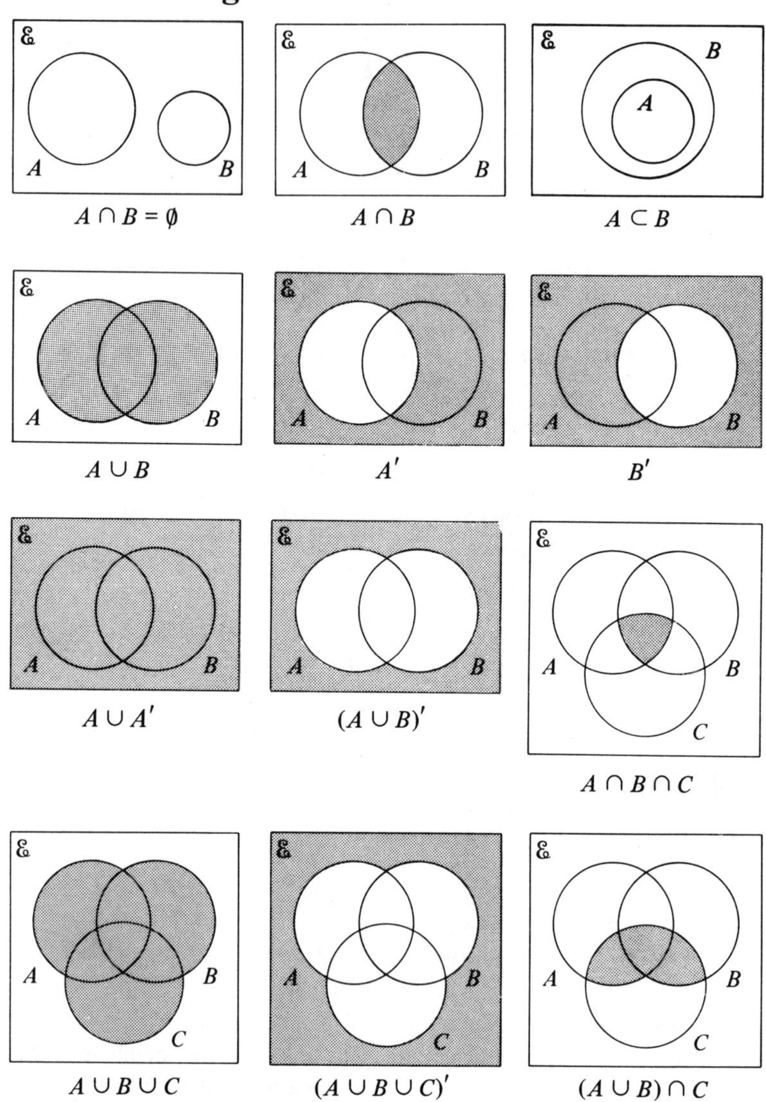

6. Matrices

A. Operations

$$A = \begin{pmatrix} 2 & 3 \\ 1 & -1 \end{pmatrix}, \; B = \begin{pmatrix} -1 & 2 \\ 3 & 1 \end{pmatrix}, \; C = \begin{pmatrix} 3 & 5 \\ -1 & 4 \end{pmatrix}$$

(i) $A + B = \begin{pmatrix} 2 & 3 \\ 1 & -1 \end{pmatrix} + \begin{pmatrix} -1 & 2 \\ 3 & 1 \end{pmatrix} = \begin{pmatrix} 1 & 5 \\ 4 & 0 \end{pmatrix}$

(ii) $A - B = \begin{pmatrix} 2 & 3 \\ 1 & -1 \end{pmatrix} - \begin{pmatrix} -1 & 2 \\ 3 & 1 \end{pmatrix} = \begin{pmatrix} 3 & 1 \\ -2 & -2 \end{pmatrix}$

(iii) $3C = 3 \times \begin{pmatrix} 3 & 5 \\ -1 & 4 \end{pmatrix} = \begin{pmatrix} 9 & 15 \\ -3 & 12 \end{pmatrix}$

(iv) $AB = \begin{pmatrix} 2 & 3 \\ 1 & -1 \end{pmatrix} \begin{pmatrix} -1 & 2 \\ 3 & 1 \end{pmatrix} = \begin{pmatrix} 7 & 7 \\ -4 & 1 \end{pmatrix}$

B. The zero matrix: 0 (all elements are zero)

(i) $A + 0 = A$ or $0 + A = A$

$$\begin{pmatrix} 2 & 3 \\ 1 & -1 \end{pmatrix} + \begin{pmatrix} 0 & 0 \\ 0 & 0 \end{pmatrix} = \begin{pmatrix} 2 & 3 \\ 1 & -1 \end{pmatrix}$$

(ii) $A + X = 0$

$$\begin{pmatrix} 2 & 3 \\ 1 & -1 \end{pmatrix} + \begin{pmatrix} -2 & -3 \\ -1 & 1 \end{pmatrix} = \begin{pmatrix} 0 & 0 \\ 0 & 0 \end{pmatrix}$$

C. The identity matrix: I

I is a square matrix, with all elements *not* on the leading diagonal equal to zero and all elements *on* the leading diagonal equal to 1,

e.g. $\begin{pmatrix} 1 & 0 \\ 0 & 1 \end{pmatrix}$ or $\begin{pmatrix} 1 & 0 & 0 \\ 0 & 1 & 0 \\ 0 & 0 & 1 \end{pmatrix}$.

If A is any matrix, then $AI = A$ or $IA = A$.

D. The inverse of a matrix

If $AB = I$ (the identity matrix) or $BA = I$, B is the multiplicative inverse of A.

The inverse of A is written as A'.

E. Standard method for calculation of the inverse of a matrix

1 Calculate the DETERMINANT: Δ — *Formulae*

Given that $A = \begin{pmatrix} 4 & 1 \\ 3 & 2 \end{pmatrix}$ — If $A = \begin{pmatrix} a & b \\ c & d \end{pmatrix}$

$\Delta = (4 \times 2) - (1 \times 3)$
$= 8 - 3$
$= 5$

$\Delta = ad - bc$

Note: If $\Delta = 0$, the matrix has no inverse and is called SINGULAR.

2 Rearrange the matrix:
(*a*) Transpose 4 and 2.
(*b*) Change the signs of 1 and 3.

$$\begin{pmatrix} 4 & 1 \\ 3 & 2 \end{pmatrix} \text{ becomes } \begin{pmatrix} 2 & -1 \\ -3 & 4 \end{pmatrix} \qquad \begin{pmatrix} a & b \\ c & d \end{pmatrix} \text{ becomes } \begin{pmatrix} d & -b \\ -c & a \end{pmatrix}$$

3 Divide each element of the rearranged matrix by Δ (in this case 5)

$$\frac{1}{5}\begin{pmatrix} 2 & -1 \\ -3 & 4 \end{pmatrix} = \begin{pmatrix} \frac{2}{5} & -\frac{1}{5} \\ -\frac{3}{5} & \frac{4}{5} \end{pmatrix} \qquad A' = \begin{pmatrix} \frac{d}{\Delta} & \frac{-b}{\Delta} \\ \frac{-c}{\Delta} & \frac{a}{\Delta} \end{pmatrix}$$

The INVERSE of A i.e. $A' = \begin{pmatrix} \frac{2}{5} & -\frac{1}{5} \\ -\frac{3}{5} & \frac{4}{5} \end{pmatrix}$

Note: $AA' = I$ (the identity matrix), therefore

$$\begin{pmatrix} 4 & 1 \\ 3 & 2 \end{pmatrix}\begin{pmatrix} \frac{3}{5} & -\frac{1}{5} \\ -\frac{3}{5} & \frac{4}{5} \end{pmatrix} = \begin{pmatrix} 1 & 0 \\ 0 & 1 \end{pmatrix}$$

When you have calculated an inverse, *always* carry out this check on your answer.

F. Matrix transformations

	(i) *Transformation*	*Matrix*
1	Identity	$\begin{pmatrix} 1 & 0 \\ 0 & 1 \end{pmatrix}$
2	Rotation about (0,0) through +90°	$\begin{pmatrix} 0 & -1 \\ 1 & 0 \end{pmatrix}$
3	Rotation about (0,0) through 180°	$\begin{pmatrix} -1 & 0 \\ 0 & -1 \end{pmatrix}$
4	Rotation about (0,0) through +270° or −90°	$\begin{pmatrix} 0 & 0 \\ -1 & 0 \end{pmatrix}$

5 Reflection in $x = 0$ (y axis) $\begin{pmatrix} -1 & 0 \\ 0 & 1 \end{pmatrix}$

6 Reflection in $y = 0$ (x axis) $\begin{pmatrix} 1 & 0 \\ 0 & -1 \end{pmatrix}$

7 Reflection in $y = x$ (or $x = y$) $\begin{pmatrix} 0 & 1 \\ 1 & 0 \end{pmatrix}$

8 Reflection in $y = -x$ ($x + y = 0$) $\begin{pmatrix} 0 & -1 \\ -1 & 0 \end{pmatrix}$

9 Enlargement, centre (0,0), Scale Factor k $\begin{pmatrix} k & 0 \\ 0 & k \end{pmatrix}$

10 Shear, $y = 0$ invariant $(0,1) \rightarrow (k, 1)$ $\begin{pmatrix} 1 & k \\ 0 & 1 \end{pmatrix}$

11 Shear, $x = 0$ invariant $(1,0) \rightarrow (1,k)$ $\begin{pmatrix} 1 & 0 \\ k & 1 \end{pmatrix}$

(ii) *Combined transformations*

If the transformation $\begin{pmatrix} x \\ y \end{pmatrix} \rightarrow A \begin{pmatrix} x \\ y \end{pmatrix}$ is followed by

the transformation $\begin{pmatrix} x \\ y \end{pmatrix} \rightarrow B \begin{pmatrix} x \\ y \end{pmatrix}$

this is equivalent to the single transformation:

$$\begin{pmatrix} x \\ y \end{pmatrix} \rightarrow BA \begin{pmatrix} x \\ y \end{pmatrix}$$

7. The circle

A. Terms and definitions

O is the centre of the circle in all the diagrams.

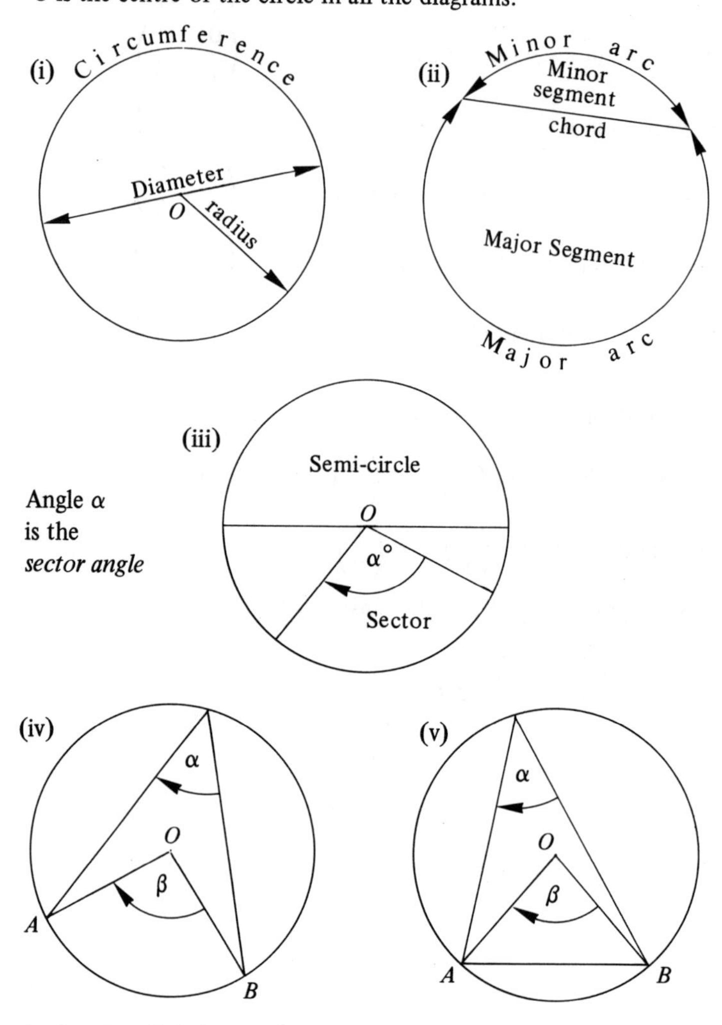

Angle α is the *sector angle*

Angle α is called the *'angle at the circumference'*.
Angle β is called the *'angle at the centre'*.
In (iv) both angles are subtended by the minor arc AB.
In (v) both angles are subtended by the chord AB.

B. Angles in a circle

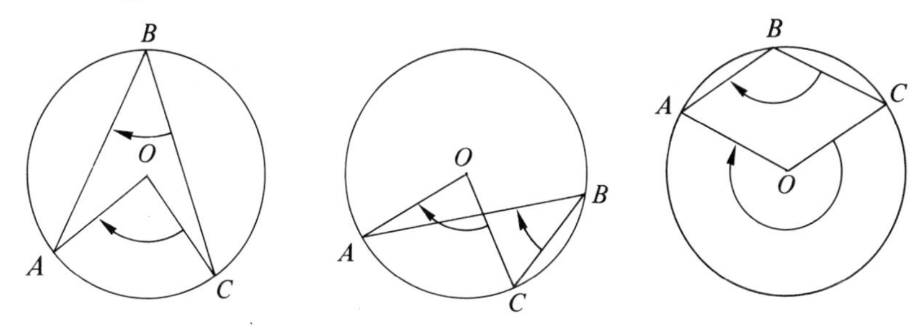

Fact: The *angle at the centre* is twice the *angle at the circumference* subtended by the same arc or chord.

The diagrams show possible positions of the two angles. In each case:

angle AOC = twice angle ABC

2 *Angles in the same segment*

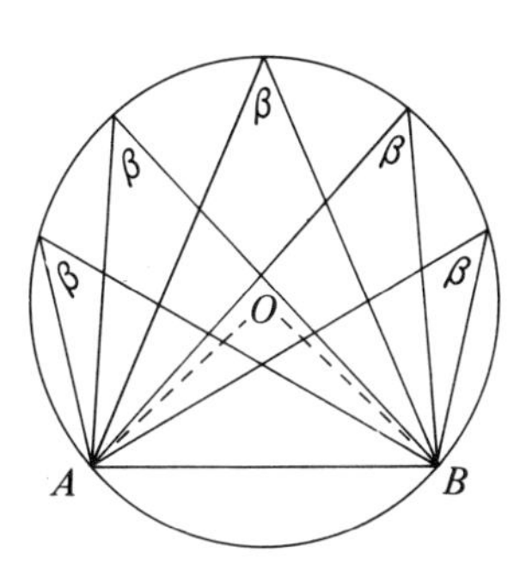

It may be seen that in each case, angle β = half angle AOB, the angle at the centre. Hence all angles at the circumference subtended by the chord AB (or the arc AB) are equal.

Fact: angles in the same segment are equal.

3 *The angle in a semi-circle*

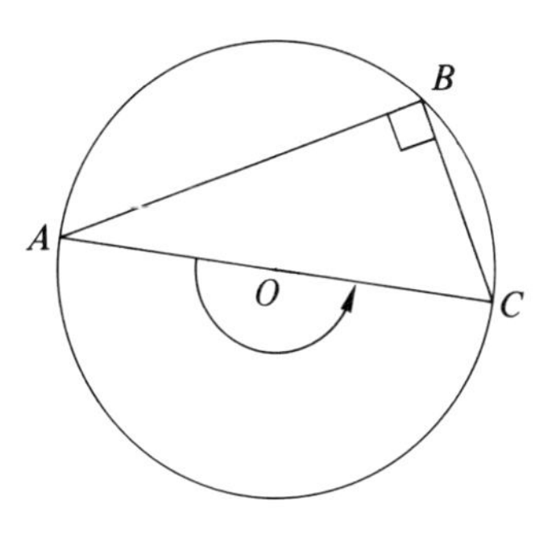

If AC is a diameter, then angle AOC (the angle at the centre) = 2 right angles.

Angle ABC (the angle at the circumference) = half angle AOC, hence:

angle ABC = 1 right angle.

Fact: the angle in a semi-circle is a right angle

4 *The opposite angles of a cyclic quadrilateral*

In the diagram: angle α = twice angle BCD and angle β = twice angle DAB.

But $\alpha + \beta$ = 4 right angles hence angle DAB + angle BCD = 2 right angles

Similarly: angle ADC + angle ABC = 2 right angles

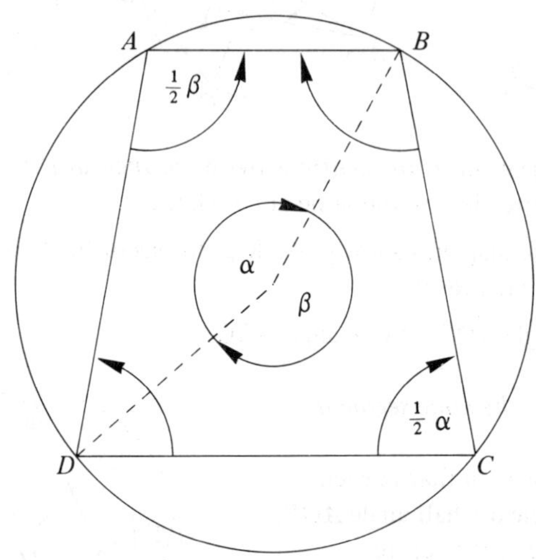

Fact: The *opposite* angles of a cyclic quadrilateral are *supplementary*.

5 *The exterior angle of a cyclic quadrilateral*

$ABCD$ is a cyclic quadrilateral, therefore: angle α + angle θ = 2 right angles, but angle β + angle θ also = 2 right angles. Therefore angle α = angle β.

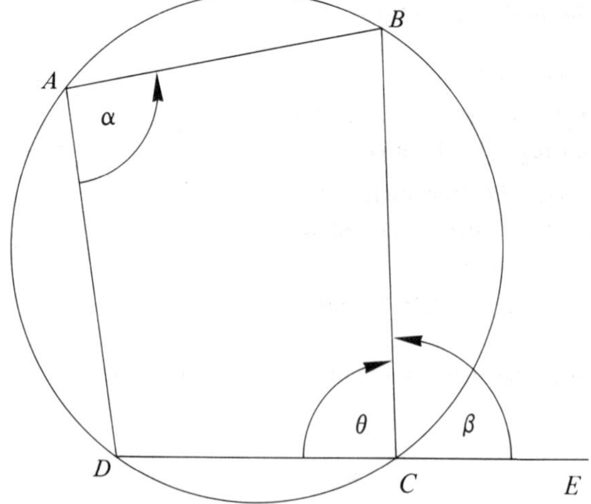

Fact: the *exterior* angle of a cyclic quadrilateral is equal to the *interior opposite angle*.

C. Tangents to a circle

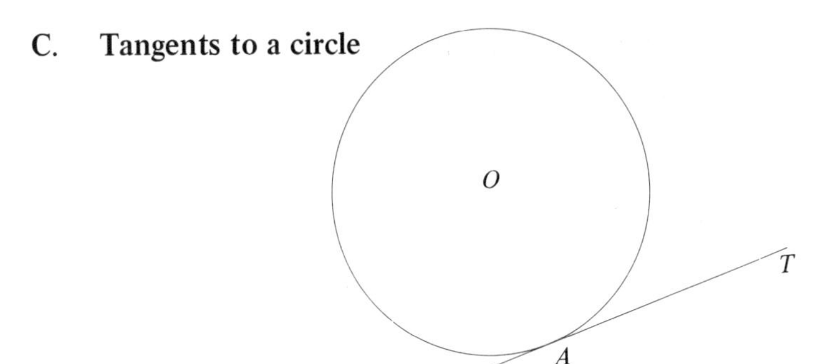

A *tangent* is a straight line which has only 1 point of contact with a circle, no matter how far the line is produced.

In the diagram, the straight line TA touches the circle at A. TA is the tangent to the circle at A.

1 *The angle between a radius and a tangent at the point of contact with the circle*

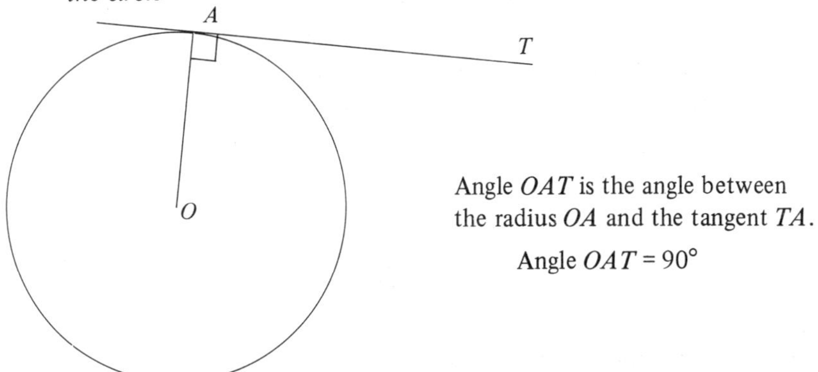

Angle OAT is the angle between the radius OA and the tangent TA.

Angle $OAT = 90°$

Fact: the angle between *the radius* and a *tangent* to the circle at the point of contact is a right angle.

2 *Two tangents to a circle, drawn from an external point*

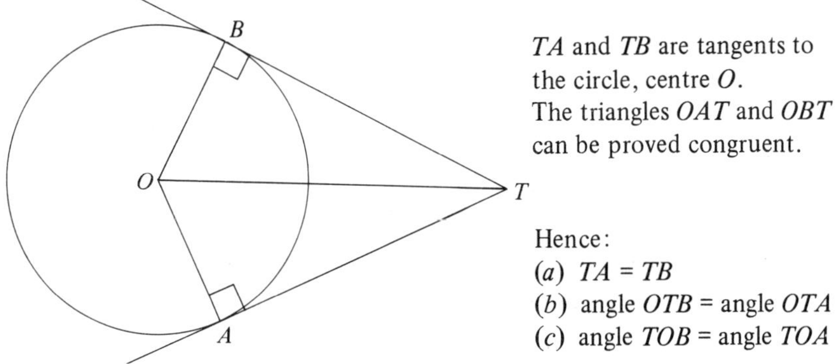

TA and TB are tangents to the circle, centre O.
The triangles OAT and OBT can be proved congruent.

Hence:
(*a*) $TA = TB$
(*b*) angle OTB = angle OTA
(*c*) angle TOB = angle TOA

Fact: Tangents drawn to a circle from an external point are equal.

3 *The angle in the 'alternate segment'*

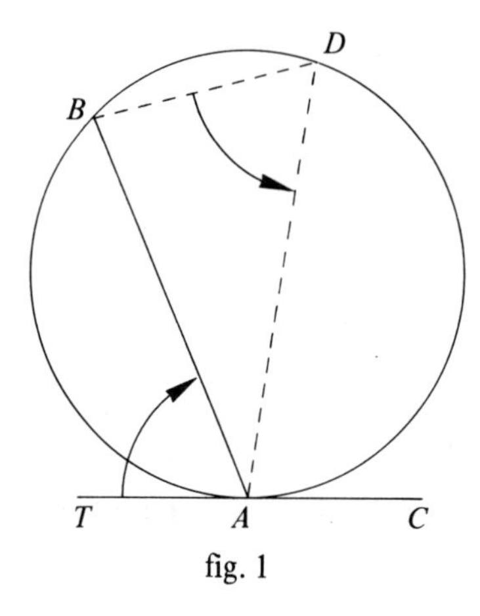

fig. 1

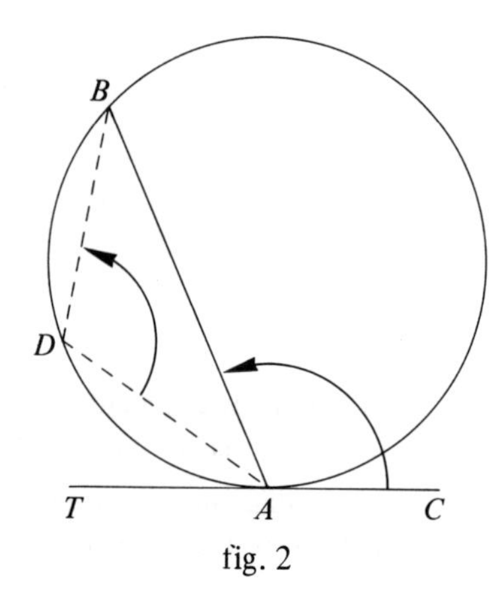

fig. 2

In the diagrams, *TAC* is a tangent to the circles at *A* and *AB* is a chord.

In fig. 1, angle *BDA* is the angle in the alternate segment (major) and angle *TAB* is the acute angle which *BA* makes with the tangent *TAC*. It may be proved that: angle *TAB* = angle *BDA*.

In fig. 2, angle *BDA* is the angle in the alternate segment (minor) and angle *CAB* is the obtuse angle which *BA* makes with the tangent *TAC*. It may be proved that: angle *CAB* = angle *BDA*.

Fact: If a chord is drawn from the point of contact of a tangent to a circle, the angles which the chord makes with the tangent are equal to the angles in the alternate segments.

D. The properties of chords

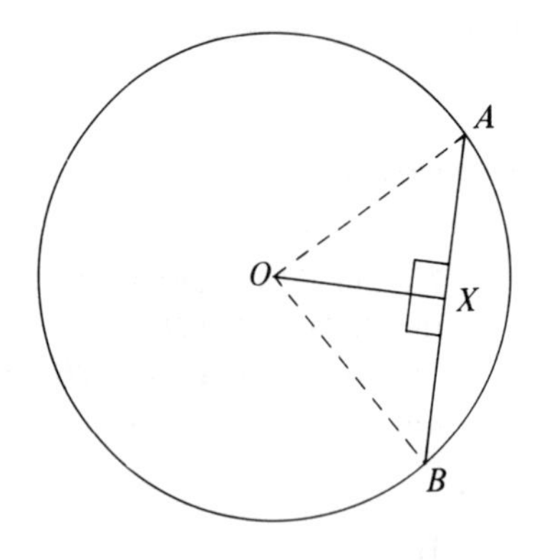

In the diagram *AB* is a chord.
O is the centre of the circle
and *x* is the mid point of *AB*.

Angle *OXA* = angle *OXB* = 90°
(triangle *OXA* ≡ triangle *OXB*).

Fact: A straight line from the centre of a circle to the mid point of a chord is perpendicular to the chord.

2

Two other facts follow from this property of chords:

Fact: A perpendicular from the centre of a circle to a chord, *bisects the chord*.

Fact: The perpendicular bisector of a chord passes through the centre of a circle. (This fact is used in the construction to find the centre of a circle.)

3

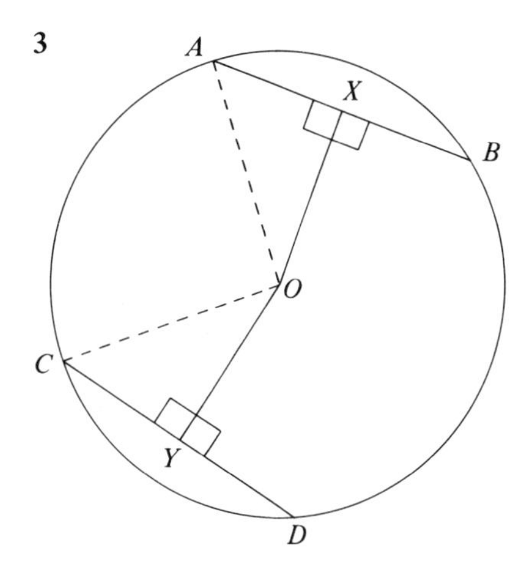

AB and *CD* are *equal* chords. *X* and *Y* are the mid points of *AB* and *CD* respectively.

It may be proved that:
triangle $AOX \equiv$ triangle COY,
hence OX = OY

Fact: Equal chords are equidistant from the centre of a circle.

4

It follows from the last fact, that chords which are equidistant from the centre of a circle are equal.

8. Pythagoras' theorem

1

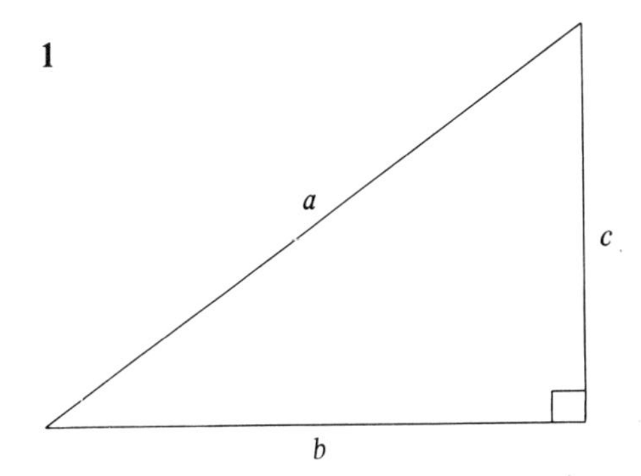

'In a right-angled triangle, the square on the hypotenuse is equal to the sum of the squares on the two shorter sides.'
With the notation of the diagram:

$$a^2 = b^2 + c^2$$

2 The basic right-angled triangle has sides in the ratio 3 : 4 : 5

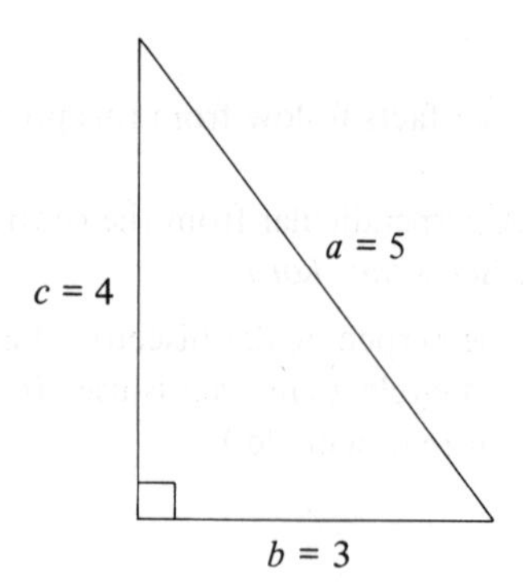

$$a^2 = 5^2 = 25$$
$$b^2 + c^2 = 3^2 + 4^2 = 9 + 16 = 25$$

$$a^2 = b^2 + c^2$$

Note: There are other ratios, sometimes called triads, which result in a right-angled triangle: e.g. 5, 12, 13 and 8, 15, 17.

3 *Using the theorem*. If we know any two of the sides of a right-angled triangle, it is possible to calculate the third.

(i) Finding the hypotenuse (the side opposite the right angle and, of course, the longest side).

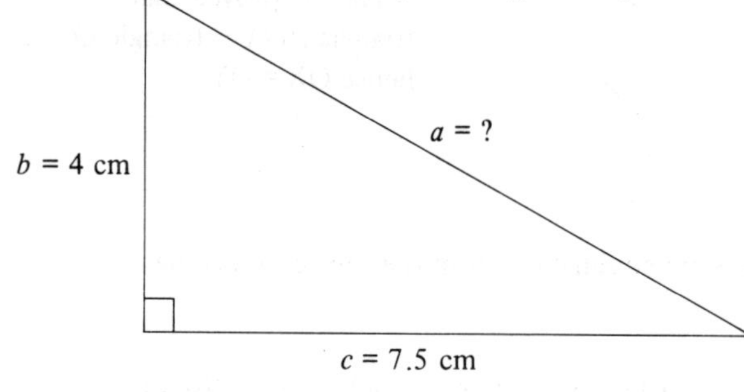

$$a^2 = b^2 + c^2$$
$$a^2 = 4^2 + 7.5^2$$
$$a^2 = 16 + 56.3$$
$$a^2 = 72.3$$
$$a = \sqrt{72.3}$$
$$a = 8.5 \text{ cm}$$

(ii) Finding a shorter side.

(*a*) By re-arranging the theorem, we get:

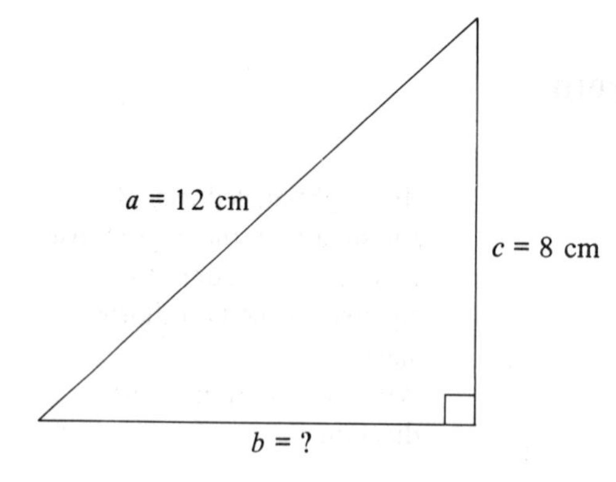

$$b^2 = a^2 - c^2$$
$$b^2 = 12^2 - 8^2$$
$$b^2 = 144 - 64$$
$$b^2 = 80$$
$$b = \sqrt{80}$$
$$b = 8.94 \text{ cm}$$

N.B. When finding a shorter side, the squares are *subtracted.*

(*b*) The side c may be similarly evaluated.

Note: When finding a shorter side and evaluating the subtraction of the squares, it is often useful to remember the fact that $a^2 - b^2 = (a + b)(a - b)$. This fact may be used as shown:

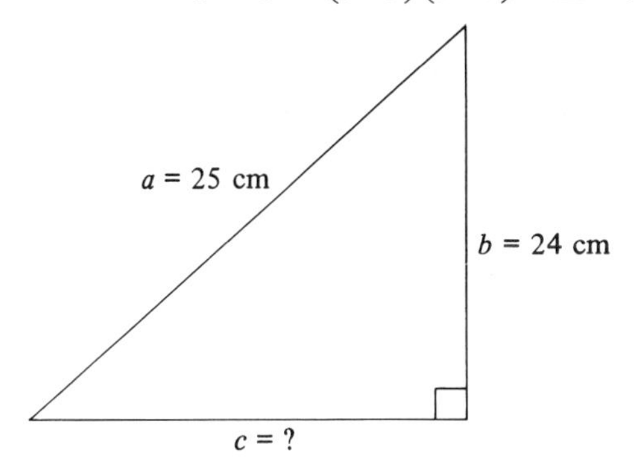

$$c^2 = a^2 - b^2$$
$$c^2 = (a + b)(a - b)$$
$$c^2 = (25 + 24)(25 - 24)$$
$$c^2 = 49 \times 1$$
$$c^2 = 49$$
$$c = \sqrt{49}$$
$$c = 7 \text{ cm}$$

9. Number bases and operations

A. Basics

When the number 3764 is expressed in decimal or denary notation, each digit represents a specific value as shown:

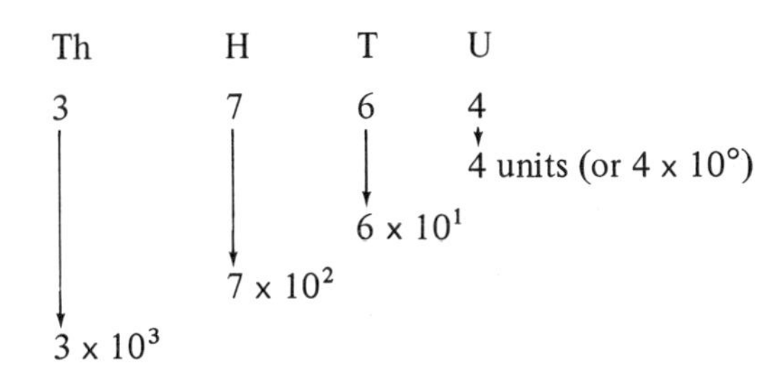

We can express numbers in other bases by raising the power of the base number as the columns proceed from the units on the right towards the left.

For example: 576_8 which is in base 8 or octal base could be evaluated in base 10 as follows:

$$\begin{aligned} &\quad 5 \qquad 7 \qquad 6 \\ &= (5 \times 8^2) + (7 \times 8^1) + (6 \text{ units}) \\ &= 320 + 56 + 6 \\ &= 382_{10} \end{aligned}$$

In the binary system or base 2, the only digits to be used are 1 and 0.

For example: $11\,011_2$ is evaluated as follows:

$$\begin{array}{ccccc} 1 & 1 & 0 & 1 & 1 \end{array}$$

$$= (1 \times 2^4) + (1 \times 2^3) + (0 \times 2^2) + (1 \times 2^1) + (1 \text{ unit})$$

$$= 32 + 8 + 0 + 2 + 1$$

$$= 43_{10}$$

Thus any number base may be developed beginning from the units column on the right e.g. base 5:

5^4	5^3	5^2	5^1	units
625	125	25	5	*u*

As mentioned previously, the only digits used in the binary system are 0 and 1. Similarly, for any other base, the only digits used are those which are less in value than the base itself, e.g. in the octal system, the digits used are: 0,1,2,3,4,5,6, and 7.

B. Addition

When adding, be careful to remember which base is being used because this will be your 'carrying' figure, e.g. carry out the following addition in base 6:

$$\begin{array}{r} 235 \\ +\ 144 \\ \hline 423_6 \end{array}$$

C. Subtraction

As in addition, remember the base you are using because this will be your 'borrowing' figure. e.g. carry out the following subtraction in base 5.

$$\begin{array}{r} 232 \\ -\ 144 \\ \hline 33_5 \end{array}$$

Remember to check your subtractions by addition (in the base being used in the question) of the last two numbers which should produce the first number, i.e. $144_5 + 33_5 = 232_5$

D. Multiplication

When multiplying numbers, use the same basic method as with base 10 numbers. The base you are using will indicate which digits may be used and what the 'carrying' figure will be, e.g. the following multiplications are in Base 7.

(*a*)
```
    34
 x   5
  236₇
```

(*b*)
```
    235
 x   26
   2112   (x 6)
   5030   (x 20)
 10142₇   (x 26)
```

E. Division

When dividing you must remember that you are working within the number base, whatever operation you perform to aid your division. If you wish, however, you could convert both numbers into base 10, carry out the division in base 10 and then convert your answer from base 10 to the required base, e.g. carry out the following division in base 4: 13 002 divided by 22.

Method 1

```
       231
22)13002
    110
     200
     132
       22
       22
```

Method 2

$13\,002_4$ =

256	64	16	4	*u*
1	3	0	0	2

= 256 + 192 + 0 + 0 + 2

= 450_{10}

22_4 = 10_{10}

$450_{10} \div 10_{10} = 45_{10}$
and $45_{10} = 231_4$

F. Converting

1. *To convert any number base into base 10*

Write out the place value of each digit and proceed as follows, e.g. convert 234_5 into base 10.

		25	5	*u*
234_5	=	(2 x 25)	+ (3 x 5)	+ 4

= 50 + 15 + 4

= 69_{10}

2 *To convert base 10 to any other base*

Whatever base you are converting your decimal number into use that as your dividing number (divisor). Set out your division clearly so that the remainders for each division may be read off for your answer, e.g. convert 79_{10} into base 5.

5	79	*r*
5	15	4
5	3	0
	0	3

↑

By reading off the remainders in the direction shown by the arrow:

$$79_{10} = 304_5$$

3 *To convert any base, other than 10, to any other base*

Convert 76_8 into base 3.

Stage 1: convert 76_8 into base 10

8	*u*
7	6

$$= (7 \times 8) + 6$$
$$= 56 + 6$$
$$= 62_{10}$$

Stage 2: convert 62_{10} into base 3

3	62	*r*
3	20	2
3	6	2
3	2	0
3	0	2

↑

$$62_{10} = 2022_3$$

therefore $76_8 = 2022_3$

10. Standard Form notation

A. Basics

A convenient method of writing very large or very small numbers. It is sometimes called Scientific Notation.

In this notation, a number is written in the form of $A \times 10^n$, where A lies between 1 and 10, i.e. $1 \leqslant A < 10$ and n is an integer.

Study the following examples of numbers written in Standard Form notation:

725 000.0	=	7.25×10^5	
72 500.0	=	7.25×10^4	
7 250.0	=	7.25×10^3	
725.0	=	7.25×10^2	
72.5	=	7.25×10^1	
7.25	=	7.25×10^0	($10^0 = 1$)
0.725	=	7.25×10^{-1}	
0.072 5	=	7.25×10^{-2}	
0.007 25	=	7.25×10^{-3}	
0.000 725	=	7.25×10^{-4}	
0.000 072 5	=	7.25×10^{-5}	

B. To write a number in Standard Form notation.

1 Move the digits across the point until the number lies between 1 and 10 ($1 \leqslant A < 10$).

2 Count the number of places the digits have been moved and this will give you the power of 10 you should multiply by.

N.B. Moving the digits to the right gives a *positive index*.
Moving digits to the left gives *a negative index*.

If you are given a number written in Standard Form notation, to convert to a normal decimal notation, use the reverse process:

Example 1: write 7.25×10^3 in decimal notation.
Move the digits 3 places to the left across the point, giving: 7250.0.

Example 2: write 7.25×10^{-6} in decimal notation.
Move the digits 6 places to the right across the point, giving: 0.000 007 25

C. Operations with numbers written in Standard Form notation.

1. *Multiplication*

$(A \times 10^x) \times (B \times 10^y) = AB \times 10^{x+y}$ (laws of indices)

Note: If AB results in a number $\geqslant 10$, the power of 10 must be adjusted.

Example (i) $(3.0 \times 10^4) \times (2.0 \times 10^6) = 6.0 \times 10^{(4+6)} = 6.0 \times 10^{10}$

Example (ii) $(6.0 \times 10^5) \times (7.0 \times 10^4) = 42.0 \times 10^9$. This cannot be left as 42.0×10^9 as A must be less than 10. To adjust this, divide 42.0 by 10 and compensate by increasing the value of n. This gives:

4.2×10^{10}

Example (iii)

$$(8.0 \times 10^{-4}) \times (9.0 \times 10^{-6}) = 72.0 \times 10^{-10}$$
$$= 7.2 \times 10^{(-10+1)}$$
$$= 7.2 \times 10^{-9}$$

2 *Division*

$(A \times 10^x) \div (B \times 10^y) = \frac{A}{B} \times 10^{x-y}$ (laws of indices).

Examples

(i) $(8.0 \times 10^7) \div (4.0 \times 10^3) = \frac{8.0}{4.0} \times 10^{7-3} = 2.0 \times 10^4$

(ii) $(7.0 \times 10^4) \div (2.0 \times 10^7) = \frac{7.0}{2.0} \times 10^{4-7} = 3.5 \times 10^{-3}$

(iii) $(6.0 \times 10^{-5}) \div (3.0 \times 10^{-10}) = \frac{6.0}{3.0} \times 10^{-5-(-10)}$

$$= 2.0 \times 10^{-5+10}$$
$$= 2.0 \times 10^5$$

(iv) $(3.0 \times 10^5) \div (4.0 \times 10^{-2}) = \frac{3.0}{4.0} \times 10^{5-(-2)}$

$$= 0.75 \times 10^{(5+2)}$$
$$= 0.75 \times 10^7$$
$$= 7.5 \times 10^6$$

11. Logarithms

(Using 3-figure tables.)

1 Logarithms (usually abbreviated to logs) are numbers written as powers of 10.

$100 = 10^{2.0}$ therefore $\log_{\text{base } 10} 100 = 2.0$

Although logs are indices, they are written as full size numbers when working with them.

2 $\log_{10} 1 = 0.0$, $\log_{10} 10 = 1.0$, $\log_{10} 100 = 2.0$, $\log_{10} 1000 = 3.0$
Numbers less than 1 may be written as negative powers of 10, e.g. $0.1 = 10^{-1}$ and as a log this is written $\bar{1}.0$ because the decimal part of a log is always positive. $\bar{1}.0$ is read as 'bar one'.

3 The number before the point in a log is called the CHARACTERISTIC and the decimal part of a log is called the MANTISSA. To find the mantissa, a table of logs is used.

4 *Writing a number as a base $_{10}$ log*

Example 1: Write 473.0 as a log.

(i) write the number in Standard Form notation (Reference Section 10)

$473.0 = 4.73 \times 10^2$

(ii) Look up 4.73 in the log tables, which gives 0.675 (the mantissa)

thus $473.0 = 10^{0.675} \times 10^2$

and $473.0 = 10^{2.675}$ (using the laws of indices)

which we write as $\log_{10} 473 = 2.675$

Example 2: Write 0.0473 as a base $_{10}$ log.

$0.0473 = 4.73 \times 10^{-2}$ (note the negative index)

$= 10^{0.675} \times 10^{-2}$

$= 10^{-2.675}$

$\log_{10} 0.0473 = \bar{2}.675$ (-2 is written as $\bar{2}$)

With practice the log of a number may be written down without showing any working. Decide the characteristic first, write it down and then add the mantissa from the tables.

5 *Writing a number from a given log*

Example 1: Write down the number whose log is 2.687 ($10^{2.687}$).

$10^{2.687} = 10^{0.687} \times 10^{2}$ (use the tables in reverse to find the number whose log is 0.687)

$= 4.86 \times 10^{2}$

$= 486.0$ (moving the digits 2 places to the left).

Example 2: Write down the number whose log is $\bar{2}.687$.

$10^{\bar{2}.687} = 10^{0.687} \times 10^{-2}$

$= 4.86 \times 10^{-2}$

$= 0.0486$ (moving the digits 2 places to the right).

6 *The operations of multiplication and division*

(i) *Multiplication*:

$a^{m} \times a^{n} = a^{(m+n)}$.

thus $10^{2.0} \times 10^{3.0} = 10^{5.0}$ *Rule*: ADD the logs

(ii) *Division*:

$a^{m} \div a^{n} = a^{m-n}$

thus $10^{7.0} \div 10^{4.0} = 10^{3.0}$ *Rule*: SUBTRACT the logs

The following worked example illustrates the rules and also provides a suggested method of setting the work out.

Evaluate: $\dfrac{3.71 \times 16.8 \times 13.2}{5.62}$

$= 146.5$

	Number	Log	
	3.71	0.569	
	16.8	1.225	Add
	13.2	+ 1.121	the logs
		2.915	Subtract
	5.62	- 0.750	the logs
Answer:	146.5	← 2.165	

7 *To calculate a power of a number*

$$n^3 = n \times n \times n = \log n + \log n + \log n = 3 \times \log n$$

Rule: Multiply the log of the number by the required power.

Example: Calculate $(71.2)^3$:

$(71.2)^3 = 360\,000$

	Number	Log
	71.2	1.852
	$(71.2)^3$	× 3
Answer:	360 000.0	← 5.556

8 *To calculate a root of a number*

$$\sqrt{n} = n^{\frac{1}{2}} = \frac{\log n}{2} \qquad \sqrt[3]{n} = n^{\frac{1}{3}} = \frac{\log n}{3}$$

Rule: Divide the log of the number by the required root.

Example: Calculate: $\sqrt[3]{29.3}$

$\sqrt[3]{29.3} = 3.08$

	Number	Log
	29.3	1.467
	$\sqrt[3]{29.3}$	3)1.467
Answer:	3.08	← 0.489

12. Laws of indices

1. *Multiplication of powers*

$$a^2 \times a^3 = (a \times a) \times (a \times a \times a) = a^5$$
$$a^3 \times a^4 = (a \times a \times a) \times (a \times a \times a \times a) = a^7$$

From these examples it may be seen that the law for multiplication is:

$$a^m \times a^n = a^{(m+n)}$$

2 *Division of powers*

$$a^4 \div a^2 = \frac{\overset{1}{\not a} \times \overset{1}{\not a} \times a \times a}{\underset{1}{\not a} \times \underset{1}{\not a}} = a^2, \quad a^5 \div a^2 = \frac{\overset{1}{\not a} \times \overset{1}{\not a} \times a \times a \times a}{\underset{1}{\not a} \times \underset{1}{\not a}} = a^3$$

From these examples it may be seen that the law for division is:

$$a^m \div a^n = a^{(m-n)}$$

Using the law in the following example produces a negative index:

$$a^2 \div a^4 = a^{(2-4)} = a^{-2}$$

3 *Multiplying a power by a power*

$$(a^2)^3 = a^2 \times a^2 \times a^2 = a^6 \quad \text{(using the first law of indices)}$$

$$(a^3)^4 = a^3 \times a^3 \times a^3 \times a^3 = a^{12}$$

In general: $(a^m)^n = a^{(mn)}$

4 *Fractional indices*

(i) $a^{\frac{1}{3}} \times a^{\frac{1}{3}} \times a^{\frac{1}{3}} = a^1 = a$ (ii) $a^{\frac{2}{3}} \times a^{\frac{2}{3}} \times a^{\frac{2}{3}} = a^{\frac{6}{3}} = a^2$

$a^{\frac{1}{3}} = \sqrt[3]{a}$ $a^{\frac{2}{3}} = (\sqrt[3]{a})^2$

e.g. $8^{\frac{1}{3}} = \sqrt[3]{8} = 2$ e.g. $27^{\frac{2}{3}} = (\sqrt[3]{27})^2 = 3^2 = 9$

5 *Negative indices*

Using the law for division: $a^2 \div a^4 = a^{(2-4)} = a^{-2}$

But $a^2 \div a^4 = \dfrac{\overset{1}{\not a} \times \overset{1}{\not a}}{\underset{1}{\not a} \times \underset{1}{\not a} \times a \times a} = \dfrac{1}{a^2}$

Therefore: $a^{-2} = \dfrac{1}{a^2}$

Similarly: $a^{-\frac{1}{2}} = \dfrac{1}{a^{\frac{1}{2}}} = \dfrac{1}{\sqrt{a}}$ and $a^{-\frac{3}{4}} = \dfrac{1}{a^{\frac{3}{4}}} = \dfrac{1}{(\sqrt[4]{a})^3}$

Numerical examples:

Evaluate: 4^{-2} : $4^{-2} = \dfrac{1}{4^2} = \dfrac{1}{16}$

Evaluate: $16^{-\frac{1}{2}}$: $16^{-\frac{1}{2}} = \dfrac{1}{\sqrt{16}} = \dfrac{1}{4}$

Evaluate: $16^{-\frac{3}{4}}$: $16^{-\frac{3}{4}} = \dfrac{1}{16^{\frac{3}{4}}} = \dfrac{1}{(\sqrt[4]{16})^3} = \dfrac{1}{2^3} = \dfrac{1}{8}$

13. Angles in a triangle

1 *The angle sum*

In any triangle, the sum of the angles is 180° (2 right angles):

angle A + angle B + angle C

= 180°

2 *Equilateral triangles*

In an equilateral triangle, the angles are each 60°:

angle A = angle B = angle C

$$= \frac{180°}{3} = 60°$$

3 *Right-angled triangles*

In a right-angled triangle the two smaller angles are complementary, i.e. add up to 90°.

angle A + angle B = 90°

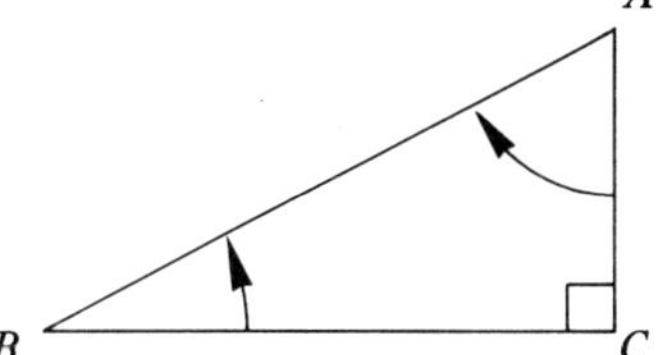

4 *Isosceles triangles*

In an isosceles triangle, the angles opposite the equal sides are equal.

If $AB = AC$

then angle B = angle C

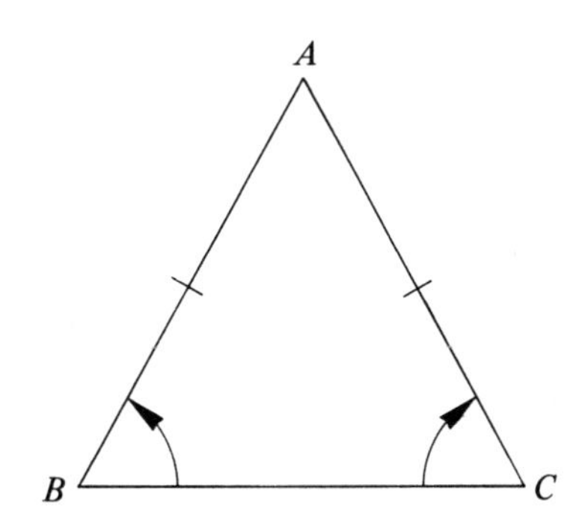